SECOND EDITION

ARM ASSEMBLY LANGUAGE

Fundamentals and Techniques

SECOND EDITION

ARM ASSEMBLY LANGUAGE

Fundamentals and Techniques

William Hohl

Christopher Hinds
ARM, Inc., Austin, Texas

CRC Press is an imprint of the
Taylor & Francis Group, an **informa** business

CRC Press
Taylor & Francis Group
6000 Broken Sound Parkway NW, Suite 300
Boca Raton, FL 33487-2742

Printed on acid-free paper
Version Date: 20140915

International Standard Book Number-13: 978-1-4822-2985-1 (Hardback)

Library of Congress Cataloging-in-Publication Data

Hohl, William.
ARM assembly language : fundamentals and techniques / William Hohl and Christopher Hinds. -- Second edition.
pages cm
Includes bibliographical references and index.
ISBN 978-1-4822-2985-1 (hardback)
1. Assembly languages (Electronic computers) 2. Embedded computer systems--Programming. I. Hinds, Christopher. II. Title.

QA76.73.A8H637 2014
006.2'2--dc23 2014024230

Visit the Taylor & Francis Web site at
http://www.taylorandfrancis.com

and the CRC Press Web site at
http://www.crcpress.com

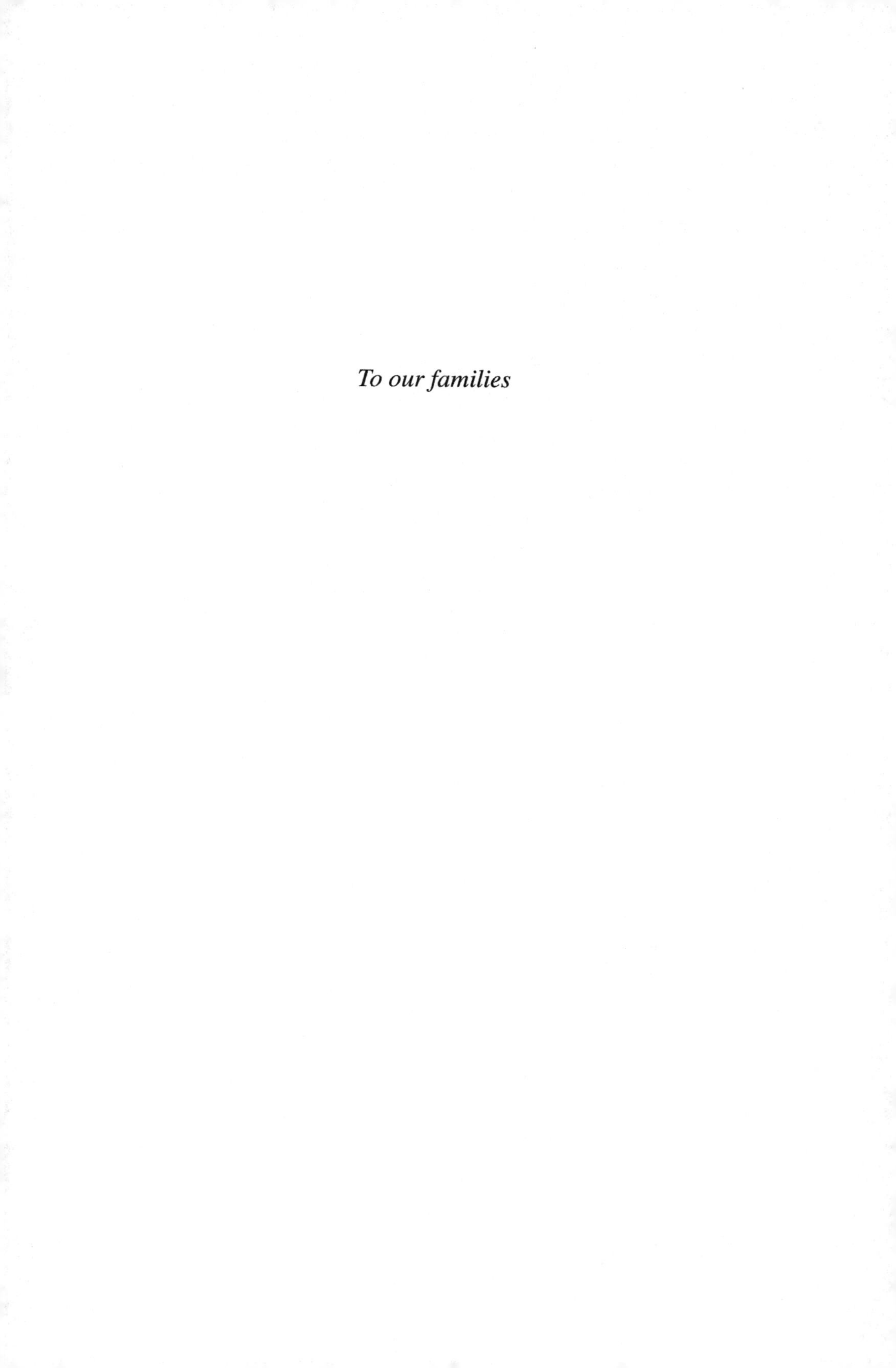

To our families

Contents

Preface

Few industries are as quick to change as those based on technology, and computer technology is no exception. Since the First Edition of *ARM Assembly Language: Fundamentals and Techniques* was published in 2009, ARM Limited and its many partners have introduced a new family of embedded processors known as the Cortex-M family. ARM is well known for applications processors, such as the ARM11, Cortex-A9, and the recently announced Cortex-5x families, which provide the processing power to modern cell phones, tablets, and home entertainment devices. ARM is also known for real-time processors, such as the Cortex-R4, Cortex-R5, and Cortex-R7, used extensively in deeply embedded applications, such as gaming consoles, routers and modems, and automotive control systems. These applications are often characterized by the presence of a real-time operating system (RTOS). However, the Cortex-M family focuses on a well-established market space historically occupied by 8-bit and 16-bit processors. These applications differ from real-time in that they rarely require an operating system, instead performing one or only a few functions over their lifetime. Such applications include game controllers, music players, automotive safety systems, smart lighting, connected metering, and consumer white goods, to name only a few. These processors are frequently referred to as microcontrollers, and a very successful processor in this space was the ubiquitous 8051, introduced by Intel but followed for decades by offerings from numerous vendors. The 68HC11, 68HC12, and 68HC16 families of microcontrollers from Motorola were used extensively in the 1980s and 1990s, with a plethora of offerings including a wide range of peripherals, memory, and packaging options. The ease of programming, availability, and low cost is partly responsible for the addition of smart functionality to such common goods as refrigerators and washers/dryers, the introduction of airbags to automobiles, and ultimately to the cell phone.

In early applications, a microcontroller operating at 1 MHz would have provided more than sufficient processing power for many applications. As product designers added more features, the computational requirements increased and the need for greater processing power was answered by higher clock rates and more powerful processors. By the early 2000s, the ARM7 was a key part of this evolution. The early Nokia cell phones and Apple iPods were all examples of systems that performed several tasks and required greater processing power than was available in microcontrollers of that era. In the case of the cell phone, the processor was controlling the user interface (keyboard and screen), the cellular radio, and monitoring the battery levels. Oh, and the Snake game was run on the ARM7 as well! In the case of the iPod, the ARM7 controlled the user interface and battery monitoring, as with the cell phone, and handled the decoding of the MP3 music for playing through headphones. With these two devices our world changed forever—ultimately phones would play music and music players would make phone calls, and each would have better games and applications than Snake!

In keeping with this trend, the mix of ARM's processor shipments is changing rapidly. In 2009 the ARM7 accounted for 55% of the processor shipments, with all Cortex processors contributing only 1%.* By 2012 the ARM7 shipments had dropped to 36%, with the Cortex-M family shipments contributing 22%.† This trend is expected to continue throughout the decade, as more of the applications that historically required only the processing power of an 8-bit or 16-bit system move to the greater capability and interoperability of 32-bit systems. This evolution is empowering more features in today's products over those of yesterday. Consider the capabilities of today's smart phone to those of the early cell phones! This increase is made possible by the significantly greater computing power available in roughly the same size and power consumption of the earlier devices. Much of the increase comes through the use of multiple processors. While early devices were capable of including one processor in the system, today's systems include between 2 and 8 processors, often different classes of processors from different processor families, each performing tasks specific to that processor's capabilities or as needed by the system at that time. In today's System-on-Chip (SoC) environment, it is common to include both application processors and microcontrollers in the same device. As an example, the Texas Instruments OMAP5 contains a dual-core Cortex-A15 application processor and two Cortex-M4 microcontrollers. Development on such a system involves a single software development system for both the Cortex-A15 and the Cortex-M4 processors. Having multiple chips from different processor families and vendors adds to the complexity, while developing with processors all speaking the same language and from the same source greatly simplifies the development.

All this brings us back to the issue raised in the first edition of this book. Why should engineers and programmers spend time learning to program in assembly language? The reasons presented in the first edition are as valid today as in 2009, perhaps even more so. The complexity of the modern SoCs presents challenges in communications between the multiple processors and peripheral devices, challenges in optimization of the sub-systems for performance and power consumption, and challenges in reducing costs by efficient use of memory. Knowledge of the assembly language of the processors, and the insight into the operation of the processors that such knowledge provides, is often the key to the timely and successful completion of these tasks and launch of the product. Further, in the drive for performance, both in speed of the product to the user and in a long battery life, augmenting the high-level language development with targeted use of hand-crafted assembly language will prove highly valuable—but we don't stop here. Processor design remains a highly skilled art in which a thorough knowledge of assembly language is essential. The same is true for those tasked with compiler design, creating device drivers for the peripheral subsystems, and those producing optimized library routines. High quality compilers, drivers, and libraries contribute directly to performance and development time. Here a skilled programmer or system designer with a knowledge of assembly language is a valuable asset.

* ARM 2009 Annual Report, www.arm.com/annualreport09/business-review

† ARM 2012 Annual Report, see www.arm.com

In the second edition, we focus on the Cortex-M4 microcontroller in addition to the ARM7TDMI. While the ARM7TDMI still outsells the Cortex-M family, we believe the Cortex-M family will soon overtake it, and in new designs this is certainly true. The Cortex-M4 family is the first ARM microcontroller to incorporate optional hardware floating-point. Chapter 9 introduces floating-point computation and contrasts it with integer computation. We present the floating-point standard of 1985, IEEE 754-1985, and the recent revision to the standard, the IEEE 754-2008, and discuss some of the issues in the use of floating-point which are not present in integer computation. In many of the chapters, floating-point instructions will be included where their usage would present a difference from that of integer usage. As an example, the floating-point instructions use a separate register file from the integer register file, and the instructions which move data between memory and these registers will be discussed in Chapters 3, 9, and 12. Example programs are repeated with floating-point instructions to show differences in usage, and new programs are added which focus on specific aspects of floating-point computation. While we will discuss floating-point at some length, we will not exhaust the subject, and where useful we will point the reader to other references.

The focus of the book remains on second- or third-year undergraduate students in the field of computer science, computer engineering, or electrical engineering. As with the first edition, some background in digital logic and arithmetic, high-level programming, and basic computer operation is valuable, but not necessary. We retain the aim of providing not only a textbook for those interested in assembly language, but a reference for coding in ARM assembly language, which ultimately helps in using any assembly language.

In this edition we also include an introduction to Code Composer Studio (from Texas Instruments) alongside the Keil RealView Microcontroller Development Kit. Appendices A and B cover the steps involved in violating just about every programming rule, so that simple assembly programs can be run in an otherwise advanced simulation environment. Some of the examples will be simulated using one of the two tools, but many can be executed on an actual hardware platform, such as a Tiva™ Launchpad from TI. Code specifically for the Tiva Launchpad will be covered in Chapter 16. In the first edition, we included a copy of the ARM v4T Instruction Set as Appendix A. To do so and include the ARM Thumb-2 and ARM FPv4-SP instruction sets of the Cortex-M4 would simply make the book too large. Appropriate references are highlighted in Section 1.7.4, all of which can be found on ARM's and TI's websites.

The first part of the book introduces students to some of the most basic ideas about computing. Chapter 1 is a very brief overview of computing systems in general, with a brief history of ARM included in the discussion of RISC architecture. This chapter also includes an overview of number systems, which should be stressed heavily before moving on to any further sections. Floating-point notation is mentioned here, but there are three later chapters dedicated to floating-point details. Chapter 2 gives a shortened description of the programmer's model for the ARM7TDMI and the Cortex-M4—a bit like introducing a new driver to the clutch, gas pedal, and steering wheel, so it's difficult to do much more than simply present it and move on. Some simple programs are presented in Chapter 3, mostly to get code running with the tools, introduce a few directives, and show what ARM and Thumb-2 instructions look like. Chapter 4 presents most of the directives that students will immediately need if they use either the Keil tools or Code Composer Studio. It is not intended to be memorized.

The next chapters cover topics that need to be learned thoroughly to write any meaningful assembly programs. The bulk of the load and store instructions are examined in Chapter 5, with the exception of load and store multiple instructions, which are held until Chapter 13. Chapter 6 discusses the creation of constants in code, and how to create and deal with literal pools. One of the bigger chapters is Chapter 7, Logic and Arithmetic, which covers all the arithmetic operations, including an optional section on fractional notation. As this is almost never taught to undergraduates, it's worth introducing the concepts now, particularly if you plan to cover floating-point. If the course is tight for time, you may choose to skip this section; however, the subject is mentioned in other chapters, particularly Chapter 12 when a sine table is created and throughout the floating-point chapters. Chapter 8 highlights the whole issue of branching and looks at conditional execution in detail. Now that the Cortex-M4 has been added to the mix, the IF-THEN constructs found in the Thumb-2 instruction set are also described.

Having covered the basics, Chapters 9 through 11 are dedicated to floating-point, particularly the formats, registers used, exception types, and instructions needed for working with single-precision and half-precision numbers found on the Cortex-M4 with floating-point hardware. Chapter 10 goes into great detail about rounding modes and exception types. Chapter 11 looks at the actual uses of floating-point in code—the data processing instructions—pointing out subtle differences between such operations as chained and fused multiply accumulate. The remaining chapters examine real uses for assembly and the situations that programmers will ultimately come across. Chapter 12 is a short look at tables and lists, both integer and floating-point. Chapter 13, which covers subroutines and stacks, introduces students to the load and store multiple instructions, along with methods for passing parameters to functions. Exceptions and service routines for the ARM7TDMI are introduced in Chapter 14, while those for v7-M processors are introduced in Chapter 15. Since the book leans toward the use of microcontroller simulation models, Chapter 16 introduces peripherals and how they're programmed, with one example specifically targeted at real hardware. Chapter 17 discusses the three different instruction sets that now exist—ARM, Thumb, and Thumb-2. The last topic, mixing C and assembly, is covered in Chapter 18 and may be added if students are interested in experimenting with this technique.

Ideally, this book would serve as both text and reference material, so Appendix A explains the use of Code Composer Studio tools in the creation of simple assembly programs. Appendix B has an introduction to the use of the RealView Microcontroller Development Kit from Keil, which can be found online at http://www.keil.com/demo. This is certainly worth covering before you begin coding. The ASCII character set is listed in Appendix C, and a complete program listing for an example found in Chapter 15 is given as Appendix D.

A one-semester (16-week) course should be able to cover all of Chapters 1 through 8. Depending on how detailed you wish to get, Chapters 12 through 16 should be enough to round out an undergraduate course. Thumb and Thumb-2 can be left off or covered as time permits. A two-semester sequence could cover the entire book, including the harder floating-point chapters (9 through 11), with more time allowed for writing code from the exercises.

Acknowledgments

To our reviewers, we wish to thank those who spent time providing feedback and suggestions, especially during the formative months of creating floating-point material (no small task): Matthew Swabey, Purdue University; Nicholas Outram, Plymouth University (UK); Joseph Camp, Southern Methodist University; Jim Garside, University of Manchester (UK); Gary Debes, Texas Instruments; and David Lutz, Neil Burgess, Kevin Welton, Chris Stephens, and Joe Bungo, ARM.

We also owe a debt of gratitude to those who helped with tools and images, answered myriad questions, and got us out of a few messy legal situations: Scott Specker at Texas Instruments, who was brave enough to take our challenge of producing five lines of assembly code in Code Composer Studio, only to spend the next three days getting the details ironed out; Ken Havens and the great FAEs at Keil; Cathy Wicks and Sue Cozart at Texas Instruments; and David Llewellyn at ARM.

As always, we would like to extend our appreciation to Nora Konopka, who believed in the book enough to produce the second edition, and Joselyn Banks-Kyle and the production team at CRC Press for publishing and typesetting the book.

William Hohl
Chris Hinds
June 2014

Authors

William Hohl held the position of Worldwide University Relations Manager for ARM, based in Austin, Texas, for 10 years. He was with ARM for nearly 15 years and began as a principal design engineer to help build the ARM1020 microprocessor. His travel and university lectures have taken him to over 40 countries on 5 continents, and he continues to lecture on low-power microcontrollers and assembly language programming. In addition to his engineering duties, he also held an adjunct faculty position in Austin from 1998 to 2004, teaching undergraduate mathematics. Before joining ARM, he worked at Motorola (now Freescale Semiconductor) in the ColdFire and 68040 design groups and at Texas Instruments as an applications engineer. He holds MSEE and BSEE degrees from Texas A&M University as well as six patents in the field of debug architectures.

Christopher Hinds has worked in the microprocessor design field for over 25 years, holding design positions at Motorola (now Freescale Semiconductor), AMD, and ARM. While at ARM he was the primary author of the ARM VFP floating-point architecture and led the design of the ARM10 VFP, the first hardware implementation of the new architecture. Most recently he has joined the Patents Group in ARM, identifying patentable inventions within the company and assisting in patent litigation. Hinds is a named inventor on over 30 US patents in the areas of floating-point implementation, instruction set design, and circuit design. He holds BSEE and MSEE degrees from Texas A&M University and an MDiv from Oral Roberts University, where he worked to establish the School of Engineering, creating and teaching the first digital logic and microprocessor courses. He has numerous published papers and presentations on the floating-point architecture of ARM processors.

1 An Overview of Computing Systems

1.1 INTRODUCTION

Most users of cellular telephones don't stop to consider the enormous amount of effort that has gone into designing an otherwise mundane object. Lurking beneath the display, below the user's background picture of his little boy holding a balloon, lies a board containing circuits and wires, algorithms that took decades to refine and implement, and software to make it all work seamlessly together. What exactly is happening in those circuits? How do such things actually work? Consider a modern tablet, considered a fictitious device only years ago, that displays live television, plays videos, provides satellite navigation, makes international Skype calls, acts as a personal computer, and contains just about every interface known to man (e.g., USB, Wi-Fi, Bluetooth, and Ethernet), as shown in Figure 1.1. Gigabytes of data arrive to be viewed, processed, or saved, and given the size of these hand-held devices, the burden of efficiency falls to the designers of the components that lie within them.

Underneath the screen lies a printed circuit board (PCB) with a number of individual components on it and probably at least two system-on-chips (SoCs). A SoC is nothing more than a combination of processors, memory, and graphics chips that have been fabricated in the same package to save space and power. If you further examine one of the SoCs, you will find that within it are two or three specialized microprocessors talking to graphics engines, floating-point units, energy management units, and a host of other devices used to move information from one device to another. The Texas Instruments (TI) TMS320DM355 is a good example of a modern SoC, shown in Figure 1.2.

System-on-chip designs are becoming increasingly sophisticated, where engineers are looking to save both money and time in their designs. Imagine having to produce the next generation of our hand-held device—would it be better to reuse some of our design, which took nine months to build, or throw it out and spend another three years building yet another, different SoC? Because the time allotted to designers for new products shortens by the increasing demand, the trend in industry is to take existing designs, especially designs that have been tested and used heavily, and build new products from them. These tested designs are examples of "intellectual property"—designs and concepts that can be licensed to other companies for use in large projects. Rather than design a microprocessor from scratch, companies will take a known design, something like a Cortex-A57 from ARM, and

FIGURE 1.1 Handheld wireless communicator.

build a complex system around it. Moreover, pieces of the project are often designed to comply with certain standards so that when one component is changed, say our newest device needs a faster microprocessor, engineers can reuse all the surrounding devices (e.g., MPEG decoders or graphics processors) that they spent years designing. Only the microprocessor is swapped out.

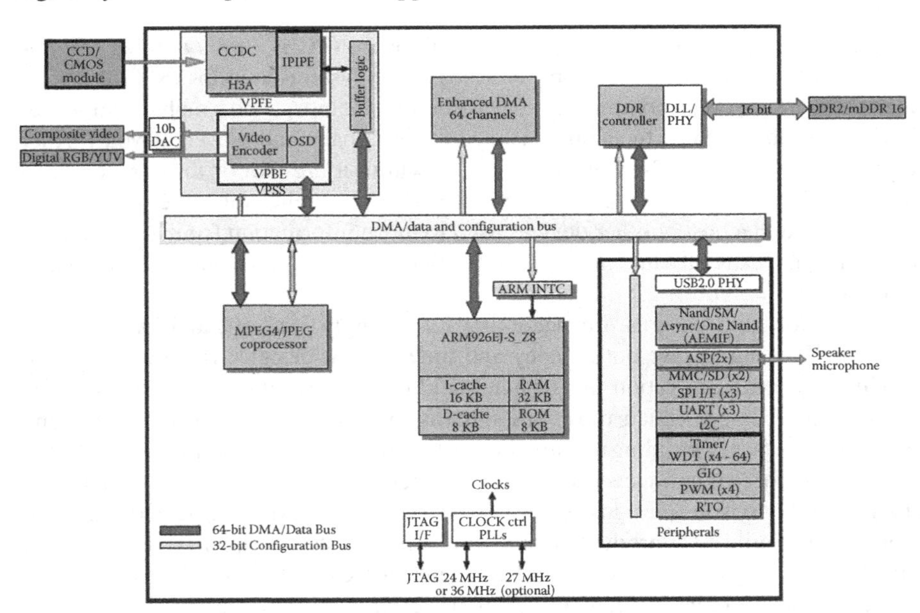

FIGURE 1.2 The TMS320DM355 System-on-Chip from Texas Instruments. (From Texas Instruments. With permission.)

This idea of building a complete system around a microprocessor has even spilled into the microcontroller industry. A microprocessor can be seen as a computing engine with no peripherals. Very simple processors can be combined with useful extras such as timers, universal asynchronous receiver/transmitters (UARTs), or analog-to-digital (A/D) converters to produce a microcontroller, which tends to be a very low-cost device for use in industrial controllers, displays, automotive applications, toys, and hundreds of other places one normally doesn't expect to find a computing engine. As these applications become more demanding, the microcontrollers in them become more sophisticated, and off-the-shelf parts today surpass those made even a decade ago by leaps and bounds. Even some of these designs are based on the notion of keeping the system the same and replacing only the microprocessor in the middle.

1.2 HISTORY OF RISC

Even before computers became as ubiquitous as they are now, they occupied a place in students' hearts and a place in engineering buildings, although it was usually under the stairs or in the basement. Before the advent of the personal computer, mainframes dominated the 1980s, with vendors like Amdahl, Honeywell, Digital Equipment Corporation (DEC), and IBM fighting it out for top billing in engineering circles. One need only stroll through the local museum these days for a glimpse at the size of these machines. Despite all the circuitry and fans, at the heart of these machines lay processor architectures that evolved from the need for faster operations and better support for more complicated operating systems. The DEC VAX series of minicomputers and superminis—not quite mainframes, but larger than minicomputers—were quite popular, but like their contemporary architectures, the IBM System/38, Motorola 68000, and the Intel iAPX-432, they had processors that were growing more complicated and more difficult to design efficiently. Teams of engineers would spend years trying to increase the processor's frequency (clock rate), add more complicated instructions, and increase the amount of data that it could use. Designers are doing the same thing today, except most modern systems also have to watch the amount of power consumed, especially in embedded designs that might run on a single battery. Back then, power wasn't as much of an issue as it is now—you simply added larger fans and even water to compensate for the extra heat!

The history of Reduced Instruction Set Computers (RISC) actually goes back quite a few years in the annals of computing research. Arguably, some early work in the field was done in the late 1960s and early 1970s by IBM, Control Data Corporation and Data General. In 1981 and 1982, David Patterson and Carlo Séquin, both at the University of California, Berkeley, investigated the possibility of building a processor with fewer instructions (Patterson and Sequin 1982; Patterson and Ditzel 1980), as did John Hennessy at Stanford (Hennessy et al. 1981) around the same time. Their goal was to create a very simple architecture, one that broke with traditional design techniques used in Complex Instruction Set Computers (CISCs), e.g., using microcode (defined below) in the processor; using instructions that had different lengths; supporting complex, multi-cycle instructions, etc. These new architectures would produce a processor that had the following characteristics:

- All instructions executed in a single cycle. This was unusual in that many instructions in processors of that time took multiple cycles. The trade-off was that an instruction such as MUL (multiply) was available without having to build it from shift/add operations, making it easier for a programmer, but it was more complicated to design the hardware. Instructions in mainframe machines were built from primitive operations internally, but they were not necessarily faster than building the operation out of simpler instructions. For example, the VAX processor actually had an instruction called INDEX that would take longer than if you were to write the operation in software out of simpler commands!
- All instructions were the same size and had a fixed format. The Motorola 68000 was a perfect example of a CISC, where the instructions themselves were of varying length and capable of containing large constants along with the actual operation. Some instructions were 2 bytes, some were 4 bytes. Some were longer. This made it very difficult for a processor to decode the instructions that got passed through it and ultimately executed.
- Instructions were very simple to decode. The register numbers needed for an operation could be found in the same place within most instructions. Having a small number of instructions also meant that fewer bits were required to encode the operation.
- The processor contained no microcode. One of the factors that complicated processor design was the use of microcode, which was a type of "software" or commands within a processor that controlled the way data moved internally. A simple instruction like MUL (multiply) could consist of dozens of lines of microcode to make the processor fetch data from registers, move this data through adders and logic, and then finally move the product into the correct register or memory location. This type of design allowed fairly complicated instructions to be created—a VAX instruction called POLY, for example, would compute the value of an nth-degree polynomial for an argument x, given the location of the coefficients in memory and a degree n. While POLY performed the work of many instructions, it only appeared as one instruction in the program code.
- It would be easier to validate these simpler machines. With each new generation of processor, features were always added for performance, but that only complicated the design. CISC architectures became very difficult to debug and validate so that manufacturers could sell them with a high degree of confidence that they worked as specified.
- The processor would access data from external memory with explicit instructions—Load and Store. All other data operations, such as adds, subtracts, and logical operations, used only registers on the processor. This differed from CISC architectures where you were allowed to tell the processor to fetch data from memory, do something to it, and then write it back to

memory using only a single instruction. This was convenient for the programmer, and especially useful to compilers, but arduous for the processor designer.

- For a typical application, the processor would execute more code. Program size was expected to increase because complicated operations in older architectures took more RISC instructions to complete the same task. In simulations using small programs, for example, the code size for the first Berkeley RISC architecture was around 30% larger than the code compiled for a VAX 11/780. The novel idea of a RISC architecture was that by making the operations simpler, you could increase the processor frequency to compensate for the growth in the instruction count. Although there were more instructions to execute, they could be completed more quickly.

Turn the clock ahead 33 years, and these same ideas live on in most all modern processor designs. But as with all commercial endeavors, there were good RISC machines that never survived. Some of the more ephemeral designs included DEC's Alpha, which was regarded as cutting-edge in its time; the 29000 family from AMD; and Motorola's 88000 family, which never did well in industry despite being a fairly powerful design. The acronym RISC has definitely evolved beyond its own moniker, where the original idea of a Reduced Instruction Set, or removing complicated instructions from a processor, has been buried underneath a mountain of new, albeit useful instructions. And all manufacturers of RISC microprocessors are guilty of doing this. More and more operations are added with each new generation of processor to support the demanding algorithms used in modern equipment. This is referred to as "feature creep" in the industry. So while most of the RISC characteristics found in early processors are still around, one only has to compare the original Berkeley RISC-1 instruction set (31 instructions) or the second ARM processor (46 operations) with a modern ARM processor (several hundred instructions) to see that the "R" in RISC is somewhat antiquated. With the introduction of Thumb-2, to be discussed throughout the book, even the idea of a fixed-length instruction set has gone out the window!

1.2.1 ARM Begins

The history of ARM Holdings PLC starts with a now-defunct company called Acorn Computers, which produced desktop PCs for a number of years, primarily adopted by the educational markets in the UK. A plan for the successor to the popular BBC Micro, as it was known, included adding a second processor alongside its 6502 microprocessor via an interface called the "Tube". While developing an entirely new machine, to be called the Acorn Business Computer, existing architectures such as the Motorola 68000 were considered, but rather than continue to use the 6502 microprocessor, it was decided that Acorn would design its own. Steve Furber, who holds the position of ICL Professor of Computer Engineering at the University of Manchester, and Sophie Wilson, who wrote the original instruction set, began working within the Acorn design team in October 1983, with VLSI Technology (bought later by Philips Semiconductor, now called NXP) as the silicon partner who produced the first samples. The ARM1 arrived back from the fab on April 26, 1985, using less than 25,000 transistors, which by today's standards would be fewer than the number found in a good integer multiplier. It's worth noting that the part worked the first time and executed code the day it arrived, which in that time frame was quite extraordinary. Unless you've lived through the evolution of computing, it's also rather important to put another metric into context, lest it be overlooked—processor speed. While today's desktop processors routinely run between 2 and 3.9 GHz in something like a 22 nanometer process, embedded processors typically run anywhere from 50 MHz to about 1 GHz, partly for power considerations. The original ARM1 was designed to run at 4 MHz (note that this is three orders of magnitude slower) in a 3 micron process! Subsequent revisions to the architecture produced the ARM2, as shown in Figure 1.3. While the processor still had no caches (on-chip, localized memory) or memory management unit (MMU), multiply and multiply-accumulate instructions were added to increase performance, along with a coprocessor interface for use with an external floating-point accelerator. More registers for handling interrupts were added to the architecture, and one of the effective address types was actually removed. This microprocessor achieved a typical clock speed of 12 MHz in a 2 micron process. Acorn used the device in the new Archimedes desktop PC, and VLSI Technology sold the device (called the VL86C010) as part of a processor chip set that also included a memory controller, a video controller, and an I/O controller.

FIGURE 1.3 ARM2 microprocessor.

1.2.2 The Creation of ARM Ltd.

In 1989, the dominant desktop architectures, the 68000 family from Motorola and the x86 family from Intel, were beginning to integrate memory management units, caches, and floating-point units on board the processor, and clock rates were going up—25 MHz in the case of the first 68040. (This is somewhat misleading, as this processor used quadrature clocks, meaning clocks that are derived from overlapping phases of two skewed clocks, so internally it was running at twice that frequency.) To compete in this space, the ARM3 was developed, complete with a 4K unified cache, also running at 25 MHz. By this point, Acorn was struggling with the dominance of the IBM PC in the market, but continued to find sales in education, specialist, and hobbyist markets. VLSI Technology, however, managed to find other companies willing to use the ARM processor in their designs, especially as an embedded processor, and just coincidentally, a company known mostly for its personal computers, Apple, was looking to enter the completely new field of personal digital assistants (PDAs).

Apple's interest in a processor for its new device led to the creation of an entirely separate company to develop it, with Apple and Acorn Group each holding a stake, and Robin Saxby (now Sir Robin Saxby) being appointed as managing director. The new company, consisting of money from Apple, twelve Acorn engineers, and free tools from VLSI Technology, moved into a new building, changed the name of the architecture from Acorn RISC Machine to Advanced RISC Machine, and developed a completely new business model. Rather than selling the processors, Advanced RISC Machines Ltd. would sell the rights to manufacture its processors to other companies, and in 1990, VLSI Technology would become the first licensee. Work began in earnest to produce a design that could act as either a standalone processor or a macrocell for larger designs, where the licensees could then add their own logic to the processor core. After making architectural extensions, the numbering skipped a few beats and moved on to the ARM6 (this was more of a marketing decision than anything else). Like its competition, this processor now included 32-bit addressing and supported both big- and little-endian memory formats. The CPU used by Apple was called the ARM610, complete with the ARM6 core, a 4K cache, a write buffer, and an MMU. Ironically, the Apple PDA (known as the Newton) was slightly ahead of its time and did quite poorly in the market, partly because of its price and partly because of its size. It wouldn't be until the late 1990s that Apple would design a device based on an ARM7 processor that would fundamentally change the way people viewed digital media—the iPod.

The ARM7 processor is where this book begins. Introduced in 1993, the design was used by Acorn for a new line of computers and by Psion for a new line of PDAs, but it still lacked some of the features that would prove to be huge selling points for its successor—the ARM7TDMI, shown in Figure 1.4. While it's difficult to imagine building a system today without the ability to examine the processor's registers, the memory system, your C++ source code, and the state of the processor all in a nice graphical interface, historically, debugging a part was often very difficult and involved adding large amounts of extra hardware to a system. The ARM7TDMI expanded the original ARM7 design to include new hardware specifically for an external debugger (the initials "D" and "I" stood for Debug and ICE, or In-Circuit

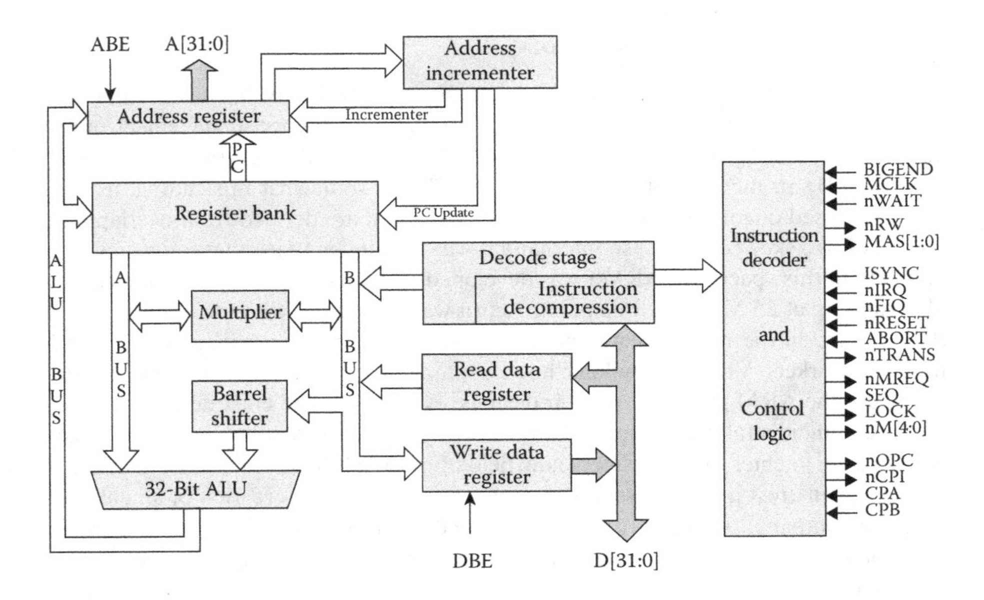

FIGURE 1.4 The ARM7TDMI.

Emulation, respectively), making it much easier and less expensive to build and test a complete system. To increase performance in embedded systems, a new, compressed instruction set was created. Thumb, as it was called, gave software designers the flexibility to either put more code into the same amount of memory or reduce the amount of memory needed for a given design. The burgeoning cell phone industry was quite keen to use this new feature, and consequently began to heavily adopt the ARM7TDMI for use in mobile handsets. The initial "M" reflected a larger hardware multiplier in the datapath of the design, making it suitable for all sorts of digital signal processing (DSP) algorithms. The combination of a small die area, very low power, and rich instruction set made the ARM7TDMI one of ARM's best-selling processors, and despite its age, continues to be used heavily in modern embedded system designs. All of these features have been used and improved upon in subsequent designs.

Throughout the 1990s, ARM continued to make improvements to the architecture, producing the ARM8, ARM9, and ARM10 processor cores, along with derivatives of these cores, and while it's tempting to elaborate on these designs, the discussion could easily fill another textbook. However, it is worth mentioning some highlights of this decade. Around the same time that the ARM9 was being developed, an agreement with Digital Equipment Corporation allowed it to produce its own version of the ARM architecture, called StrongARM, and a second version was slated to be produced alongside the design of the ARM10 (they would be the same processor). Ultimately, DEC sold its design group to Intel, who then decided to continue the architecture on its own under the brand XScale. Intel produced a second version of its design, but has since sold this design to Marvell. Finally, on a corporate note, in 1998 ARM Holdings PLC was floated on the London and New York Stock Exchanges as a publicly traded company.

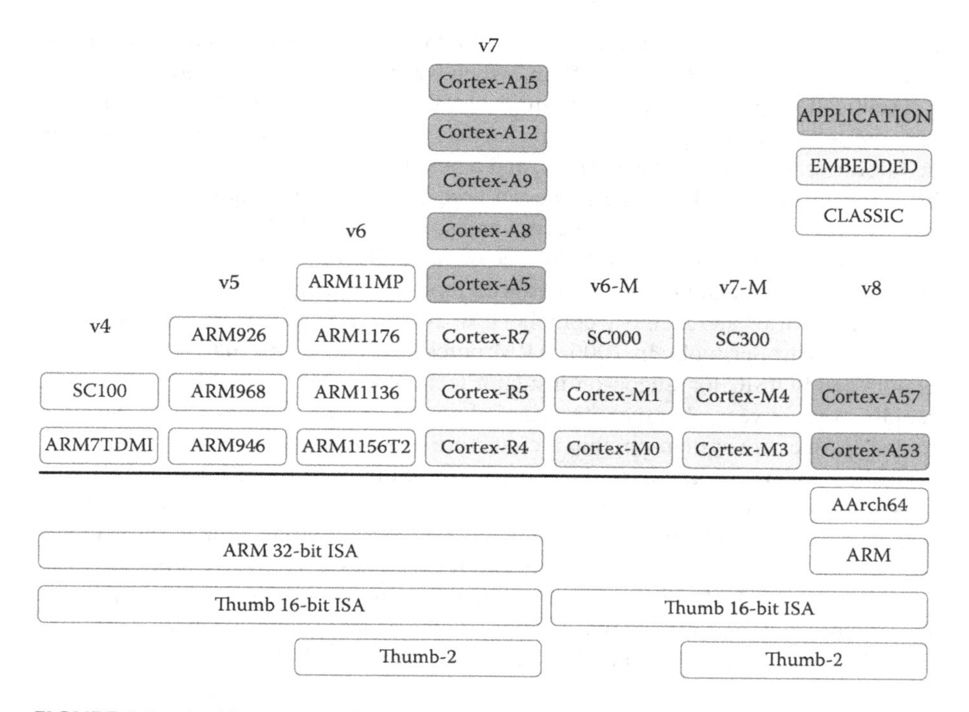

FIGURE 1.5 Architecture versions.

In the early part of the new century, ARM released several new processor lines, including the ARM11 family, the Cortex family, and processors for multi-core and secure applications. The important thing to note about all of these processors, from a programmer's viewpoint anyway, is the version. From Figure 1.5, you can see that while there are many different ARM cores, the version precisely defines the instruction set that each core executes. Other salient features such as the memory architecture, Java support, and floating-point support come mostly from the individual cores. For example, the ARM1136JF-S is a synthesizable processor, one that supports both floating-point and Java in hardware; however, it supports the version 6 instruction set, so while the implementation is based on the ARM11, the instruction set architecture (ISA) dictates which instructions the compiler is allowed to use. The focus of this book is the ARM version 4T and version 7-M instruction sets, but subsequent sets can be learned as needed.

1.2.3 ARM Today

By 2002, there were about 1.3 billion ARM-based devices in myriad products, but mostly in cell phones. By this point, Nokia had emerged as a dominant player in the mobile handset market, and ARM was the processor powering these devices. While TI supplied a large portion of the cellular market's silicon, there were other ARM partners doing the same, including Philips, Analog Devices, LSI Logic,

PrairieComm, and Qualcomm, with the ARM7 as the primary processor in the offerings (except TI's OMAP platform, which was based on the ARM9).

Application Specific Integrated Circuits (ASICs) require more than just a processor core—they require peripheral logic such as timers and USB interfaces, standard cell libraries, graphics engines, DSPs, and a bus structure to tie everything together. To move beyond just designing processor cores, ARM began acquiring other companies focusing on all of these specific areas. In 2003, ARM purchased Adelante Technologies for data engines (DSP processors, in effect). In 2004, ARM purchased Axys Design Automation for new hardware tools and Artisan Components for standard cell libraries and memory compilers. In 2005, ARM purchased Keil Software for microcontroller tools. In 2006, ARM purchased Falanx for 3D graphics accelerators and SOISIC for silicon-on-insulator technology. All in all, ARM grew quite rapidly over six years, but the ultimate goal was to make it easy for silicon partners to design an entire system-on-chip architecture using ARM technology.

Billions of ARM processors have been shipped in everything from digital cameras to smart power meters. In 2012 alone, around 8.7 billion ARM-based chips were created by ARM's partners worldwide. Average consumers probably don't realize how many devices in their pockets and their homes contain ARM-based SoCs, mostly because ARM, like the silicon vendor, does not receive much attention in the finished product. It's unlikely that a Nokia cell phone user thinks much about the fact that TI provided the silicon and that ARM provided part of the design.

1.2.4 The Cortex Family

Due to the radically different requirements of embedded systems, ARM decided to split the processor cores into three distinct families, where the end application now determines both the nature and the design of the processors, but all of them go by the trade name of Cortex. The Cortex-A, Cortex-R, and Cortex-M families continue to add new processors each year, generally based on performance requirements as well as the type of end application the cores are likely to see. A very basic cell phone doesn't have the same throughput requirements as a smartphone or a tablet, so a Cortex-A5 might work just fine, whereas an infotainment system in a car might need the ability to digitally sample and process very large blocks of data, forcing the SoC designer to build a system out of two or four Cortex-A15 processors. The controller in a washing machine wouldn't require a 3 GHz processor that costs eight dollars, so a very lightweight Cortex-M0 solves the problem for around 70 cents. As we explore the older version 4T instructions, which operate seamlessly on even the most advanced Cortex-A and Cortex-R processors, the Cortex-M architecture resembles some of the older microcontrollers in use and requires a bit of explanation, which we'll provide throughout the book.

1.2.4.1 The Cortex-A and Cortex-R Families

The Cortex-A line of cores focuses on high-end applications such as smart phones, tablets, servers, desktop processors, and other products which require significant computational horsepower. These cores generally have large caches, additional arithmetic blocks for graphics and floating-point operations, and memory management units to support large operating systems, such as Linux, Android, and Windows. At the high end of the computing spectrum, these processors are also likely to support systems containing multiple cores, such as those found in servers and wireless base stations, where you may need up to eight processors at once. The 32-bit Cortex-A family includes the Cortex-A5, A7, A8, A9, A12, and A15 cores. Newer, 64-bit architectures include the A57 and A53 processors. In many designs, equipment manufacturers build custom solutions and do not use off-the-shelf SoCs; however, there are quite a few commercial parts from the various silicon vendors, such as Freescale's i.MX line based around the Cortex-A8 and A9; TI's Davinci and Sitara lines based on the ARM9 and Cortex-A8; Atmel's SAMA5D3 products based on the Cortex-A5; and the OMAP and Keystone multi-core solutions from TI based on the Cortex-A15. Most importantly, there are very inexpensive evaluation modules for which students and instructors can write and test code, such as the Beaglebone Black board, which uses the Cortex-A8.

The Cortex-R cores (R4, R5, and R7) are designed for those applications where real-time and/or safety constraints play a major role; for example, imagine an embedded processor designed within an anti-lock brake system for automotive use. When the driver presses on the brake pedal, the system is expected to have completely deterministic behavior—there should be no guessing as to how many cycles it might take for the processor to acknowledge the fact that the brake pedal has been pressed! In complex systems, a simple operation like loading multiple registers can introduce unpredictable delays if the caches are turned on and an interrupt comes in at the just the wrong time. Safety also plays a role when considering what might happen if a processor fails or becomes corrupted in some way, and the solution involves building redundant systems with more than one processor. X-ray machines, CT scanners, pacemakers, and other medical devices might have similar requirements. These cores are also likely to be asked to work with operating systems, large memory systems, and a wide variety of peripherals and interfaces, such as Bluetooth, USB, and Ethernet. Oddly enough, there are only a handful of commercial offerings right now, along with their evaluation platforms, such as TMS570 and RM4 lines from TI.

1.2.4.2 The Cortex-M Family

Finally, the Cortex-M line is targeted specifically at the world of microcontrollers, parts which are so deeply embedded in systems that they often go unnoticed. Within this family are the Cortex-M0, M0+, M1, M3, and M4 cores, which the silicon vendors then take and use to build their own brand of off-the-shelf controllers. As the much older, 8-bit microcontroller space moves into 32-bit processing, for controlling car seats, displays, power monitoring, remote sensors, and industrial robotics, industry requires a variety of microcontrollers that cost very little, use virtually no power, and can be programmed quickly. The Cortex-M family has surfaced as a very popular product with silicon vendors: in 2013, 170 licenses were held by 130 companies, with their parts costing anywhere from two dollars to twenty cents. The Cortex-M0 is the simplest, containing only a core, a nested vectored interrupt controller (NVIC), a bus interface, and basic debug logic. Its tiny size, ultra-low gate count, and small instruction set (only 56 instructions) make it well suited for applications that only require a basic controller. Commercial parts include the LPC1100 line from NXP, and the XMC1000 line from Infineon. The Cortex-M0+ is similar to the M0, with

FIGURE 1.6 Tiva LaunchPad from Texas Instruments.

the addition of a memory protection unit (MPU), a relocatable vector table, a single-cycle I/O interface for faster control, and enhanced debug logic. The Cortex-M1 was designed specifically for FPGA implementations, and contains a core, instruction-side and data-side tightly coupled memory (TCM) interfaces, and some debug logic. For those controller applications that require fast interrupt response times, the ability to process signals quickly, and even the ability to boot a small operating system, the Cortex-M3 contains enough logic to handle such requirements. Like its smaller cousins, the M3 contains an NVIC, MPU, and debug logic, but it has a richer instruction set, an SRAM and peripheral interface, trace capability, a hardware divider, and a single-cycle multiplier array. The Cortex-M4 goes further, including additional instructions for signal processing algorithms; the Cortex-M4 with optional floating-point hardware stretches even further with additional support for single-precision floating-point arithmetic, which we'll examine in Chapters 9, 10, and 11. Some commercial parts offering the Cortex-M4 include the SAM4SD32 controllers from Atmel, the Kinetis family from Freescale, and the Tiva C series from TI, shown in its evaluation module in Figure 1.6.

1.3 THE COMPUTING DEVICE

More definitions are probably in order before we start speaking of processors, programs, and bits. At the most fundamental level, we can look at machines that are given specific instructions or commands through any number of mechanisms—paper tape, switches, or magnetic materials. The machine certainly doesn't have to be electronic to be considered. For example, in 1804 Joseph Marie Jacquard invented a way to weave designs into fabric by controlling the warp and weft threads on a silk loom with cards that had holes punched in them. Those same cards were actually modified (see Figure 1.7) and used in punch cards to feed instructions to electronic computers from the 1960s to the early 1980s. During the process of writing even short programs, these cards would fill up boxes, which were then handed to someone

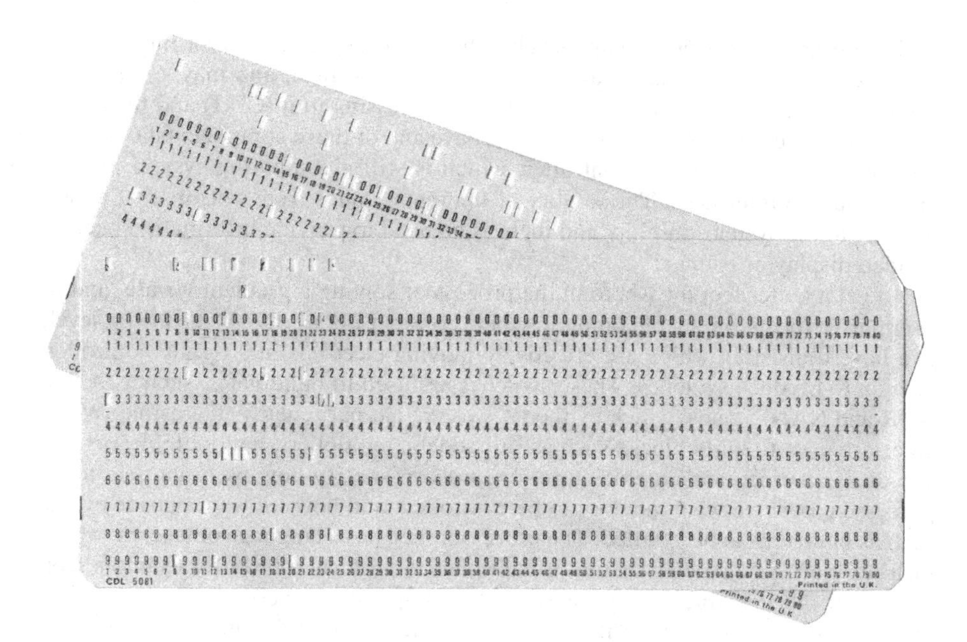

FIGURE 1.7 Hollerith cards.

behind a counter with a card reader. Woe to the person who spent days writing a program using punch cards without numbering them, since a dropped box of cards, all of which looked nearly identical, would force someone to go back and punch a whole new set in the proper order! However the machine gets its instructions, to do any computational work those instructions need to be stored somewhere; otherwise, the user must reload them for each iteration. The stored-program computer, as it is called, fetches a sequence of instructions from memory, along with data to be used for performing calculations. In essence, there are really only a few components to a computer: a processor (something to do the actual work), memory (to hold its instructions and data), and busses to transfer the data and instructions back and forth between the two, as shown in Figure 1.8. Those instructions are the focus of this book—assembly language programming is the use of the most fundamental operations of the processor, written in a way that humans can work with them easily.

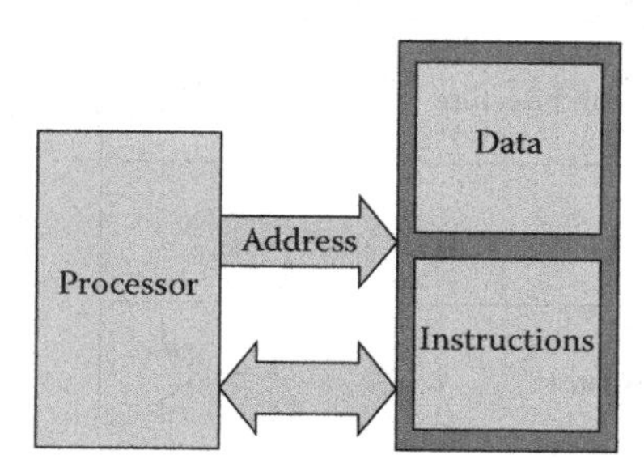

FIGURE 1.8 The stored-program computer model.

The classic model for a computer also shows typical interfaces for input/output (I/O) devices, such as a keyboard, a disk drive for storage, and maybe a printer. These interfaces connect to both the central processing unit (CPU) and the memory; however, embedded systems may not have any of these components! Consider a device such as an engine controller, which is still a computing system, only it has no human interfaces. The totality of the input comes from sensors that attach directly to the system-on-chip, and there is no need to provide information back to a video display or printer.

To get a better feel for where in the process of solving a problem we are, and to summarize the hierarchy of computing then, consider Figure 1.9. At the lowest level, you have transistors which are effectively moving electrons in a tightly controlled fashion to produce switches. These switches are used to build gates, such as AND, NOR and NAND gates, which by themselves are not particularly interesting. When gates are used to build blocks such as full adders, multipliers, and multiplexors, we can create a processor's architecture, i.e., we can specify how we want data to be processed, how we want memory to be controlled, and how we want outside events such as interrupts to be handled. The processor then has a language of its own, which instructs various elements such as a multiplier to perform a task; for example, you might tell the machine to multiply two floating-point numbers together and store the result in a register. We will spend a great deal of time learning this language and seeing the best ways to write assembly code for the ARM architecture. Beyond the scope of what is addressed in this text, certainly you could go to the next levels, where assembly code is created from a higher-level language such as C or C++, and then on to work with operating systems like Android that run tasks or applications when needed.

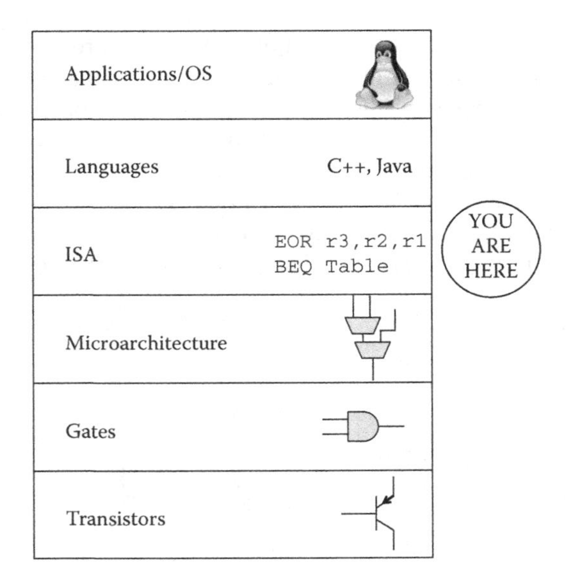

FIGURE 1.9 Hierarchy of computing.

1.4 NUMBER SYSTEMS

Since computers operate internally with transistors acting as switches, the combinational logic used to build adders, multipliers, dividers, etc., understands values of 1 or 0, either on or off. The binary number system, therefore, lends itself to use in computer systems more easily than base ten numbers. Numbers in base two are centered on the idea that each digit now represents a power of two, instead of a power of ten. In base ten, allowable numbers are 0 through 9, so if you were to count the number of sheep in a pasture, you would say 0, 1, 2, 3, 4, 5, 6, 7, 8, 9, and then run out of digits. Therefore, you place a 1 in the 10's position (see Figure 1.10), to indicate you've counted this high already, and begin using the old digits again—10, 11, 12, 13, etc.

Now imagine that you only have two digits with which to count: 0 or 1. To count that same set of sheep, you would say 0, 1 and then you're out of digits. We know the next value is 2 in base ten, but in base two, we place a 1 in the 2's position and keep counting—10, 11, and again we're out of digits to use. A marker is then placed in the 4's position, and we do this as much as we like.

EXAMPLE 1.1

Convert the binary number 110101_2 to decimal.

SOLUTION

This can be seen as

2^5	2^4	2^3	2^2	2^1	2^0
1	1	0	1	0	1

This would be equivalent to $32 + 16 + 4 + 1 = 53_{10}$.

The subscripts are normally only used when the base is not 10. You will see quickly that a number such as 101 normally doesn't raise any questions until you start using computers. At first glance, this is interpreted as a base ten number—one hundred one. However, careless notation could have us looking at this number in base two, so be careful when writing and using numbers in different bases.

After staring at 1's and 0's all day, programming would probably have people jumping out of windows, so better choices for representing numbers are base eight (octal, although you'd be hard pressed to find a machine today that mainly uses octal notation) and base sixteen (hexadecimal or hex, the preferred choice), and here the digits

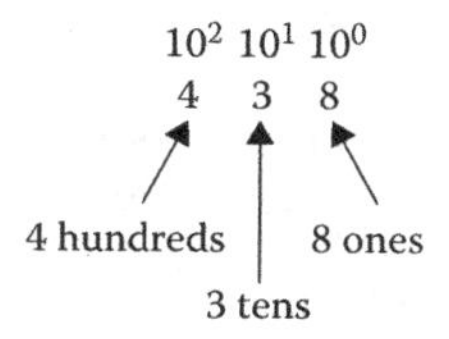

FIGURE 1.10 Base ten representation of 438.

are now a power of sixteen. These numbers pack quite a punch, and are surprisingly big when you convert them to decimal. Since counting in base ten permits the numbers 0 through 9 to indicate the number of 1's, 10's, 100's, etc., in any given position, the numbers 0 through 9 don't go far enough to indicate the number of 1's we have in base sixteen. In other words, to count our sheep in base sixteen using only one digit, we would say 0, 1, 2, 3, 4, 5, 6, 7, 8, 9, and then we can keep going since the next position represents how many 16's we have. So the first six letters of the alphabet are used as placeholders. So after 9, the counting continues—A, B, C, D, E, and then F. Once we've reached F, the next number is 10_{16}.

EXAMPLE 1.2

Find the decimal equivalent of $A5E9_{16}$.

SOLUTION

This hexadecimal number can be viewed as

16^3	16^2	16^1	16^0
A	5	E	9

So our number above would be $(10 \times 16^3) + (5 \times 16^2) + (14 \times 16^1) + (9 \times 16^0) = 42{,}473_{10}$. Notice that it's easier to mentally treat the values A, B, C, D, E, and F as numbers in base ten when doing the conversion.

EXAMPLE 1.3

Calculate the hexadecimal representation for the number 862_{10}.

SOLUTION

While nearly all handheld calculators today have a conversion function for this, it's important that you can do this by hand (this is a very common task in programming). There are tables that help, but the easiest way is to simply evaluate how many times a given power of sixteen can go into your number. Since 16^3 is 4096, there will be none of these in your answer. Therefore, the next highest power is 16^2, which is 256, and there will be

$$862/256 = 3.3672$$

or 3 of them. This leaves

$$862 - (3 \times 256) = 94.$$

The next highest power is 16^1, and this goes into 94 five times with a remainder of 14. Our number in hexadecimal is therefore

16^3	16^2	16^1	16^0
	3	5	E

TABLE 1.1
Binary and Hexadecimal Equivalents

Decimal	Binary	Hexadecimal
0	0000	0
1	0001	1
2	0010	2
3	0011	3
4	0100	4
5	0101	5
6	0110	6
7	0111	7
8	1000	8
9	1001	9
10	1010	A
11	1011	B
12	1100	C
13	1101	D
14	1110	E
15	1111	F

The good news is that conversion between binary and hexadecimal is very easy—just group the binary digits, referred to as bits, into groups of four and convert the four digits into their hexadecimal equivalent. Table 1.1 shows the binary and hexadecimal values for decimal numbers from 0 to 15.

EXAMPLE 1.4

Convert the following binary number into hexadecimal:

$$11011111000010101111_2$$

SOLUTION

By starting at the least significant bit (at the far right) and grouping four bits together at a time, the first digit would be F_{16}, as shown below.

$$1101111100001010\underbrace{1111}_{F_{16}}{}_2$$

The second group of four bits would then be 1010_2 or A_{16}, etc., giving us

$$DF0AF_{16}.$$

One comment about notation—you might see hexadecimal numbers displayed as 0xFFEE or &FFEE (depending on what's allowed by the software development tools you are using), and binary numbers displayed as 2_1101 or b1101.

1.5 REPRESENTATIONS OF NUMBERS AND CHARACTERS

All numbers and characters are simply bit patterns to a computer. It's unfortunate that something inside microprocessors cannot interpret a programmer's meaning, since this could have saved countless hours of debugging and billions of dollars in equipment. Programmers have been known to be the cause of lost space probes, mostly because the processor did exactly what the software told it to do. When you say 0x6E, the machine sees 0x6E, and that's about it. This could be a character (a lowercase "n"), the number 110 in base ten, or even a fractional value! We're going to come back to this idea over and over—computers have to be told how to treat all types of data. The programmer is ultimately responsible for interpreting the results that a processor provides and making it clear in the code. In these next three sections, we'll examine ways to represent integer numbers, floating-point numbers, and characters, and then see another way to represent fractions in Chapter 7.

1.5.1 INTEGER REPRESENTATIONS

For basic mathematical operations, it's not only important to be able to represent numbers accurately but also use as few bits as possible, since memory would be wasted to include redundant or unnecessary bits. Integers are often represented in byte (8-bit), halfword (16-bit), and word (32-bit) quantities. They can be longer depending on their use, e.g., a cryptography routine may require 128-bit integers.

Unsigned representations make the assumption that every bit signifies a positive contribution to the value of the number. For example, if the hexadecimal number 0xFE000004 were held in a register or in memory, and assuming we treat this as an unsigned number, it would have the decimal value

$$(15 \times 16^7) + (14 \times 16^6) + (4 \times 16^0) = 4{,}261{,}412{,}868.$$

Signed representations make the assumption that the most significant bit is used to create positive and negative values, and they come in three flavors: sign-magnitude, one's complement and two's complement.

Sign-magnitude is the easiest to understand, where the most significant bit in the number represents a sign bit and all other bits represent the magnitude of the number. A one in the sign bit indicates the number is negative and a zero indicates it is positive.

EXAMPLE 1.5

The numbers –18 and 25 are represented in 16 bits as

$$-18 = 1000000000010010$$
$$25 = 0000000000011001$$

To add these two numbers, it's first necessary to determine which number has the larger magnitude, and then the smaller number would be subtracted from it. The sign would be the sign of the larger number, in this case a zero. Fortunately, sign-magnitude representations are not used that much, mostly because their use implies making comparisons first, and this adds extra instructions in code just to perform basic math.

One's complement numbers are not used much in modern computing systems either, mostly because there is too much extra work necessary to perform basic arithmetic operations. To create a negative value in this representation, simply invert all the bits of its positive, binary value. The sign bit will be a 1, just like sign-magnitude representations, but there are two issues that arise when working with these numbers. The first is that you end up with two representations for 0, and the second is that it may be necessary to adjust a sum when adding two values together, causing extra work for the processor. Consider the following two examples.

EXAMPLE 1.6

Assuming that you have 16 bits to represent a number, add the values –124 to 236 in one's complement notation.

SOLUTION

To create –124 in one's complement, simply write out the binary representation for 124, and then invert all the bits:

```
 124    0000000001111100
–124    1111111110000011
```

Adding 236 gives us

```
–124          1111111110000011
+236         +0000000011101100
----         -----------------
    carry → 1 0000000001101111
```

The problem is that the answer is actually 112, or 0x70 in hex. In one's complement notation, a carry in the most significant bit forces us to add a one back into the sum, which is one extra step:

```
              0000000001101111
 + 1     +                   1
 ---      --------------------
 112          0000000001110000
```

EXAMPLE 1.7

Add the values –8 and 8 together in one's complement, assuming 8 bits are available to represent the numbers.

SOLUTION

Again, simply take the binary representation of the positive value and invert all the bits to get –8:

	8	00001000
+	−8	11110111
	0	11111111

Since there was no carry from the most significant bit, this means that 00000000 and 11111111 both represent zero. Having a +0 and a −0 means extra work for software, especially if you're testing for a zero result, leading us to the use of two's complement representations and avoiding this whole problem.

Two's complement representations are easier to work with, but it's important to interpret them correctly. As with the other two signed representations, the most significant bit represents the sign bit. However, in two's complement, the most significant bit is *weighted*, which means that it has the same magnitude as if the bit were in an unsigned representation. For example, if you have 8 bits to represent an unsigned number, then the most significant bit would have the value of 2^7, or 128. If you have 8 bits to represent a two's complement number, then the most significant bit represents the value −128. A base ten number n can be represented as an m-bit two's complement number, with b being an individual bit's value, as

$$n = -b_{m-1}2^{m-1} + \sum_{i=0}^{m-2} b_i 2^i$$

To interpret this more simply, the most significant bit can be thought of as the only negative component to the number, and all the other bits represent positive components. As an example, −114 represented as an 8-bit, two's complement number is

$$10001110_2 = -2^7 + 2^3 + 2^2 + 2^1 = -114.$$

Notice in the above calculation that the only negative value was the most significant bit. Make no mistake—you must be told in advance that this number is treated as a two's complement number; otherwise, it could just be the number 142 in decimal.

The two's complement representation provides a range of positive and negative values for a given number of bits. For example, the number 8 could not be represented in only 4 bits, since 1000_2 sets the most significant bit, and the value is now interpreted as a negative number (−8, in this case). Table 1.2 shows the range of values produced for certain bit lengths, using ARM definitions for halfword, word, and double word lengths.

EXAMPLE 1.8

Convert −9 to a two's complement representation in 8 bits.

SOLUTION

Since 9 is 1001_2, the 8-bit representation of −9 would be

00001001	9
11110110	−9 in one's complement
+ 1	
11110111	−9 in two's complement

TABLE 1.2
Two's Complement Integer Ranges

Length	Number of Bits	Range
	m	-2^{m-1} to $2^{m-1}-1$
Byte	8	–128 to 127
Halfword	16	–32,768 to 32,767
Word	32	–2,147,483,648 to 2,147,483,647
Double word	64	-2^{64} to $2^{64}-1$

Note: To calculate the two's complement representation of a negative number, simply take its magnitude, convert it to binary, invert all the bits, and then add 1.

Arithmetic operations now work as expected, without having to adjust any final values. To convert a two's complement binary number back into decimal, you can either subtract one and then invert all the bits, which in this case is the fastest way, or you can view it as -2^7 plus the sum of the remaining weighted bit values, i.e.,

$$-2^7 + 2^6 + 2^5 + 2^4 + 2^2 + 2^1 + 2^0 =$$
$$-128 + 119 =$$
$$-9$$

EXAMPLE 1.9

Add the value –384 to 2903 using 16-bit, two's complement arithmetic.

Solution

First, convert the two values to their two's complement representations:

$$384 = 0000000110000000_2$$
$$-384 = 1111111001111111_2 \quad +1 = \quad 1111111010000000_2$$
$$+2903 = \qquad + 0000101101010111_2$$
$$2519 \qquad 0000100111010111_2$$

1.5.2 Floating-Point Representations

In many applications, values larger than 2,147,483,647 may be needed, but you still have only 32 bits to represent numbers. Very large and very small values can be constructed by using a floating-point representation. While the format itself has a long history to it, with many varieties of it appearing in computers over the years, the IEEE 754 specification of 1985 (Standards Committee 1985) formally defined a 32-bit data type called single-precision, which we'll cover extensively in Chapter 9. These floating-point numbers consist of an exponent, a fraction, a sign bit, and a bias.

For "normal" numbers, and here "normal" is defined in the specification, the value of a single-precision number F is given as

$$F = -1^s \times 1.f \times 2^{e-b}$$

where s is the sign bit, and f is the fraction made up of the lower 23 bits of the format. The most significant fraction bit has the value 0.5, the next bit has the value 0.25, and so on. To ensure all exponents are positive numbers, a bias b is added to the exponent e. For single-precision numbers, the exponent bias is 127.

While the range of an unsigned, 32-bit integer is 0 to 2^{32}-1 (4.3×10^9), the positive range of a single-precision floating-point number, also represented in 32 bits, is 1.2×10^{-38} to $3.4 \times 10^{+38}$! Note that this is only the positive range; the negative range is congruent. The amazing range is a trade-off, actually. Floating-point numbers trade accuracy for range, since the delta between representable numbers gets larger as the exponent gets larger. Integer formats have a fixed precision (each increment is equal to a fixed value).

EXAMPLE 1.10

Represent the number 1.5 in a single-precision, floating-point format.
We would form the value as

$s = 0$ (a positive number)
$f =$ 100 0000 0000 0000 0000 0000 (23 fraction bits representing 0.5)
$e = 0 + 127$ (8 bits of true exponent plus the bias)

$F =$ 0 0111111 100 0000 0000 0000 0000 0000

or 0x3FC00000, as shown in Figure 1.11.

The large dynamic range of floating-point representations has made it popular for scientific and engineering computing. While we've only seen the single-precision format, the IEEE 754 standard also specifies a 64-bit, double-precision format that has a range of $\pm 2.2 \times 10^{-308}$ to $1.8 \times 10^{+308}$! Table 1.3 shows what two of the most common formats specified in the IEEE standard look like (single- and double-precision). Single precision provides typically 6–9 digits of numerical precision, while double precision gives 15–17.

Special hardware is required to handle numbers in these formats. Historically, floating-point units were separate ICs that were attached to the main processor, e.g., the Intel 80387 for the 80386 and the Motorola 68881 for the 68000. Eventually these were integrated onto the same die as the processor, but at a cost. Floating-point units are often quite large, typically as large as the rest of the processor without caches and other memories. In most applications, floating-point computations are rare and

<table>
<tr><td>S</td><td>E</td><td>E</td><td>E</td><td>E</td><td>E</td><td>E</td><td>E</td><td>E</td><td>F</td><td>F</td><td>F</td><td>F</td><td>F</td><td>F</td><td>F</td><td>F</td><td>F</td><td>F</td><td>F</td><td>F</td><td>F</td><td>F</td><td>F</td><td>F</td><td>F</td><td>F</td><td>F</td><td>F</td><td>F</td><td>F</td><td>F</td></tr>
<tr><td>0</td><td>0</td><td>1</td><td>1</td><td>1</td><td>1</td><td>1</td><td>1</td><td>1</td><td>1</td><td>0</td><td>0</td><td>0</td><td>0</td><td>0</td><td>0</td><td>0</td><td>0</td><td>0</td><td>0</td><td>0</td><td>0</td><td>0</td><td>0</td><td>0</td><td>0</td><td>0</td><td>0</td><td>0</td><td>0</td><td>0</td><td>0</td></tr>
<tr><td colspan="4">3</td><td colspan="4">F</td><td colspan="4">C</td><td colspan="4">0</td><td colspan="4">0</td><td colspan="4">0</td><td colspan="4">0</td><td colspan="4">0</td></tr>
</table>

FIGURE 1.11 Formation of 1.5 in single-precision.

TABLE 1.3
IEEE 754 Single- and Double-Precision Formats

Format	Single Precision	Double Precision
Format width in bits	32	64
Exponent width in bits	8	11
Fraction bits	23	52
Exp maximum	+127	+1023
Exp minimum	–126	–1022
Exponent bias	127	1023

not speed-critical. For these reasons, most microcontrollers do not include specialized floating-point hardware; instead, they use software routines to emulate floating-point operations. There is actually another format that can be used when working with real values, which is a fixed-point format; it doesn't require a special block of hardware to implement, but it does require careful programming practices and often complicated error and bounds analysis. Fixed-point formats will be covered in great detail in Chapter 7.

1.5.3 Character Representations

Bit patterns can represent numbers or characters, and the interpretation is based entirely on context. For example, the binary pattern 01000001 could be the number 65 in an audio codec routine, or it could be the letter "A". The program determines how the pattern is used and interpreted. Fortunately, standards for encoding character data were established long ago, such as the American Standard Code for Information Interchange, or ASCII, where each letter or control character is mapped to a binary value. Other standards include the Extended Binary-Coded-Decimal Interchange Code (EBCDIC) and Baudot, but the most commonly used today is ASCII. The ASCII table for character codes can be found in Appendix C.

While most devices may only need the basic characters, such as letters, numbers, and punctuation marks, there are some control characters that can be interpreted by the device. For example, old teletype machines used to have a bell that rang in a Pavlovian fashion, alerting the user that something exciting was about to happen. The control character to ring the bell is 0x07. Other control characters include a backspace (0x08), a carriage return (0x0D), a line feed (0x0A), and a delete character (0x7F), all of which are still commonly used.

Using character data in assembly language is not difficult, and most assemblers will let you use a character in the program without having to look up the equivalent hexadecimal value in a table. For example, instead of saying

```
MOV    r0, #0x42; move a 'B' into register r0
```

you can simply say

```
MOV    r0, #'B'; move a 'B' into register r0
```

Character data will be seen throughout the book, so it's worth spending a little time becoming familiar with the hexadecimal equivalents of the alphabet.

1.6 TRANSLATING BITS TO COMMANDS

All processors are programmed with a set of instructions, which are unique patterns of bits, or 1's and 0's. Each set is unique to that particular processor. These instructions might tell the processor to add two numbers together, move data from one place to another, or sit quietly until something wakes it up, like a key being pressed. A processor from Intel, such as the Pentium 4, has a set of bit patterns that are completely different from a SPARC processor or an ARM926EJ-S processor. However, all instruction sets have some common operations, and learning one instruction set will help you understand nearly any of them.

The instructions themselves can be of different lengths, depending on the processor architecture—8, 16, or 32 bits long, or even a combination of these. For our studies, the instructions are either 16 or 32 bits long; although, much later on, we'll examine how the ARM processors can use some shorter, 16-bit instructions in combination with 32-bit Thumb-2 instructions. Reading and writing a string of 1's and 0's can give you a headache rather quickly, so to aid in programming, a particular bit pattern is mapped onto an instruction name, or a *mnemonic*, so that instead of reading

```
E0CC31B0
1AFFFFF1
E3A0D008
```

the programmer can read

```
STRH      sum, [pointer], #16
BNE       loop_one
MOV       count, #8
```

which makes a little more sense, once you become familiar with the instructions themselves.

EXAMPLE 1.11

Consider the bit pattern for the instruction above:

```
MOV    count, #8
```

The pattern is the hex number 0xE3A0D008. From Figure 1.12, you can see that the ARM processor expects parts of our instruction in certain fields—the

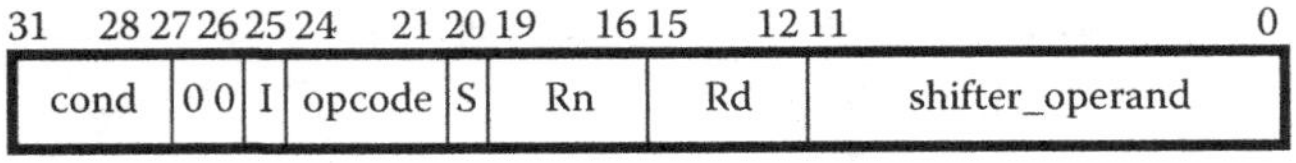

31 28	27 26	25	24 21	20	19 16	15 12	11 0
cond	0 0	I	opcode	S	Rn	Rd	shifter_operand

FIGURE 1.12 The MOV instruction.

number 8, for example, would be placed in the field called 8_bit_immediate, and the instruction itself, moving a number into a register, is encoded in the field called opcode. The parameter called count is a convenience that allows the programmer to use names instead of register numbers. So somewhere in our program, count is assigned to a real register and that register number is encoded into the field called Rd. We will see the uses of MOV again in Chapter 6.

Most mnemonics are just a few letters long, such as B for branch or ADD for, well, add. Microprocessor designers usually try and make the mnemonics as clear as possible, but every once in a while you come across something like RSCNE (from ARM), DCBZ (from IBM) or the really unpronounceable AOBLSS (from DEC) and you just have to look it up. Despite the occasionally obtuse name, it is still much easier to remember RSCNE than its binary or hex equivalent, as it would make programming nearly impossible if you had to remember each command's pattern. We could do this mapping or translation by hand, taking the individual mnemonics, looking them up in a table, then writing out the corresponding bit pattern of 1's and 0's, but this would take hours, and the likelihood of an error is very high. Therefore, we rely on tools to do this mapping for us.

To complicate matters, reading assembly language commands is not always trivial, even for advanced programmers. Consider the sequence of mnemonics for the IA-32 architecture from Intel:

```
mov         eax, DWORD PTR _c
add         eax, DWORD PTR _b
mov         DWORD PTR _a, eax
cmp         DWORD PTR _a, 4
```

This is actually seen as pretty typical code, really—not so exotic. Even more intimidating to a new learner, mnemonics for the ColdFire microprocessor look like this:

```
mov.l       (0,%a2,%d2.l*4),%d6
mac.w       %d6:u,%d5:u<<1
mac.w       %d6:l,%d3:u<<1
```

Where an experienced programmer would immediately recognize these commands as just variations on basic operations, along with extra characters to make the software tools happy, someone just learning assembly would probably close the book and go home. The message here is that coding in assembly language takes practice and time to learn. Each processor's instruction set looks different, and tools sometimes force a few changes to the syntax, producing something like the above code; however, nearly all assembly formats follow some basic rules. We will begin learning assembly using ARM instructions, which are very readable.

1.7 THE TOOLS

At some point in the history of computing, it became easier to work with high-level languages instead of coding in 1's and 0's, or machine code, and programmers described loops and variables using statements and symbols. The earlier languages include COBOL, FORTRAN, ALGOL, Forth, and Ada. FORTRAN was required knowledge for an undergraduate electrical engineering student in the 1970s and 1980s, and that has largely been replaced with C, C++, Java, and even Python. All of these languages still have one thing in common: they all contain near-English descriptions of code that are then translated into the native instruction set of the microprocessor. The program that does this translation is called a *compiler*, and while compilers get more and more sophisticated, their basic purpose remains the same, taking something like an "if…then" statement and converting it into assembly language. Modern systems are programmed in high-level languages much of the time to allow code portability and to reduce design time.

As with most programming tasks, we also need an automated way of translating our assembly language instructions into bit patterns, and this is precisely what an *assembler* does, producing a file that a piece of hardware (or a software simulator) can understand, in machine code using only 1's and 0's. To help out even further, we can give the assembler some pseudo-instructions, or directives (either in the code or with options in the tools), that tell it how to do its job, provided that we follow the particular assembler's rules such as spacing, syntax, the use of certain markers like commas, etc. If you follow the tools flow in Figure 1.13, you can see that an *object file* is produced by the assembler from our *source file,* or the file that contains our assembly language program. Note that a compiler will also use a source file, but the code might be C or C++. Object files differ from executable files in that they often contain debugging information, such as program symbols (names of variables and functions) for linking or debugging, and are usually used to build a larger executable. Object files, which can be produced in different formats, also contain relocation

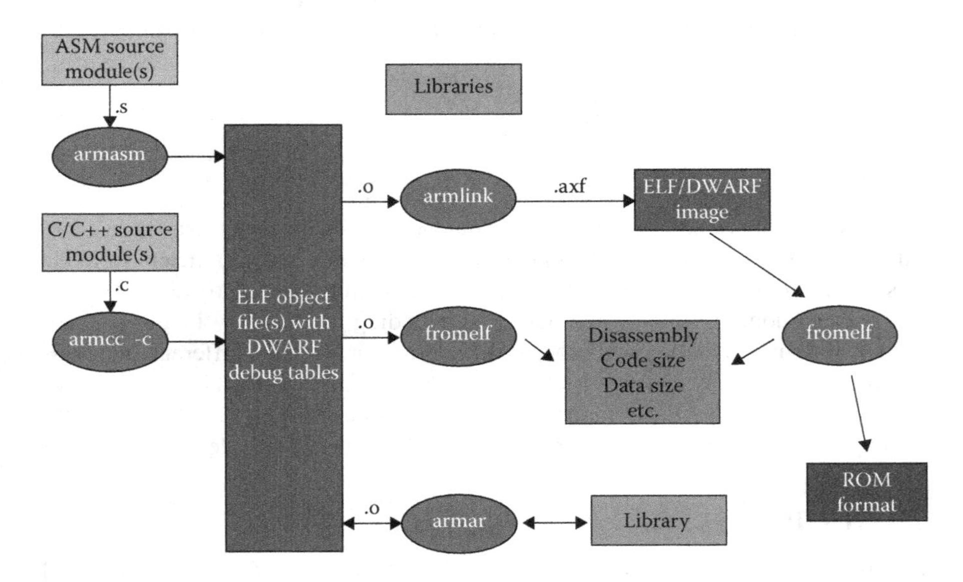

FIGURE 1.13 Tools flow.

information. Once you've assembled your source files, a *linker* can then be used to combine them into an executable program, even including other object files, say from customized libraries. Under test conditions, you might choose to run these files in a debugger (as we'll do for the majority of examples), but usually these executables are run by hardware in the final embedded application. The *debugger* provides access to registers on the chip, views of memory, and the ability to set and clear breakpoints and watchpoints, which are methods of stopping the processor on instruction or memory accesses, respectively. It also provides views of code in both high-level languages and assembly.

1.7.1 Open Source Tools

Many students and professors steer clear of commercial software simply to avoid licensing issues; most software companies don't make it a policy to give away their tools, but there are non-profits that do provide free toolchains. Linaro, a not-for-profit engineering organization, focuses on optimizing open source software for the ARM architecture, including the GCC toolchain and the Linux kernel, and providing regular releases of the various tools and operating systems. You can find downloads on their website (www.linaro.org). What they define as "bare-metal" builds for the tools can also be found if you intend on working with gcc (the gnu compiler) and gdb (the gnu debugger). Clicking on the links take you to prebuilt gnu toolchains for Cortex-M and Cortex-R controllers located at https://launchpad.net/gcc-arm-embedded. There are dozens of other open source sites for ARM tools, found with a quick Web search.

1.7.2 Keil (ARM)

ARM's C and C++ compilers generate optimized code for all of the instruction sets, ARM, Thumb, and Thumb-2, and support full ISO standard C and C++. Modern tool sets, like ARM's RealView Microcontroller Development Kit (RVMDK), which is found at http://www.keil.com/demo, can display both the high-level code and its assembly language equivalent together on the screen, as shown in Figure 1.14. Students have found that the Keil tools are relatively easy to use, and they support hundreds of popular microcontrollers. A limitation appears when a larger microprocessor, such as a Cortex-A9, is used in a project, since the Keil tools are designed specifically for microcontrollers. Otherwise, the tools provide everything that is needed:

- C and C++ compilers
- Macro assembler
- Linker
- True integrated source-level debugger with a high-speed CPU and peripheral simulator for popular ARM-based microcontrollers
- µVision4 Integrated Development Environment (IDE), which includes a full-featured source code editor, a project manager for creating and maintaining projects, and an integrated make facility for assembling, compiling, and linking embedded applications

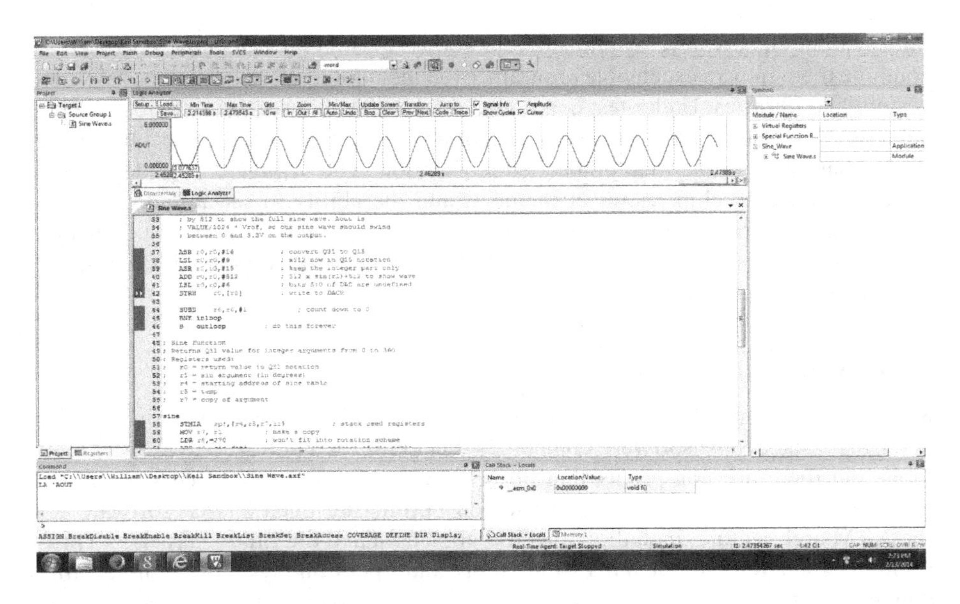

FIGURE 1.14 Keil simulation tools.

- Execution profiler and performance analyzer
- File conversion utility (to convert an executable file to a HEX file, for example)
- Links to development tools manuals, device datasheets, and user's guides

It turns out that you don't always choose either a high-level language or assembly language for a particular project—sometimes, you do both. Before you progress through the book, read the *Getting Started User's Guide* in the RVMDK tools' documentation.

1.7.3 Code Composer Studio

Texas Instruments has a long history of building ARM-based products, and as a leading supplier, makes their own tools. Code Composer Studio (CCS) actually supports all of their product lines, not just ARM processors. As a result, they include some rather nice features, such as a free operating system (SYS/BIOS), in their tool suite. The CCS tools support microcontrollers, e.g., the Cortex-M4 products, as well as very large SoCs like those in their Davinci and Sitara lines, so there is some advantage in starting with a more comprehensive software package, provided that you are aware of the learning curve associated with it. The front end to the tools is based on the Eclipse open source software framework, shown in Figure 1.15, so if you have used another development tool for Java or C++ based on Eclipse, the CCS tools might look familiar. Briefly, the CCS tools include:

- Compilers for each of TI's device families
- Source code editor

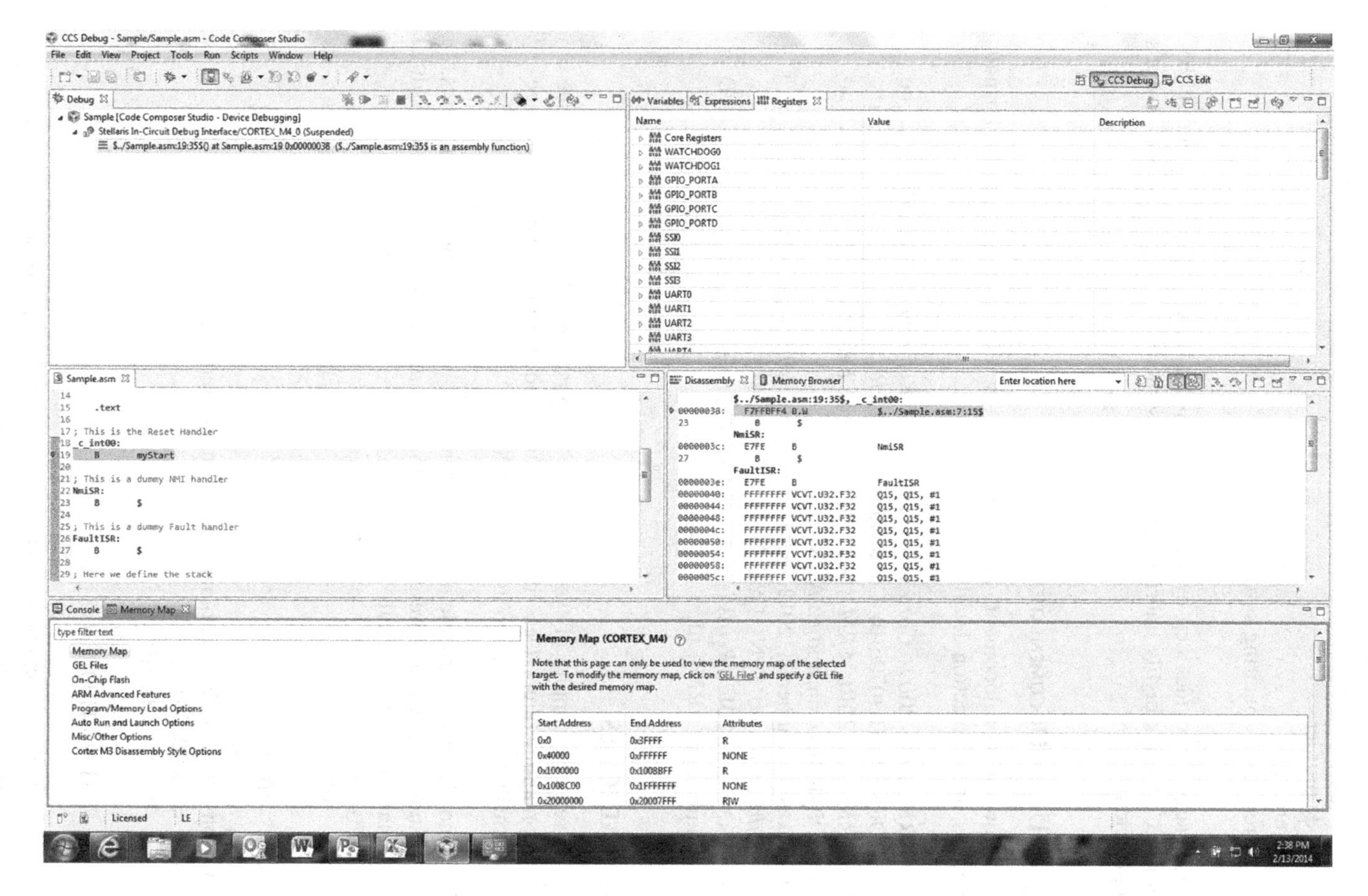

FIGURE 1.15 Code Composer Studio development tools.

- Project build environment
- Debugger
- Code profiler
- Simulators
- A real-time operating system

Appendix A provides step-by-step instructions for running a small assembly program in CCS—it's highly unorthodox and not something done in industry, but it's simple and it works!

1.7.4 Useful Documentation

The following free documents are likely to be used often for looking at formats, examples, and instruction details:

- ARM Ltd. 2009. *Cortex-M4 Technical Reference Manual.* Doc. no. DDI0439C (ID070610). Cambridge: ARM Ltd.
- ARM Ltd. 2010. *ARM v7-M Architectural Reference Manual.* Doc. no. DDI0403D. Cambridge: ARM Ltd.
- Texas Instruments. 2012. *ARM Assembly Language Tools v5.0 User's Guide.* Doc. no. SPNU118K. Dallas: Texas Instruments.
- ARM Ltd. 2012. *RealView Assembler User Guide* (online), Revision D. Cambridge: ARM Ltd.

1.8 EXERCISES

1. Give two examples of system-on-chip designs available from semiconductor manufacturers. Describe their features and interfaces. They do not necessarily have to contain an ARM processor.

2. Find the two's complement representation for the following numbers, assuming they are represented as a 16-bit number. Write the value in both binary and hexadecimal.
   a. –93
   b. 1034
   c. 492
   d. –1094

3. Convert the following binary values into hexadecimal:
   a. 10001010101111
   b. 10101110000110
   c. 1011101010111110
   d. 1111101011001110

4. Write the 8-bit representation of –14 in one's complement, two's complement, and sign-magnitude representations.

5. Convert the following hexadecimal values to base ten:
   a. 0xFE98
   b. 0xFEED
   c. 0xB00
   d. 0xDEAF

6. Convert the following base ten numbers to base four:
   a. 812
   b. 101
   c. 96
   d. 3640

7. Using the smallest data size possible, either a byte, a halfword (16 bits), or a word (32 bits), convert the following values into two's complement representations:
   a. –18,304
   b. –20
   c. 114
   d. –128

8. Indicate whether each value could be represented by a byte, a halfword, or a word-length two's complement representation:
   a. –32,765
   b. 254
   c. –1,000,000
   d. –128

9. Using the information from the *ARM v7-M Architectural Reference Manual*, write out the 16-bit binary value for the instruction `SMULBB r5, r4, r3`.

10. Describe all the ways of interpreting the hexadecimal number 0xE1A02081 (hint: it might not be data).

11. If the hexadecimal value 0xFFE3 is a two's complement, halfword value, what would it be in base ten? What if it were a word-length value (i.e., 32 bits long)?

12. How do you think you could quickly compute values in octal (base eight) given a value in binary?

13. Convert the following decimal numbers into hexadecimal:
    a. 256
    b. 1000
    c. 4095
    d. 42

14. Write the 32-bit representation of –247 in sign-magnitude, one's complement, and two's complement notations. Write the answer using 8 hex digits.

15. Write the binary pattern for the letter "Q" using the ASCII representation.

16. Multiply the following binary values. Notice that binary multiplication works exactly like decimal multiplication, except you are either adding 0 to the final product or a scaled multiplicand. For example:

$$
\begin{array}{rrl}
 & 100 & \text{(multiplicand)} \\
\times & \underline{110} & \text{(multiplier)} \\
 & 0 & \\
 & 1000 & \text{(scaled multiplicand – by 2)} \\
 & \underline{10000} & \text{(scaled multiplicand – by 4)} \\
 & 11000 &
\end{array}
$$

a. $\begin{array}{r} 1100 \\ \times\ 1111 \\ \hline \end{array}$

b. $\begin{array}{r} 1010 \\ \times\ 1011 \\ \hline \end{array}$

c. $\begin{array}{r} 1000 \\ \times\ 1001 \\ \hline \end{array}$

d. $\begin{array}{r} 11100 \\ \times\ \ 111 \\ \hline \end{array}$

17. How many bits would the following C data types use by the ARM7TDMI?
   a. int
   b. long
   c. char
   d. short
   e. long long

18. Write the decimal number 1.75 in the IEEE single-precision floating-point format. Use one of the tools given in the References to check your answer.

2 The Programmer's Model

2.1 INTRODUCTION

All microprocessors have a set of features that programmers use. In most instances, a programmer will not need an understanding of how the processor is actually constructed, meaning that the wires, transistors, and/or logic boards that were used to build the machine are not typically known. From a programmer's perspective, what *is* necessary is a model of the device, something that describes not only the way the processor is controlled but also the features available to you from a high level, such as where data can be stored, what happens when you give the machine an invalid instruction, where your registers are stacked during an exception, and so forth. This description is called the programmer's model. We'll begin by examining the basic parts of the ARM7TDMI and Cortex-M4 programmer's models, but come back to certain elements of them again in Chapters 8, 13, 14, and 15, where we cover branching, stacks, and exceptions in more detail. For now, a brief treatment of the topic will provide some definition, just enough to let us begin writing programs.

2.2 DATA TYPES

Data in machines is represented as *b*inary dig*its*, or *bits*, where one binary digit can be seen as either on or off, a one or a zero. A collection of bits are often grouped together into units of eight, called bytes, or larger units whose sizes depend on the maker of the device, oddly enough. For example, a 16-bit data value for a processor such as the Intel 8086 or MC68040 is called a word, where a 32-bit data value is a word for the ARM cores. When describing both instructions and data, normally the length is factored in, so that we often speak of 16-bit instructions or 32-bit instructions, 8-bit data or 16-bit data, etc. Specifically for data, the ARM7TDMI and Cortex-M4 processors support the following data types:

Byte, or 8 bits
Halfword, or 16 bits
Word, or 32 bits

For the moment, the length of the instructions is immaterial, but we'll see later than they can be either 16 or 32 bits long, so you will need two bytes to create a Thumb instruction and four bytes to create either an ARM instruction or a Thumb-2 instruction. For the ARM7TDMI, when reading or writing data, halfwords must be aligned to two-byte boundaries, which means that the address in memory must end in an even number. Words must be aligned to four-byte boundaries, i.e., addresses ending in 0, 4, 8, or C. The Cortex-M4 allows unaligned accesses under certain conditions, so it is actually possible to read or write a word of data located at an odd address. Don't worry, we'll cover memory accesses in much more detail when we get to addressing modes in Chapter 5. Most data operations, e.g., ADD, are performed on word quantities, but we'll also work with smaller, 16-bit values later on.

2.3 ARM7TDMI

The motivation behind examining an older programmer's model is to show its similarity to the more advanced cores—the Cortex-A and Cortex-R processors, for example, look very much like the ARM7TDMI, only with myriad new features and more modes, but everything here applies. Even though the ARM7TDMI appears simple (only three stages in its pipeline) when compared against the brobdingnagian Cortex-A15 (highly out-of-order pipeline with fifteen stages), there are still enough details to warrant a more cautious introduction to modes and exceptions, omitting some details for now. It is also noteworthy to point out features that are common to all ARM processors but differ by number, use, and limitations, for example, the size of the integer register file on the Cortex-M4. The registers look and act the same as those on an ARM7TDMI, but there are just fewer of them. Our tour of the programmer's model starts with the processor modes.

2.3.1 Processor Modes

Version 4T cores support seven processor modes: User, FIQ, IRQ, Supervisor, Abort, Undefined, and System, as shown in Figure 2.1. It is possible to make mode changes under software control, but most are normally caused by external conditions or exceptions. Most application programs will execute in User mode. The other modes are known as privileged modes, and they provide a way to service exceptions or to access protected resources, such as bits that disable sections of the core, e.g., a branch predictor or the caches, should the processor have either of these.

	Mode	Description	
Exception modes	Supervisor (SVC)	Entered on reset and when a Software Interrupt (SWI) instruction is executed	Privileged modes
	FIQ	Entered when a high priority (fast) interrupt is raised	
	IRQ	Entered when a low priority (normal) interrupt is raised	
	Abort	Used to handle memory access violations	
	Undef	Used to handle undefined instructions	
	System	Privileged mode using the same registers as User mode	
	User	Mode under which most applications/OS tasks run	Unprivileged mode

FIGURE 2.1 Processor modes.

A simple way to look at this is to view a mode as an indication of what the processor is actually doing. Under normal circumstances, the machine will probably be in either User mode or Supervisor mode, happily executing code. Consider a device such as a cell phone, where not much happens (aside from polling) until either a signal comes in or the user has pressed a key. Until that time, the processor has probably powered itself down to some degree, waiting for an event to wake it again, and these external events could be seen as *interrupts*. Processors generally have differing numbers of interrupts, but the ARM7TDMI has two types: a fast interrupt and a lower priority interrupt. Consequently, there are two modes to reflect activities around them: FIQ mode and IRQ mode. Think of the fast interrupt as one that might be used to indicate that the machine is about to lose power in a few milliseconds! Lower priority interrupts might be used for indicating that a peripheral needs to be serviced, a user has touched a screen, or a mouse has been moved.

Abort mode allows the processor to recover from exceptional conditions such as a memory access to an address that doesn't physically exist, for either an instruction or data. This mode can also be used to support virtual memory systems, often a requirement of operating systems such as Linux. The processor will switch to Undefined mode when it sees an instruction in the pipeline that it does not recognize; it is now the programmer's (or the operating system's) responsibility to determine how the machine should recover from such as error. Historically, this mode could be used to support valid floating-point instructions on machines without actual floating-point hardware; however, modern systems rarely rely on Undefined mode for such support, if at all. For the most part, our efforts will focus on working in either User mode or Supervisor mode, with special attention paid to interrupts and other exceptions in Chapter 14.

2.3.2 Registers

The register is the most fundamental storage area on the chip. You can put most anything you like in one—data values, such as a timer value, a counter, or a coefficient for an FIR filter; or addresses, such as the address of a list, a table, or a stack in memory. Some registers are used for specific purposes. The ARM7TDMI processor has a total of 37 registers, shown in Figure 2.2. They include

- 30 general-purpose registers, i.e., registers which can hold any value
- 6 status registers
- A Program Counter register

The general-purpose registers are 32 bits wide, and are named r0, r1, etc. The registers are arranged in partially overlapping banks, meaning that you as a programmer see a different register bank for each processor mode. This is a source of confusion sometimes, but it shouldn't be. At any one time, 15 general-purpose registers (r0 to r14), one or two status registers, and the Program Counter (PC or r15) are visible. You always call the registers the same thing, but depending on which mode you are in, you are simply looking at different registers. Looking at Figure 2.2, you

Mode					
User/System	Supervisor	Abort	Undefined	Interrupt	Fast interrupt
R0	R0	R0	R0	R0	R0
R1	R1	R1	R1	R1	R1
R2	R2	R2	R2	R2	R2
R3	R3	R3	R3	R3	R3
R4	R4	R4	R4	R4	R4
R5	R5	R5	R5	R5	R5
R6	R6	R6	R6	R6	R6
R7	R7	R7	R7	R7	R7
R8	R8	R8	R8	R8	R8_FIQ
R9	R9	R9	R9	R9	R9_FIQ
R10	R10	R10	R10	R10	R10_FIQ
R11	R11	R11	R11	R11	R11_FIQ
R12	R12	R12	R12	R12	R12_FIQ
R13	R13_SVC	R13_ABORT	R13_UNDEF	R13_IRQ	R13_FIQ
R14	R14_SVC	R14_ABORT	R14_UNDEF	R14_IRQ	R14_FIQ
PC	PC	PC	PC	PC	PC
CPSR	CPSR	CPSR	CPSR	CPSR	CPSR
	SPSR_SVC	SPSR_ABORT	SPSR_UNDEF	SPSR_IRQ	SPSR_FIQ

= *banked register*

FIGURE 2.2 Register organization.

can see that in User/System mode, you have registers r0 to r14, a Program Counter, and a Current Program Status Register (CPSR) available to you. If the processor were to suddenly change to Abort mode for whatever reason, it would swap, or bank out, registers r13 and r14 with different r13 and r14 registers. Notice that the largest number of registers swapped occurs when the processor changes to FIQ mode. The reason becomes apparent when you consider what the processor is trying to do very quickly: save the state of the machine. During an interrupt, it is normally necessary to drop everything you're doing and begin to work on one task: namely, saving the state of the machine and transition to handling the interrupt code quickly. Rather than moving data from all the registers on the processor to external memory, the machine simply swaps certain registers with new ones to allow the programmer access to fresh registers. This may seem a bit unusual until we come to the chapter on exception handling. The banked registers are shaded in the diagram.

While most of the registers can be used for any purpose, there are a few registers that are normally reserved for special uses. Register r13 (the stack pointer or SP) holds the address of the stack in memory, and a unique stack pointer exists in each mode (except System mode which shares the User mode stack pointer). We'll examine this register much more in Chapter 13. Register r14 (the Link Register or LR) is

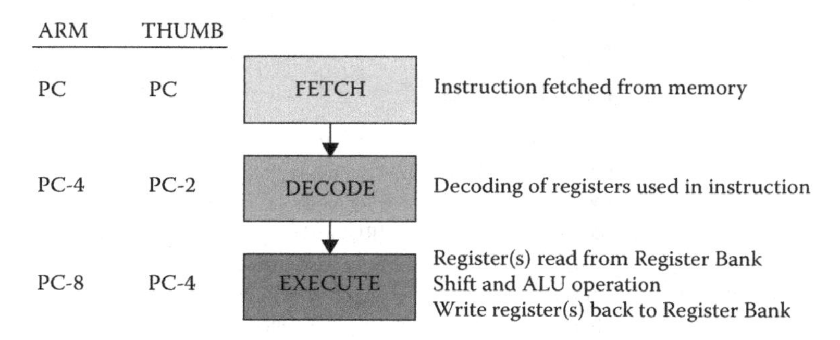

FIGURE 2.3 ARM7TDMI pipeline diagram.

used to hold subroutine and exception return addresses. As with the stack pointers, a unique r14 exists in all modes (except System mode which shares the User mode r14). In Chapters 8 and 13, we will begin to work with branches and subroutines, and this register will hold the address to which we need to return should our program jump to a small routine or a new address in memory. Register r15 holds the Program Counter (PC). The ARM7TDMI is a pipelined architecture, as shown in Figure 2.3, meaning that while one instruction is being fetched, another is being decoded, and yet another is being executed. The address of the instruction that is being *fetched* (not the one being executed) is contained in the Program Counter. This register is not normally accessed by the programmer unless certain specific actions are needed, such as jumping long distances in memory or recovering from an exception. You can read a thorough treatment of pipelined architectures in Patterson and Hennessy (2007).

The Current Program Status Register (CPSR) can be seen as the state of the machine, allowing programs to recover from exceptions or branch on the results of an operation. It contains condition code flags, interrupt enable flags, the current mode, and the current state (more on the differences between ARM and Thumb state is discussed in Chapter 17). Each privileged mode (except System mode) has a Saved Program Status Register (SPSR) that is used to preserve the value of the CPSR when an exception occurs. Since User mode and System mode are not entered on any exception, they do not have an SPSR, and a register to preserve the CPSR is not required. In User mode or System mode, if you attempt to read the SPSR, you will get an unpredictable value back, meaning the data cannot be used in any further operations. If you attempt to write to the SPSR in one of these modes, the data will be ignored.

The format of the Current Program Status Register and the Saved Program Status Register is shown in Figure 2.4. You can see that it contains four bits at the top,

31	30	29	28	27 26 25 24 23 22 21 20 19 18 17 16 15 14 13 12 11 10 9 8	7	6	5	4	3	2	1	0
N	Z	C	V	Do not modify/Read as zero	I	F	T	M 4	M 3	M 2	M 1	M 0

FIGURE 2.4 Format of the program status registers.

TABLE 2.1
The Mode Bits

xPSR[4:0]	Mode
10000	User mode
10001	FIQ mode
10010	IRQ mode
10011	Supervisor mode
10111	Abort mode
11011	Undefined mode
11111	System mode

collectively known as the condition code flags, and eight bits at the bottom. The condition code flags in the CPSR can be altered by arithmetic and logical instructions, such as subtractions, logical shifts, and rotations. Furthermore, by allowing these bits to be used with all the instructions on the ARM7TDMI, the processor can conditionally execute an instruction, providing improvements in code density and speed. Conditional execution and branching are covered in detail in Chapter 8.

The bottom eight bits of a status register (the mode bits M[4:0], I, F, and T) are known as the control bits. The I and F bits are the interrupt disable bits, which disable interrupts in the processor if they are set. The I bit controls the IRQ interrupts, and the F bit controls the FIQ interrupts. The T bit is a status bit, meant only to indicate the state of the machine, so as a programmer you would only read this bit, not write to it. If the bit is set to 1, the core is executing Thumb code, which consists of 16-bit instructions. The processor changes between ARM and Thumb state via a special instruction that we'll examine much later on. Note that these control bits can be altered by software only when the processor is in a privileged mode.

Table 2.1 shows the interpretation of the least significant bits in the PSRs, which determine the mode in which the processor operates. Note that while there are five bits that determine the processor's mode, not all of the configurations are valid (there's a historical reason behind this). If any value not listed here is programmed into the mode bits, the result is unpredictable, which by ARM's definition means that the fields do not contain valid data, and a value may vary from moment to moment, instruction to instruction, and implementation to implementation.

2.3.3 The Vector Table

There is one last component of the programmer's model that is common in nearly all processors—the vector table, shown in Table 2.2. While it is presented here for reference, there is actually only one part of it that's needed for the introductory work in the next few chapters. The exception vector table consists of designated addresses in external memory that hold information necessary to handle an exception, an interrupt, or other atypical event such as a reset. For example, when an interrupt (IRQ) comes along, the processor will change the Program Counter to 0x18 and begin fetching instructions from there. The data values that are located at these addresses

TABLE 2.2
ARM7TDMI Exception Vectors

Exception Type	Mode	Vector Address
Reset	SVC	0x00000000
Undefined instruction	UNDEF	0x00000004
Software Interrupt (SVC)	SVC	0x00000008
Prefetch abort (instruction fetch memory abort)	ABORT	0x0000000C
Data abort (data access memory abort)	ABORT	0x00000010
IRQ (interrupt)	IRQ	0x00000018
FIQ (fast interrupt)	FIQ	0x0000001C

are actual ARM instructions, so the next instruction that the machine will likely fetch is a branch (B) instruction, assuming the programmer put such an instruction at address 0x18. Once this branch instruction is executed, the processor will begin fetching instructions for the interrupt handler that resides at the target address, also specified with the branch instruction, somewhere in memory. It is worth noting here that many processors, including the Cortex-M4, have *addresses* at these vector locations in memory. The ARM7TDMI processor places *instructions* here. You can use the fact that instructions reside at these vectors for a clever shortcut, but it will have to wait until Chapter 14.

The one exception vector with which we do need to concern ourselves before writing some code is the Reset exception vector, which is at address 0x0 in memory. Since the machine will fetch from this address immediately as it comes out of reset, we either need to provide a reset exception handler (to provide an initialization routine for turning on parts of the device and setting bits the way we like) or we can begin coding at this address, assuming we have a rather unusual system with no errors, exceptions, or interrupts. Many modern development tools provide a startup file for specific microcontrollers, complete with startup code, initialization routines, exception vector assignments, etc., so that when we begin programming, the first instruction in your code isn't *really* the first instruction the machine executes. However, to concentrate on the simpler instructions, we will bend the rules a bit and ignore exceptional conditions for the time being.

2.4 CORTEX-M4

The Cortex-M family differs significantly from earlier ARM designs, but the programmer's model is remarkably similar. The cores are very small. They may only implement a subset of instructions. The memory models are relatively simple. In some ways the Cortex-M3 and M4 processors resemble much older microcontrollers used in the 1970s and 1980s, and the nod to these earlier designs is justified by the markets that they target. These cores are designed to be used in applications that require 32-bit processors to achieve high code density, fast interrupt response times, and now even the ability to handle signal processing algorithms, but the final product produced by silicon vendors may cost only a few dollars. The line between the world of microcontrollers and the world of high-end microprocessors is beginning to blur a bit, as we see features like IEEE floating-point units, real-time operating system support, and advanced trace capabilities in an inexpensive device like the Tiva microcontrollers from TI. There is no substitute for actually writing code, so for now, we will learn enough detail of the programmer's model to bring the processor out of reset, play with some of the registers in the Cortex-M4 and its floating-point unit, and then stop a simulation. Again, we begin with the processor modes.

2.4.1 Processor Modes

The Cortex-M4 has only two modes: Handler mode and Thread mode. As shown in Figure 2.5, there are also two access levels to go along with the modes, Privileged and User, and depending on what the system is doing, it will switch between the two using a bit in the CONTROL register. For very simple applications, the processor may only stay in a single access level—there might not be any User-level code running at all. In situations where you have an embedded operating system, such as SYS/BIOS controlling everything, security may play a role by partitioning the kernel's stack memory from any user stack memory to avoid problems. In Chapter 15, we will examine the way the Cortex-M4 handles exceptions more closely.

2.4.2 Registers

There appear to be far fewer physical registers on a Cortex-M4 than an ARM7TDMI, as shown in Figure 2.6, but the same 16 registers appear as those in User mode on the ARM7TDMI. If you have a Cortex-M4 that includes a floating-point unit, there are actually more. Excluding peripherals, the Cortex-M4 with floating-point hardware contains the following registers as part of the programmer's model:

- 17 general purpose registers, i.e., registers than can hold any value
- A status register than can be viewed in its entirety or in three specialized views

	Privileged	User
Handler mode	Use: Exception handling Stack: Main	
Thread mode	Use: Applications Stack: Main or Process	Use: Applications Stack: Main or Process

FIGURE 2.5 Cortex-M4 modes.

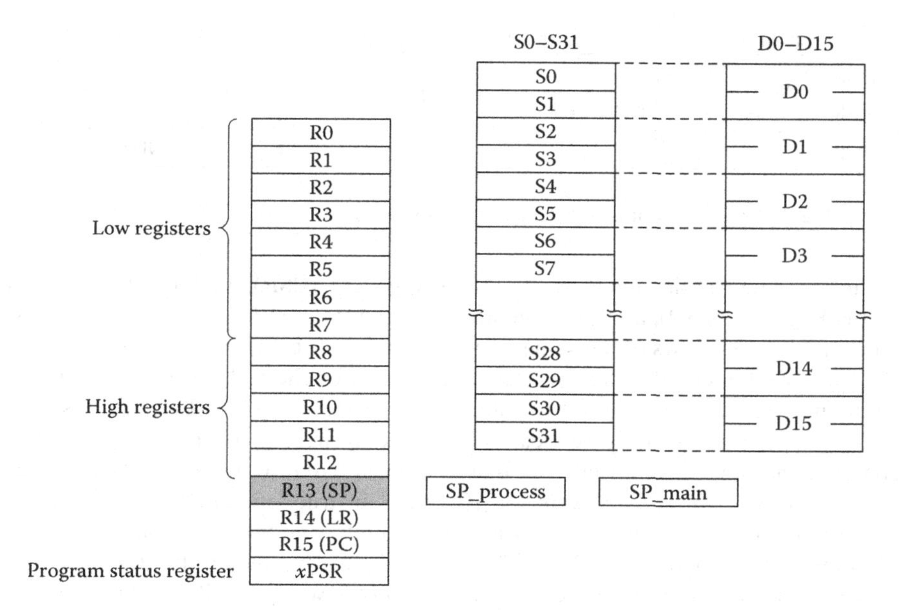

FIGURE 2.6 Cortex-M4 with floating-point register organization.

- 3 interrupt mask registers
- A control register
- 32 single-precision floating-point registers (s0–s31) *or* 16 double-precision registers (d0–d15) *or* a mix
- 4 floating-point control registers (although these are memory-mapped, not physical registers)

As described in the previous section, registers r0 through r12 are general purpose registers, and the registers hold 32-bit values that can be anything you like—addresses, data, packed data, fractional data values, anything. There are some special purpose registers, such as register r13, the stack pointer (and there are two of them, giving you the ability to work with separate stacks); register r14, the Link Register; and register r15, which is the Program Counter. Like the ARM7TDMI, register r13 (the stack pointer or SP) holds the address of the stack in memory, only there are just two of them in the Cortex-M4, the Main Stack Pointer (MSP) and the Process Stack Pointer (PSP). We'll examine these registers much more in Chapter 15. Register r14 (the Link Register or LR) is used to hold subroutine and exception return addresses. Unlike the ARM7TDMI, there is only one Link Register. Register r15, the Program Counter or PC, points to the instruction being fetched, but due to pipelining, there are enough corner cases to make hard and fast rules about its value difficult, so details can be safely tabled for now.

The Program Status Register, or xPSR, performs the same function that the ARM7TDMI's CPSR does, but with different fields. The entire register can be accessed all at once, or you can examine it in three different ways, as shown in

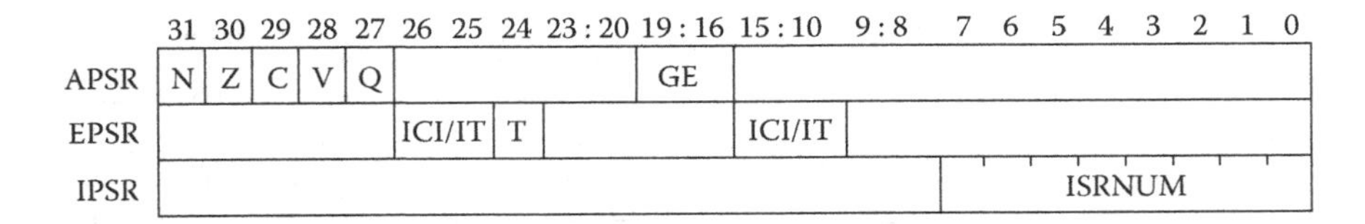

FIGURE 2.7 Program status registers on the Cortex-M4.

Figure 2.7. The Application Program Status Register (APSR), the Interrupt Program Status Register (IPSR), and the Execution Program Status Register (EPSR) are just three specialized views of the same register. The APSR contains the status flags (N, C, V, and Z), the Greater Than or Equal flags (used by the SEL instruction), and an additional "sticky" Q flag used in saturation arithmetic, where sticky in this case means that the bit can only be cleared by explicitly writing a zero to it. The IPSR contains only an exception number that is used in handling faults and other types of exceptions. Two fields contain the IF-THEN instruction status bits overlapped with the Interrupt-Continuable Instruction (ICI) bits, and when combined with the Thumb (T) bit, produce the EPSR. The IF-THEN instruction will be seen when we begin loops and conditional execution in Chapter 8; however, the ICI/IT bits are used for recovering from exceptions, which will not be covered. See the *ARM Cortex-M4 Devices Generic User Guide* (ARM 2010b) for more details.

The interrupt mask registers, PRIMASK, FAULTMASK, and BASEPRI are use to mask certain types of interrupts and exceptions. PRIMASK and FAULTMASK are actually just single-bit registers. BASEPRI can be up to eight bits wide, and the value contained in this register sets the priority level of allowable interrupts that the processor will acknowledge. In Chapter 15, we'll see examples of interrupt handling, but for more complex interrupt situations, see Yiu (2014), where the use of interrupt mask registers is illustrated in more detail.

The last special purpose register is the CONTROL register, which consists of only three bits. The least significant bit, CONTROL[0], changes the access level while in Thread mode to either a Privileged or User level. The next most significant bit, CONTROL[1], selects which stack the processor is to use, either the Main Stack Pointer (MSP) or the Process Stack Pointer (PSP). The most significant bit, CONTROL[2], indicates whether or not to preserve the floating-point state during exception processing. We'll work with this register a bit more in Chapter 15.

2.4.3 The Vector Table

The Cortex-M4 vector table is probably one of the larger departures from all previous ARM processor designs. Returning to the idea that *addresses* are stored in the vector table, rather than instructions, the Cortex-M model looks very much like older microcontrollers such as the 8051 and MC6800 in this respect. From Table 2.3, you can see how the various exception types have their own type number and address in memory. An important point here, not normally too prominent if you are coding in C, since a compiler will take care of this issue for you, is that the least significant bit of these exception vectors (addresses) should be set to a 1. When we cover instructions

TABLE 2.3
Cortex-M4 Exception Vectors

Exception Type	Exception No.	Vector Address
(Top of Stack)	—	0x00000000
Reset	1	0x00000004
NMI	2	0x00000008
Hard fault	3	0x0000000C
Memory management fault	4	0x00000010
Bus fault	5	0x00000014
Usage fault	6	0x00000018
SVcall	11	0x0000002C
Debug monitor	12	0x00000030
PendSV	14	0x00000038
SysTick	15	0x0000003C
Interrupts	16 and above	0x00000040 and above

over the next few chapters, we'll discover that the Cortex-M4 only executes Thumb-2 instructions, rather than ARM instructions as the ARM7TDMI does, and the protocol requires it. This vector table is relocatable after the processor comes out of reset; however, our focus for now is to write short blocks of code without any exceptions or errors, covering procedural details first and worrying about all of the variations later.

2.5 EXERCISES

1. How many modes does the ARM7TDMI processor have? How many states does it have? How many modes does the Cortex-M4 have?

2. What do you think would happen if the instruction SMULTT (an instruction that runs fine on a Cortex-M4) were issued to an ARM7TDMI? Which mode do you think it would be in after this instruction entered the execute stage of its pipeline?

3. What is the standard use of register r14? Register r13? Register r15?

4. On an ARM7TDMI, in any given mode, how many registers does a programmer see at one time?

5. Which bits of the ARM7TDMI status registers contain the flags? Which register on the Cortex-M4 holds the status flags?

6. If an ARM7TDMI processor encounters an undefined instruction, from what address will it begin fetching instructions after it changes to Undefined mode? What about a reset?

7. What is the purpose of FIQ mode?

8. Which mode on an ARM7TDMI can assist in supporting operating systems, especially for supporting virtual memory systems?

9. How do you enable interrupts on the ARM7TDMI?

10. How many stages does the ARM7TDMI pipeline have? Name them.

11. Suppose that the Program Counter, register r15, contained the hex value 0x8000. From what address would an ARM7TDMI fetch an instruction (assume you are in ARM state)?

12. What is the function of the Saved Program Status Register?

13. On an ARM7TDMI, is it permitted to put the instruction

    ```
    SUB r0, r2, r3
    ```

    at address 0x4? How about at address 0x0? Can you put that same bit pattern at address 0x4 in a system using a Cortex-M4?

14. Describe the exception vector table for any other microprocessor. How does it differ from the ARM7TDMI processor? How does it differ from the Cortex-M4?

15. Give an example of an instruction that would typically be placed at address 0x0 on an ARM7TDMI. What value is typically placed at address 0x0 on a Cortex-M4?

16. Explain the current program state of an ARM7TDMI if the CPSR had the value 0xF00000D3.

3 Introduction to Instruction Sets

v4T and v7-M

3.1 INTRODUCTION

This chapter introduces basic program structure and a few easy instructions to show how directives and code create an assembly program. What *are* directives? How is the code stored in memory? What *is* memory? It's unfortunate that the tools and the mechanics of writing assembly have to be learned simultaneously. Without software tools, the best assembly ever written is virtually useless, difficult to simulate in your head, and even harder to debug. You might find reading sections with unfamiliar instructions while using new tools akin to learning to swim by being thrown into a pool. It is. However, after going through the exercise of running a short block of code, the remaining chapters take time to look at all of the details: directives, memory, arithmetic, and putting it all together. This chapter is meant to provide a gentle introduction to the concepts behind, and rules for writing, assembly programs.

First, we need tools. While the ideas behind assemblers haven't changed over the years, the way that programmers work with an assembler has, in that command-line assemblers aren't really the first tool that you want to use. Integrated Development Environments (IDEs) have made learning assembly much easier, as the assembler can be driven graphically. Gone are the days of having paper tape as an input to the machine, punch cards have been relegated to museums, and errors are reported in milliseconds instead of hours. More importantly, the countless options available with command-line assemblers are difficult to remember, so our introduction starts the easy way, graphically. Graphical user interfaces display not only the source code, but memory, registers, flags, the binary listings, and assembler output all at once. Tools such as the Keil MDK and Code Composer Studio will set up most of the essential parameters for us.

If you haven't already installed and familiarized yourself with the tools you plan to use, you should do so now. By using tools that support integrated development, such as those from Keil, ARM, IAR, and Texas Instruments, you can enter, assemble, and test your code all in the same environment. Refer to Appendices A and B for instructions on creating new projects and running the code samples in the book. You may also choose to use other tools, either open-source (like gnu) or commercial, but note there might be subtle changes to the syntax presented throughout this book, and you will want to consult your software's documentation for those details. Either way, today's tools are vastly more helpful than those used 20 years ago—no clumsy breakpoint exceptions are needed; debugging aids are already provided; and everything is visual!

3.2 ARM, THUMB, AND THUMB-2 INSTRUCTIONS

There is no clean way to avoid the subject of instruction length once you begin writing code, since the instructions chosen for your program will depend on the processor. Even more daunting, there are options on the length of the instruction—you can choose a 32-bit instruction or let the assembler optimize it for you if a smaller one exists. So some background on the instructions themselves will guide us in making sense of these differences. ARM instructions are 32 bits wide, and they were the first to be used on older architectures such as the ARM7TDMI, ARM9, ARM10, and ARM11. Thumb instructions, which are a subset of ARM instructions, also work on 32-bit data; however, they are 16 bits wide. For example, adding two 32-bit numbers together can be done one of two ways:

```
ARM instruction      ADD    r0, r0, r2
Thumb instruction    ADD    r0, r2
```

The first example takes registers r0 and r2, adds them together, then stores the result back in register r0. The data contained in those registers as well as the ARM instruction itself is 32 bits wide. The second example does the exact same thing, only the *instruction* is 16 bits wide. Notice there are only two operands in the second example, so one of the operands, register r0, acts as both the source and destination of the data. Thumb instructions are supported in older processors such as the ARM7TDMI, ARM9, and ARM11, and all of the Cortex-A and Cortex-R families.

Thumb-2 is a superset of Thumb instructions, including new 32-bit instructions for more complex operations. In other words, Thumb-2 is a combination of both 16-bit and 32-bit instructions. Generally, it is left to the compiler or assembler to choose the optimal size, but a programmer can force the issue if necessary. Some cores, such as the Cortex-M3 and M4, *only* execute Thumb-2 instructions—there are no ARM instructions at all. The good news is that Thumb-2 code looks very similar to ARM code, so the Cortex-M4 examples below resemble those for the ARM7TDMI, allowing us to concentrate more on getting code to actually run. In Chapter 17, Thumb and Thumb-2 are discussed in detail, especially in the context of optimizing code, but for now, only a few basic operations will be needed.

3.3 PROGRAM 1: SHIFTING DATA

Finally, we get around to writing up and describing a real, albeit small, program using a few simple instructions, some directives, and the tools to watch everything in action. The code below takes a simple value (0x11), loads it into a register, and then shifts it one bit to the left, twice. The code could be written identically for either the Cortex-M4 or an ARM7TDMI, but we'll look at the first example using only the ARM7TDMI using Keil directives, shown below.

```
        AREA Prog1, CODE, READONLY
        ENTRY

        MOV     r0, #0x11       ; load initial value
        LSL     r1, r0, #1      ; shift 1 bit left
        LSL     r2, r1, #1      ; shift 1 bit left

stop    B       stop            ; stop program
        END
```

For the assembler to create a block of code, we need the AREA declaration, along with the type of data we have—in this case, we are creating instructions, not just data (hence the CODE option), and we specify the block to be read-only. Since all programs need at least one ENTRY declaration, we place it in the only file that we have, with the only section of code that we have. The only other directive we have for the assembler in this file is the END statement, which is needed to tell the assembler there are no further instructions beyond the B (branch) instruction.

For most of the instructions (there are a few exceptions), the general format is

instruction *destination*, *source*, *source*

with data going from the source to the destination. Our first MOV instruction has register r0 as its destination register, with an immediate value, a hex number, as the source operand. We'll find throughout the book that instructions have a variety of source types, including numbers, registers, registers with a shift or rotate, etc. The MOV command is normally used to shuffle data from one *register* to another *register*. It is not used to load data from external memory into a register, and we will see that there are dedicated load and store instructions for doing that. The LSL instruction takes the value in register r0, shifts it one bit to the left, and moves the result to register r1. In Chapter 6, we will look at the datapaths of the ARM7TDMI and the Cortex-M4 in more detail, but for now, note that we can also modify other instructions for performing simple shifts, such as an ADD, using two registers as the source operands in the instruction, and then providing a shift count. The second LSL instruction is the same as the first, shifting the value of register r1 one bit to the left and moving the result to register r2. We expect to have the values 0x11, 0x22, and 0x44 in registers r0, r1, and r2, respectively, after the program completes.

The last instruction in the program tells the processor to branch to the branch instruction itself, which puts the code into an infinite loop. This is hardly a graceful exit from a program, but for the purpose of trying out code, it allows us to terminate the simulation easily by choosing Start/Stop Debug Session from the Debug menu or clicking the Halt button in our tools.

3.3.1 Running the Code

Learning assembly requires an adventurous programmer, so you should try each code sample (and write your own). The best way to hone your skills is to assemble and run these short routines, study their effects on registers and memory, and make improvements as needed. Following the examples provided in Appendices A and B, create a project and a new assembly file. You may wish to choose a simple microcontroller, such as the LPC2104 from NXP, as your ARM7TDMI target, and the TM4C1233H6PM from TI as your Cortex-M4 target (NB: this part is listed as LM4F120H5QR in the Keil tools). Once you've started the debugger, you can single-step through the code, executing one instruction at a time until you come to the last instruction (the branch). You may also wish to view the assembly listing as it appears in memory. If you're using the MDK tools, choose Disassembly Window from the View menu, and your code will appear as in Figure 3.1. You can see the mnemonics in the sample program alongside their equivalent binary representations. Code Composer Studio has a similar Disassembly window, found in its View menu.

Recall from Chapter 1 that a stored program computer holds instructions in memory, and in this first exercise for the ARM7TDMI, memory begins at address 0x00000000 and the last instruction of our program can be found at address 0x0000000C. Notice that the branch instruction at this address has been changed, and that our label called stop has been replaced with its numerical equivalent, so that the line reads

```
0x0000000C  EAFFFFFE  B    0x0000000C
```

The label `stop` in this case is the address of the B instruction, which is 0x0000000C. In Chapter 8, we'll explore how branches work in detail, but it's worth noting here that the mnemonic has been translated into the binary number 0xEAFFFFFE. Referring to Figure 3.2 we can see that a 32-bit (ARM) branch instruction consists of four bits to indicate the instruction itself, bits 24 to 27, along with twenty-four bits to be used as an offset. When a program uses the B instruction

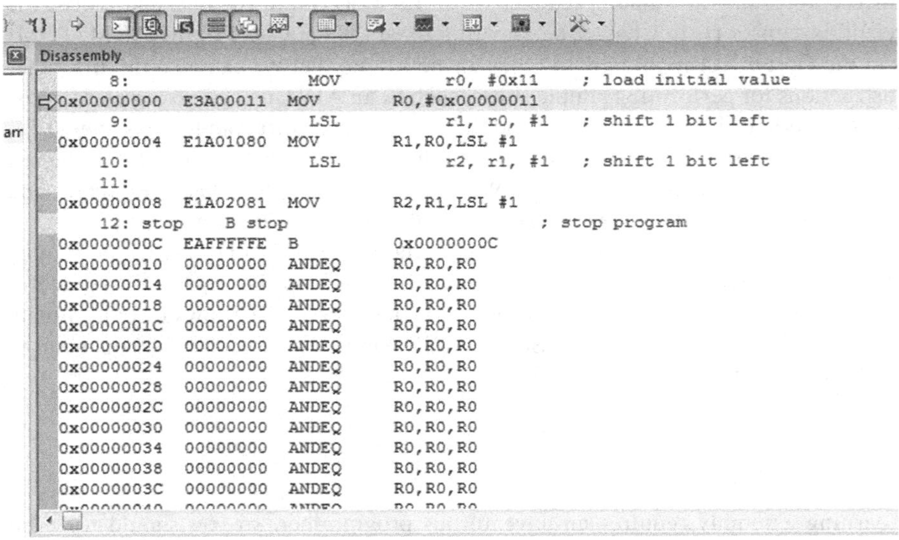

```
Disassembly
     8:                    MOV         r0, #0x11      ; load initial value
0x00000000  E3A00011  MOV         R0,#0x00000011
     9:                    LSL         r1, r0, #1     ; shift 1 bit left
0x00000004  E1A01080  MOV         R1,R0,LSL #1
    10:                    LSL         r2, r1, #1     ; shift 1 bit left
    11:
0x00000008  E1A02081  MOV         R2,R1,LSL #1
    12: stop     B stop                               ; stop program
0x0000000C  EAFFFFFE  B           0x0000000C
0x00000010  00000000  ANDEQ       R0,R0,R0
0x00000014  00000000  ANDEQ       R0,R0,R0
0x00000018  00000000  ANDEQ       R0,R0,R0
0x0000001C  00000000  ANDEQ       R0,R0,R0
0x00000020  00000000  ANDEQ       R0,R0,R0
0x00000024  00000000  ANDEQ       R0,R0,R0
0x00000028  00000000  ANDEQ       R0,R0,R0
0x0000002C  00000000  ANDEQ       R0,R0,R0
0x00000030  00000000  ANDEQ       R0,R0,R0
0x00000034  00000000  ANDEQ       R0,R0,R0
0x00000038  00000000  ANDEQ       R0,R0,R0
0x0000003C  00000000  ANDEQ       R0,R0,R0
```

FIGURE 3.1 Disassembly window.

31 28	27 26 25	24	23 0
cond	1 0 1	L	24_bit_signed_offset

FIGURE 3.2 Bit pattern for a branch instruction.

to jump or branch to some new place in memory, it uses the Program Counter to create an address. For our case, the Program Counter contains the value 0x00000014 when the branch instruction is in the execute stage of the ARM7TDMI's pipeline. Remember that the Program Counter points to the address of the instruction being *fetched*, not executed. Our branch instruction sits at address 0x0000000C, and in order to create this address, the machine needs merely to subtract 8 from the Program Counter. It turns out that the branch instruction takes its twenty-four-bit offset and shifts it two bits to the left first, effectively multiplying the value by four. Therefore, the two's complement representation of –2, which is 0xFFFFFE, is placed in the instruction, producing a binary encoding of 0xEAFFFFFE.

Examining memory beyond our small program shows a seemingly endless series of ANDEQ instructions. A quick examination of the bit pattern with all bits clear will show that this translates into the AND instruction. The source and destination registers are register r0, and the conditional field, to be explained in Chapter 8, translates to "if equal to zero." The processor will fetch these instructions but never execute them, since the branch instruction will always force the processor to jump back to itself.

3.3.2 Examining Register and Memory Contents

Again referring back to the stored program computer in Chapter 1, we know that both registers and memory can hold data. While you write and debug code, it can be extremely helpful to monitor the changes that occur to registers *and* memory contents. The upper left-hand corner of Figure 3.3 shows the register window in the Keil tools, where the entire register bank can be viewed and altered. Changing values during debugging sessions can often save time, especially if you just want to test the effect of a single instruction on data. The lower right-hand corner of Figure 3.3 shows a memory window that will display the contents of memory locations given a starting address. Code Composer Studio has these windows, too, shown in Figure 3.4. For now, just note that our ARM7TDMI program starts at address 0x00000000 in memory, and the instructions can be seen in the following 16 bytes. For the next few chapters, we'll see examples of moving data to and from memory before unleashing all the details about memory in Chapter 5.

Breakpoints can also be quite useful for debugging purposes. A *breakpoint* is an instruction that has been tagged in such a way that the processor stops just before its execution. To set a breakpoint on an instruction, simply double-click the instruction in the gray bar area. You can use either the source window or the disassembly window. You should notice a red box beside the breakpointed instruction. When you run your code, the processor will stop automatically upon hitting the breakpoint. For larger programs, when you need to examine memory and register contents, set

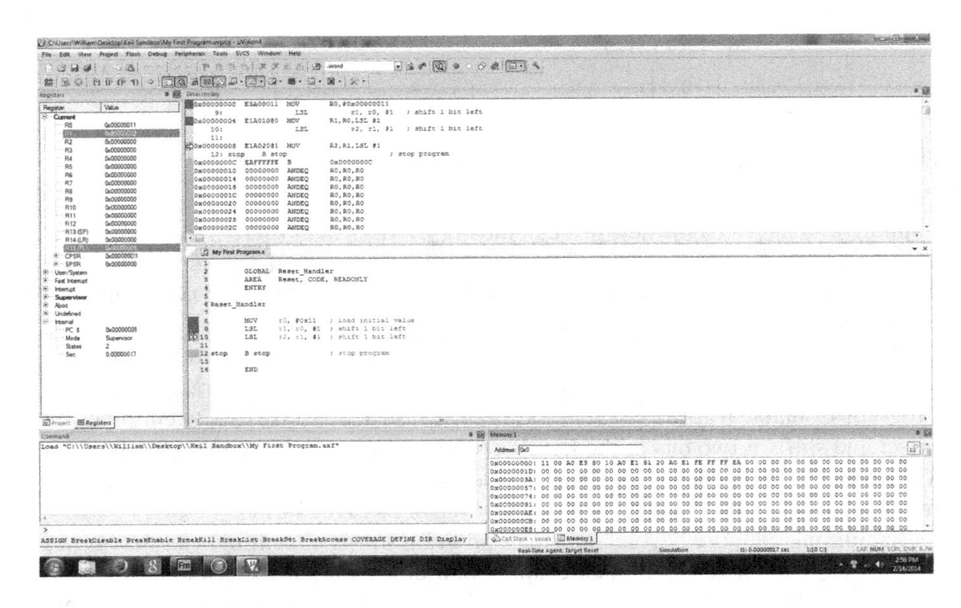

FIGURE 3.3 Register and memory windows in the Keil tools.

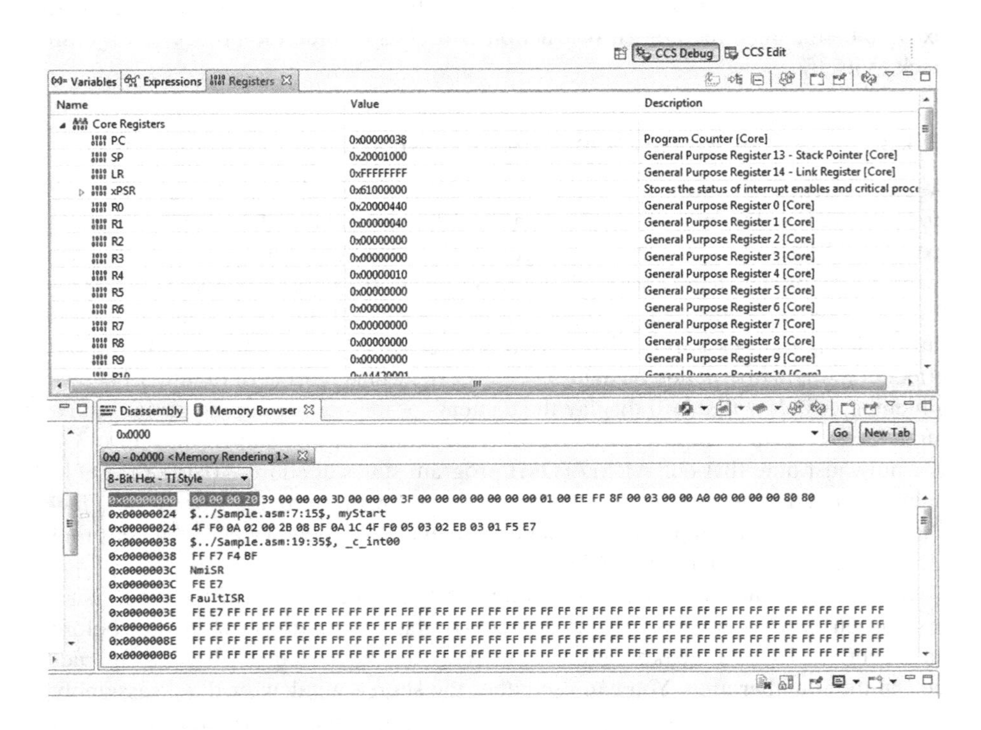

FIGURE 3.4 Register and memory windows in CCS.

a breakpoint at strategic points in the code, especially in areas where you want to single-step through complex instruction sequences.

3.4 PROGRAM 2: FACTORIAL CALCULATION

The next simple programs we look at for both the ARM7TDMI and the Cortex-M4 are ones that calculate the value of $n!$, which is a relatively short loop using only a few instructions. Recall that $n!$ is defined as

$$n! = \prod_{i=1}^{n} i = n(n-1)(n-2)\ldots(1)$$

For a given value of n, the algorithm iteratively multiplies a current product by a number that is one less than the number it used in the previous multiplication. The code continues to loop until it is no longer necessary to perform a multiplication, that is, when the multiplier is equal to zero.

For the ARM7TDMI code below, we can introduce the topics of

Conditional execution—The multiplication, subtraction, and branch may or may not be performed, depending on the result of another instruction.

Setting flags—The CMP instruction directs the processor to update the flags in the Current Program Status Register based on the result of the comparison.

Change-of-flow instructions—A branch will load a new address, called a branch target, into the Program Counter, and execution will resume from this new address.

Flags, in particular their use and meaning, are covered in detail in Chapters 7 and 8, but one condition that is quite easy to understand is greater-than, which simply tells you whether a value is greater than another or not. After a comparison instruction (CMP), flags in the CPSR are set and can be combined so that we might say one value is less than another, greater than another, etc. In order for one signed value to be greater than another, the Z flag must be clear, and the N and V flags must be equal. From a programmer's viewpoint, you simply write the condition in the code, e.g., GE for greater-than-or-equal, LT for less-than, or EQ for equal.

```
        AREA Prog2, CODE, READONLY
        ENTRY
        MOV     r6,#10          ; load n into r6
        MOV     r7,#1           ; if n=0, at least n!=1
loop    CMP     r6, #0
        MULGT   r7, r6, r7
        SUBGT   r6, r6, #1      ; decrement n
        BGT     loop            ; do another mul if counter!= 0
stop    B       stop            ; stop program
        END
```

As in the first program, we have directives for the Keil assembler to create an area with code in it, and we have an ENTRY point to mark the start of our code. The first MOV instruction places the decimal value 10, our initial value, into register r6. The second MOV instruction moves a default value of one into register r7, our result register, in the event the value of *n* equals zero. The next instruction simply subtracts zero from register r6, setting the condition code flags. We will cover this in much more detail in the next few chapters, but for now, note that if we want to make a decision based on an arithmetic operation, say if we are subtracting one from a counter until the counter expires (and then branching when finished), we must tell the instructions to save the condition codes by appending the "S" to the instruction. The CMP instruction does not need one—setting the condition codes is the only function of CMP.

The bulk of the arithmetic work rests with the only multiplication instruction in the code, MULGT, or multiply conditionally. The MULGT instruction is executed based on the results of that comparison we just did—if the subtraction ended up with a result of zero, then the zero (Z) flag in the Current Program Status Register (CPSR) will be set, and the condition greater-than does not exist. The multiply instruction reads "multiply register r6 times register r7, putting the results in register r7, but only if r6 is greater than zero," meaning if the previous comparison produced a result greater than zero. If the condition fails, then this instruction proceeds through the pipeline without doing anything. It's a no-operation instruction, or a nop (pronounced no op).

The next SUB instruction decrements the value of *n* during each pass of the loop, counting down until we get to where *n* equals zero. Like the multiplier instruction, the conditional subtract (SUBGT) instruction only executes if the result from the comparison is greater than zero. There are two points here that are important. The first is that we have not modified the flag results of the earlier CMP instruction. In other words, once the flags were set or cleared by the CMP instruction, they stay that way until something else comes along to modify them. There are explicit commands to modify the flags, such as CMP, TST, etc., or you can also append the "S" to an instruction to set the flags, which we'll do later. The second thing to point out is that we could have two, three, five, or more instructions all with this GT suffix on them to avoid having to make another branch instruction. Notice that we don't have to branch around certain instructions when the subtraction finally produces a value of zero in our counter—each instruction that fails the comparison will simply be ignored by the processor, including the branch (BGT), and the code is finished.

As before, the last branch instruction just branches to itself so that we have a stopping point. Run this code with different values for *n* to verify that it works, including the case where *n* equals zero.

The factorial algorithm can be written in a similar fashion for the Cortex-M4 as

```
        MOV     r6,#10          ; load 10 into r6
        MOV     r7,#1           ; if n=0, at least n!=1
loop    CMP     r6, #0
        ITTT    GT              ; start of our IF-THEN block
        MULGT   r7, r6, r7
```

```
        SUBGT   r6, r6, #1
        BGT     loop        ; end of IF-THEN block

stop    B       stop        ; stop program
```

The code above looks a bit like ARM7TDMI code, only these are Thumb-2 instructions (technically, a combination of 16-bit Thumb instructions and some new 32-bit Thumb-2 instructions, but since we're not looking at the code produced by the assembler just yet, we won't split hairs).

The first two MOV instructions load our value for *n* and our default product into registers r6 and r7, respectively. The comparison tests our counter against zero, just like the ARM7TDMI code, except the Cortex-M4 cannot conditionally execute instructions in the same way. Since Thumb instructions do not have a 4-bit conditional field (there are simply too few bits to include one), Thumb-2 provides an IF-THEN structure that can be used to build small loops efficiently. The format will be covered in more detail in Chapter 8, but the ITTT instruction indicates that there are three instructions following an IF condition that are treated as THEN operations. In other words, we read this as "if register r6 is greater than zero, perform the multiply, the subtraction, and the branch; otherwise, do not execute any of these instructions."

3.5 PROGRAM 3: SWAPPING REGISTER CONTENTS

This next program is actually a useful way to shuffle data around, and a good exercise in Boolean arithmetic. A fast way to swap the contents of two registers without using an intermediate storage location (such as memory or another register) is to use the exclusive OR operator. Suppose two values A and B are to be exchanged. The following algorithm could be used:

$$A = A \oplus B$$
$$B = A \oplus B$$
$$A = A \oplus B$$

The ARM7TDMI code below implements this algorithm using the Keil assembler, where the values of A = 0xF631024C and B = 0x17539ABD are stored in registers r0 and r1, respectively.

```
      AREA Prog3, CODE, READONLY
      ENTRY
      LDR   r0, =0xF631024C     ; load some data
      LDR   r1, =0x17539ABD     ; load some data
      EOR   r0, r0, r1          ; r0 XOR r1
      EOR   r1, r0, r1          ; r1 XOR r0
      EOR   r0, r0, r1          ; r0 XOR r1
stop  B     stop                ; stop program
      END
```

After execution, r0 = 0x17539ABD and r1 = 0xF631024C. Exclusive OR statements work on register data only, so we perform three EOR operations using our preloaded values. There are two funny-looking LDR (load) instructions, and in fact, they are not legal instructions. Rather, they are pseudo-instructions that we put in the code to make it easier on us, the programmer. While LDR instructions are normally used to bring data from memory into a register, here they are used to load the hexadecimal values 0xF631024C and 0x17539ABD into registers. This pseudo-instruction is not supported by all tools, so in Chapter 6, we investigate all the different ways of loading constants into a register.

3.6 PROGRAM 4: PLAYING WITH FLOATING-POINT NUMBERS

The Cortex-M4 is the first Cortex-M processor to offer an optional floating-point unit, allowing real values to be used in microcontroller routines more easily. This is no small block of logic; consequently, it is worth examining a short program to introduce the subject, as well as the format of the numbers themselves. The following code adds 1.0 and 1.0 together, which is not at all obvious:

```
LDR      r0, =0xE000ED88    ; Read-modify-write
LDR      r1, [r0]
ORR      r1, r1, #(0xF << 20) ; Enable CP10, CP11
STR      r1, [r0]
VMOV.F   s0, #0x3F800000    ; single-precision 1.0
VMOV.F   s1, s0
VADD.F   s2, s1, s0         ; 1.0 + 1.0 = ??
```

The first instruction, LDR, is actually the same pseudo-instruction we saw in Program 3 above, placing a 32-bit constant into register r0. We then use a real load instruction, LDR, to perform a read-modify-write operation, first reading a value at address 0xE000ED88 into register r1. This is actually the address of the Coprocessor Access Control Register, one of the memory-mapped registers used for controlling the floating-point unit. We then use a logical-OR instruction to set bits r1[23:20] to give us full access to coprocessors 10 and 11 (covered in Chapter 9). The final store instruction (STR) writes the value into the memory-mapped register, turning on the floating-point unit.

If you run the code using the Keil tools, you will see all of the registers for the processor, including the floating-point registers, in the Register window, shown in Figure 3.5. As you single-step through the code, notice that the first floating-point register, s0, eventually gets loaded with the value 0x3F800000, which is the decimal value 1.0 represented as a single-precision floating-point number. The second move operation (VMOV.F) copies that value from register s0 to s1. The VADD.F instruction adds the two numbers together, but the resulting 32-bit value, 0x40000000, definitely feels a little odd—that's 2.0 as a single-precision floating-point value! Run the code again, replacing the value in register s0 with 0x40000000. You anticipate that the value is 4.0, but the result requires a bit of interpretation.

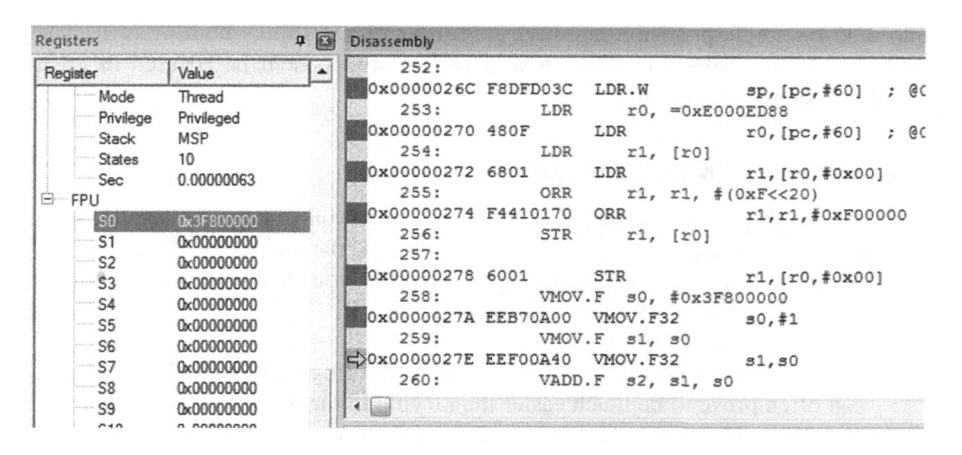

FIGURE 3.5 Register window in the Keil tools.

3.7 PROGRAM 5: MOVING VALUES BETWEEN INTEGER AND FLOATING-POINT REGISTERS

It's worth exploring one more short example. Here data is transferred between the ARM integer processor and the floating-point unit. Type in and run the following code on a Cortex-M4 microcontroller with floating-point hardware, single-stepping through each instruction to see the register values change.

```
LDR      r0,  =0xE000ED88         ; Read-modify-write
LDR      r1,  [r0]
ORR      r1,  r1, #(0xF << 20)    ; Enable CP10, CP11
STR      r1,  [r0]
LDR      r3,  =0x3F800000         ; single precision 1.0
VMOV.F   s3,  r3                  ; transfer contents from ARM to FPU
VLDR.F   s4,  =6.0221415e23       ; Avogadro's constant
VMOV.F   r4,  s4                  ; transfer contents from FPU to ARM
```

The first four instructions are those that we saw in the previous example to enable the floating-point unit. In line five, the LDR instruction loads register r3 with the representation of 1.0 in single precision. The VMOV.F instruction then takes the value stored in an integer register and transfers it to a floating-point register, register s3. Notice that the VMOV instruction was also used earlier to transfer data between two floating-point registers. Finally, Avogadro's constant is loaded into a floating-point register directly with the VLDR pseudo-instruction, which works just like the LDR pseudo-instruction in Programs 3 and 4. The VMOV.F instruction transfers the 32-bit value into the integer register r4. As you step through the code, watch the values move between integer and floating-point registers. Remember that the microcontroller really has little control over what these 32-bit values mean, and while there are some special values that do get treated differently in the floating-point logic, the integer logic just sees the value 0x66FF0C30 (Avogadro's constant now converted into a 32-bit single-precision number) in register r4 and thinks nothing of it. The exotic world of IEEE-compatible floating-point numbers will be covered in great detail in Chapters 9 through 11.

3.8 PROGRAMMING GUIDELINES

Writing assembly code is generally not difficult once you've become familiar with the processor's abilities, the instructions available, and the problem you are trying to solve. When writing code for the first time, however, you should keep a few things in mind:

- Break your problem down into small pieces. Writing smaller blocks of code can often prove to be much easier than trying to tackle a large problem all at one go. The trade-off, of course, is that you must now ensure that the smaller blocks of code can share information and work together without introducing bugs in the final routine.
- Always run a test case through your finished code, even if the code looks like it will "obviously" work. Often you will find a corner case that you haven't anticipated, and spending some time trying to break your own code is time well spent.
- Use the software tools to their fullest when writing a block of code. For example, the Keil MDK and Code Composer Studio tools provide a nice interface for setting breakpoints on instructions and watchpoints on data so that you can track the changes in registers, memory, and the condition code flags. As you step through your code, watch the changes carefully to ensure your code is doing exactly what you expect.
- Always make the assumption that someone else will be reading your code, so don't use obscure names or labels. A frequent complaint of programmers, even experienced ones, is that they can't understand their own code at certain points because they didn't write down what they were thinking at the time they wrote it. Years may pass before you examine your software again, so it's important to notate as much as possible, as carefully as possible, while you're writing the code and it's fresh in your mind.
- While it's tempting to make a program look very sophisticated and clever, especially if it's being evaluated by a teacher or supervisor, this often leads to errors. Simplicity is usually the best bet for beginning programs.
- Your first programs will probably not be optimal and efficient. This is normal. As you gain experience coding, you will learn about optimization techniques and pipeline effects later, so focus on getting the code running without errors first. Optimal code will come with practice.
- Don't be afraid to make mistakes or try something out. The software tools that you have available make it very easy to test code sections or instructions without doing any permanent damage to anything. Write some code, run it, watch the effects on the registers and memory, and if it doesn't work, find out why and try again!

- Using flowcharts may be useful in describing algorithms. Some programmers don't use them, so the choice is ultimately left to the writer.
- Pay attention to initialization. When your programs or modules begin, make a note of what values you expect to find in various registers—are they to be clear? Do you need to reset certain parameters at the start of a loop? Check for constants and fixed values that can be stored in memory or in the program itself. Before using variables (register or memory contents), it's always a good idea to set them to a known value. In some cases, this may not be necessary, e.g., if you subtracted two numbers and stored the result in a register that had not been initialized, the operation itself will set the register to a known value. However, if you use a register assuming the contents are clear, even a memory-mapped register, you can easily introduce errors in your code since some memory-mapped registers are described as undefined coming out of reset and may not be set to zero. Memory-mapped registers are examined in more detail in Chapter 16.

3.9 EXERCISES

1. Change Program 1, replacing the last LSL instruction with

```
ADD    r2, r1, r1, LSL #2
```

   and rerun the simulation. What value is in register r2 when the code reaches the infinite loop (the B instruction)? What is the ADD instruction actually doing?

2. Using a Disassembly window, write out the seven machine codes (32-bit instructions) for Program 2.

3. How many bytes does the code for Program 2 occupy? What about Program 3?

4. Change the value in register r6 at the start of Program 2 to 12. What value is in register r7 when the code terminates? Verify that this hex number is correct.

5. Run Program 3. After the first EOR instruction, what is the value in register r0? After the second EOR instruction, what is the value in register r1?

6. Using the instructions in Program 2 as a guide, write a program for both the ARM7TDMI and the Cortex-M4 that computes $6x^2 - 9x + 2$ and leaves the result in register r2. You can assume x is in register r3. For the syntax of the instructions, such as addition and subtraction, see the *ARM Architectural Reference Manual* and the *ARM v7-M Architectural Reference Manual.*

7. Show two different ways to clear all the bits in register r12 to zero. You may not use any registers other than r12.

8. Using Program 3 as a guide, write a program that adds the 32-bit two's complement representations of −149 and −4321. Place the result in register r7. Show your code and the resulting value in register r7.

9. Using Program 2 as a guide, execute the following instructions on an ARM7TDMI. Place small values in the registers beforehand. What do the instructions actually do?
   a. `MOVS r6, r6, LSL #5`
   b. `ADD r9, r8, r8, LSL #2`
   c. `RSB r10, r9, r9, LSL #3`
   d. (b) Followed by (c)

10. Suppose a branch instruction is located at address 0x0000FF00 in memory. What ARM instruction (32-bit binary pattern) do you think would be needed so that this B instruction could branch to itself?

11. Translate the following machine code into ARM mnemonics. What does the machine code do? What is the final value in register r2? You will want to compare these bit patterns with instructions found in the *ARM Architectural Reference Manual.*

Address	Machine code
00000000	E3A00019
00000004	E3A01011
00000008	E0811000
0000000C	E1A02001

12. Using the VLDR pseudo-instruction shown in Program 5, change Program 4 so that it adds the value of pi (3.1415926) to 2.0. Verify that the answer is correct using one of the floating-point conversion tools given in the References.

13. The floating-point instruction VMUL.F works very much like a VADD.F instruction. Using Programs 4 and 5 as a guide, multiply the floating-point representation for Avogadro's constant and 4.0 together. Verify that the result is correct using a floating-point conversion tool.

4 Assembler Rules and Directives

4.1 INTRODUCTION

The ARM assembler included with the RealView Microcontroller Development Kit contains an extensive set of features found on most assemblers—essential for experienced programmers, but somewhat unnerving if you are forced to wade through volumes of documentation as a beginner. Code Composer Studio also has a nice assembler with myriad features, but the details in the *ARM Assembly Language Tools User's Guide* run on for more than three hundred pages. In an attempt to cut right to the heart of programming, we now look at rules for the assembler, the structure of a program, and directives, which are instructions to the assembler for creating areas of code, aligning data, marking the end of your code, and so forth. These are unlike processor instructions, which tell the processor to add two numbers or jump somewhere in your code, since they never turn into actual machine instructions. Although both the ARM and TI assemblers are easy to learn, be aware that other assemblers have slightly different rules; e.g., gnu tools have directives that are preceded with a period and labels that are followed by a colon. It's a Catch-22 situation really, as you cannot learn assembly without knowing how to use directives, but it's difficult to learn directives without seeing a little assembly. Fortunately, it is unlikely that you will use every directive or every assembler option immediately, so for now, we start with what is essential. Read this chapter to get an overview of what's possible, but don't panic. As we proceed through more chapters of the book, you may find yourself flipping back to this chapter quite often, which is normal. You can, of course, refer back to the *RealView Assembler User's Guide* found in the RVMDK tools or the Code Composer Studio documentation for the complete specifications of the assemblers if you need even more detail.

4.2 STRUCTURE OF ASSEMBLY LANGUAGE MODULES

We begin by examining a very simple module as a starting point. Consider the following code:

```
      AREA ARMex, CODE, READONLY
      ; Name this block of code ARMex
      ENTRY                  ; Mark first instruction to execute
start MOV    r0, #10         ; Set up parameters
      MOV    r1, #3
```

```
        ADD     r0, r0, r1      ; r0 = r0 + r1
stop    B       stop            ; infinite loop
        END                     ; Mark end of file
```

While the routine may appear a little cryptic, it only does one thing: it adds the numbers 10 and 3 together. The rest of the code consists of directives for the assembler and an instruction at the end to put the processor in an infinite loop. You can see that there is some structure to the lines of code, and the general form of source lines in your assembly files is

```
{label} {instruction|directive|pseudo-instruction} {;comment}
```

where each field in braces is optional. Labels are names that you choose to represent an address somewhere in memory, and while they eventually do need to be translated into a numeric value, as a programmer you simply work with the name throughout your code. The linker will calculate the correct address during the linkage process that follows assembly. Note that a label name can only be defined once in your code, and *labels must start at the beginning of the line* (there are some assemblers that will allow you to place the label at any point, but they require delimiters such as a colon).

The instructions, directives, and pseudo-instructions (such as ADR that we will see in Chapter 6) must be preceded by a white space, either a tab or any number of spaces, even if you don't have a label at the beginning. One of the most common mistakes new programmers make is starting an instruction in column one. To make your code more readable, you may use blank lines, since all three sections of the source line are optional. ARM and Thumb instructions available on the ARM7TDMI are from the ARM version 4T instruction set; the Thumb-2 instructions used on the Cortex-M4 are from the v7-M instruction set. All of these can be found in the respective Architectural Reference Manuals, along with their mnemonics and uses. Just to start us off, the ARM instructions for the ARM7TDMI are also listed in Table 4.1, and we'll slowly introduce the v7-M instructions throughout the text. There are many directives and pseudo-instructions, but we will cover only a handful throughout this chapter to get a sense of what is possible.

The current ARM/Thumb assembler language, called Unified Assembler Language (UAL), has superseded earlier versions of both the ARM and Thumb assembler languages (we saw a few Thumb instructions in Chapter 3, and we'll see more throughout the book, particularly in Chapter 17). To give you some idea of the subtle changes involved, compare the two formats for performing a shift operation:

Old ARM format	**UAL format**
`MOV <Rd>, <Rn>, LSL shift`	`LSL <Rd>, <Rn>, shift`
`LDR{cond}SB`	`LDRSB{cond}`
`LDMFD sp!,{reglist}`	`PUSH {reglist}`

Code written using UAL can be assembled for ARM, Thumb, or Thumb-2, which is an extension of the Thumb instruction set found on the more recent ARM

TABLE 4.1
ARM Version 4T Instruction Set

ADC	ADD	AND	B	BL
BX	CDP	CMN	CMP	EOR
LDC	LDM	LDR	LDRB	LDRBT
LDRH	LDRSB	LDRSH	LDRT	MCR
MLA	MOV	MRC	MRS	MSR
MUL	MVN	ORR	RSB	RSC
SBC	SMLAL	SMULL	STC	STM
STR	STRB	STRBT	STRH	STRT
SUB	SWI[a]	SWP	SWPB	TEQ
TST	UMLAL	UMULL		

[a] The SWI instruction was deprecated in the latest version of the ARM *Architectural Reference Manual* (2007c), so while you should use the SVC instruction, you may still see this instruction in some older code.

processors, e.g., Cortex-A8. However, you're likely to find a great a deal of code written using the older format, so be mindful of the changes when you review older programs. Also be aware that a disassembly of your code will show the UAL notations if you are using the RealView tools or Code Composer Studio. You can find more details on UAL formats in the *RealView Assembler User's Guide* located in the RVMDK tools.

We'll examine commented code throughout the book, but in general it is a good idea to document your code as much as possible, with clear statements about the operation of certain lines. Remember that on large projects, you will probably not be the only one reading your code. Guidelines for good comments include the following:

- Don't comment the obvious. If you're adding one to a register, don't write "Register r3 + 1."
- Use concise language when describing what registers hold or how a function behaves.
- Comment the sections of code where you think another programmer might have a difficult time following your reasoning. Complicated algorithms usually require a deep understanding of the code, and a bug may take days to find without adequate documentation.
- In addition to commenting individual instructions, include a short description of functions, subroutines, or long segments of code.
- Do not abbreviate, if possible.
- Acronyms should be avoided, but this can be difficult sometimes, since peripheral register names tend to be shortened. For example, VIC0_VA7R might not mean much in a comment, so if you use the name in the instruction, describe what the register does.

If you are using the Keil tools, the first semicolon on a line indicates the beginning of a comment, unless you have the semicolon inside of a string constant, for example,

```
abc     SETS    "This is a semicolon;"
```

Here, a string is assigned to the variable abc, but since the semicolon lies within quotes, there is no comment on this line. The end of the line is the end of the comment, and a comment can occupy the entire line if you wish. The TI assembler will allow you to place either an asterisk (*) or a semicolon in column 1 to denote a comment, or a semicolon anywhere else on the line.

At some point, you will begin using constants in your assembly, and they are allowed in a handful of formats:

- Decimal, for example, 123
- Hexadecimal, for example, 0x3F
- n_xxx (Keil only) where:
  - n is a base between 2 and 9
  - xxx is a number in that base

Character constants consist of opening and closing *single* quotes, enclosing either a single character or an escaped character, using the standard C escape characters (recall that escape characters are those that act as nonprinting characters, such as \n for creating a new line). String constants are contained within *double* quotes. The standard C escape sequences can be used within string constants, but they are done differently by assemblers. For example, in the Keil tools, you could say something like

```
     MOV    r3, #'A'  ; single character constant
     GBLS   str1      ; set the value of global string variable
str1 SETS   "Hello world!\n"
```

In the Code Composer Studio tools, you might say

```
   .string "Hello world!"
```

which places 8-bit characters in the string into a section of code, but the .string directive neither adds a NUL character at the end of the characters nor interprets escape characters. Instead, you could say

```
   .cstring "Hello world!\n"
```

which both adds the NUL character for you and correctly interprets the \n escape character at the end.

Before we move into directives, we need to cover a few housekeeping rules. For the Keil tools, there are case rules associated with your commands, so while you can write the instruction mnemonics, directives, and symbolic register names in either uppercase or lowercase, you cannot mix them. For example ADD or add are acceptable, but not Add. When it comes to mnemonics, the TI assembler is case-insensitive.

To make the source file easier to read, the Keil tools allow you to split up a single line into several lines by placing a backslash character (\) at the end of a line. If you had a long string, you might write

```
ISR_Stack_Size EQU (UND_Stack_Size + SVC_Stack_Size + ABT_Stack_Size + \
                    FIQ_Stack_Size + IRQ_Stack_Size)
```

There must not be any other characters following the backslash, such as a space or a tab. The end-of-line sequence is treated as a white space by the assembler. Using the Keil tools, you may have up to 4095 characters for any given line, including any extensions using backslashes. The TI tools only allow 400 characters per line—anything longer is truncated. For either tool, keep the lines relatively short for easier reading!

4.3 PREDEFINED REGISTER NAMES

Most assemblers have a set of register names that can be used interchangeably in your code, mostly to make it easier to read. The ARM assembler is no different, and includes a set of predefined, case-sensitive names that are synonymous with registers. While the tools recognize predeclared names for basic registers, status registers, floating-point registers, and coprocessors, only the following are of immediate use to us:

r0-r15 or R0-R15
s0-s31 or S0-S31
a1-a4 (argument, result, or scratch registers, synonyms for r0 to r3)
sp or SP (stack pointer, r13)
lr or LR (Link Register, r14)
pc or PC (Program Counter, r15)
cpsr or CPSR (current program status register)
spsr or SPSR (saved program status register)
apsr or APSR (application program status register)

4.4 FREQUENTLY USED DIRECTIVES

A complete description of the assembler directives can be found in Section 4.3 of the *RealView Assembler User's Guide* or Chapter 4 of *ARM Assembly Language Tools User's Guide*; however, in order to start coding, you only need a few. We'll examine the more frequently used directives first, shown in Table 4.2, and leave the others as reference material should you require them. Then we'll move on to macros in the next section.

4.4.1 DEFINING A BLOCK OF DATA OR CODE

As you create code, particularly compiled code from C programs, the tools will need to be told how to treat all the different parts of it—data sections, program sections,

TABLE 4.2
Frequently Used Directives

Keil Directive	CCS Directive	Uses
AREA	.sect	Defines a block of code or data
RN	.asg	Can be used to associate a register with a name
EQU	.equ	Equates a symbol to a numeric constant
ENTRY		Declares an entry point to your program
DCB, DCW, DCD	.byte, .half, .word	Allocates memory and specifies initial runtime contents
ALIGN	.align	Aligns data or code to a particular memory boundary
SPACE	.space	Reserves a zeroed block of memory of a particular size
LTORG		Assigns the starting point of a literal pool
END	.end	Designates the end of a source file

blocks of coefficients, etc. These sections, which are indivisible and named, then get manipulated by the linker and ultimately end up in the correct type of memory in a system. For example, data, which could be read-write information, could get stored in RAM, as opposed to the program code which might end up in Flash memory. Normally you will have separate sections for your program and your data, especially in larger programs. Blocks of coefficients or tables can be placed in a section of their own. Since the two main tool sets that we'll use throughout the book do things in very different ways, both formats are presented below.

4.4.1.1 Keil Tools

You tell the assembler to begin a new code or data section using the AREA directive, which has the following syntax:

AREA *sectionname*{,*attr*}{,*attr*}...

where *sectionname* is the name that the section is to be given. Sections can be given almost any name, but if you start a section name with a digit, it must be enclosed in bars, e.g., |1_DataArea|; otherwise, the assembler reports a missing section name error. There are some names you cannot use, such as |.text|, since this is used by the C compiler (but it would be a rather odd name to pick at random). Your code must have at least one AREA directive in it, which you'll usually find in the first few lines of a program. Table 4.3 shows some of the attributes that are available, but a full list can be found in the *RealView Assembler User Guide* in the Keil tools.

EXAMPLE 4.1

The following example defines a read-only code section named Example.

```
AREA Example,CODE,READONLY ; An example code section.
; code
```

TABLE 4.3
Valid Section Attributes (Keil Tools)

ALIGN = *expr*	This aligns a section on a 2^{expr}-byte boundary (note that this is different from the ALIGN directive); e.g., if *expr* = 10, then the section is aligned to a 1KB boundary.
CODE	The section is machine code (READONLY is the default)
DATA	The section is data (READWRITE is the default)
READONLY	The section can be placed in read-only memory (default for sections of CODE)
READWRITE	The section can be placed in read-write memory (default for sections of DATA)

4.4.1.2 Code Composer Studio Tools

It's often helpful to break up large assembly files into sections, e.g., creating a separate section for large data sets or blocks of coefficients. In fact, the TI assembler has directives to address similar concepts. Table 4.4 shows some of the directives used to create sections.

The .sect directive is similar to the AREA directive in that you use it to create an initialized section, to put either your code or some initialized data there. Sections can be made read-only or read-write, just as with Keil tools. You can make as many sections as you like; however, it is usually best to make only as many as needed. An example of a section of data called Coefficients might look like

```
.sect "Coefficients"
.float 0.05
.float 2.03
.word 0AAh
```

The default section is the .text section, which is where your assembly program will normally sit, and in fact, you can create it either by saying

```
.sect ".text"
```

TABLE 4.4
TI Assembler Section Directives

	Directive	Use
Uninitialized sections	.bss	Reserves space in the .bss section
	.usect	Reserves space in a specified uninitialized named section
Initialized sections	.text	The default section where the compiler places code
	.data	Normally used for pre-initialized variables or tables
	.sect	Defines a named section similar to the default .text and .data sections

or by simply typing

```
.text
```

Anything after this will be placed in the .text section. As we'll see in Chapter 5 for both the Keil tools and the Code Composer Studio tools, there is a linker command file and a memory map that determines where all of these sections ultimately end up in memory. As with most silicon vendors, TI ships a default linker command file for their MCUs, so you shouldn't need to modify anything to get up and running.

4.4.2 Register Name Definition

4.4.2.1 Keil Tools

In the ARM assembler that comes with the Keil tools, there is a directive RN that defines a register name for a specified register. It's not mandatory to use such a directive, but it can help in code readability. The syntax is

name RN *expr*

where *name* is the name to be assigned to the register. Obviously *name* cannot be the same as any of the predefined names listed in Section 4.3. The *expr* parameter takes on values from 0 to 15. Mind that you do not assign two or more names to the same register.

EXAMPLE 4.2

The following registers have been given names that can be used throughout further code:

```
coeff1      RN      8       ; coefficient 1
coeff2      RN      9       ; coefficient 2
dest        RN      0       ; register 0 holds the pointer to
                            ; destination matrix
```

4.4.2.2 Code Composer Studio

You can assign names to registers using the .asg directive. The syntax is

.asg "character string", substitution symbol

For example, you might say

```
.asg    R13, STACKPTR
ADD     STACKPTR, STACKPTR, #3
```

4.4.3 Equating a Symbol to a Numeric Constant

It is frequently useful to give a symbolic name to a numeric constant, a register-relative value, or a program-relative value. Such a directive is similar to the use of #define to define a constant in C. Note that the assembler doesn't actually place anything at a particular memory location. It merely equates a label with an operand, either a value or another label, for example.

4.4.3.1 Keil Tools

The syntax for the EQU directive is

name EQU *expr*{,*type*}

where *name* is the symbolic name to assign to the value, *expr* is a register-relative address, a program-relative address, an absolute address, or a 32-bit integer constant. The parameter *type* is optional and can be any one of

ARM
THUMB
CODE16
CODE32
DATA

EXAMPLE 4.3

```
SRAM_BASE   EQU    0x04000000    ; assigns SRAM a base address
abc         EQU    2             ; assigns the value 2 to the symbol abc
xyz         EQU    label+8       ; assigns the address (label+8)
                                 ; to the symbol xyz
fiq         EQU    0x1C, CODE32  ; assigns the absolute address
                                 ; 0x1C to the symbol fiq, and marks it
                                 ; as code
```

4.4.3.2 Code Composer Studio

There are two identical (and interchangeable) directives for equating names with constants and other values: .set and .equ. Notice that registers can be given names using these directives as well as values. Their syntax is

symbol .set *value*
symbol .equ *value*

EXAMPLE 4.4

```
AUX_R4    .set    R4              ; equate symbol AUX_R4 to register R4
OFFSET    .equ    50/2 + 3        ; equate OFFSET to a numeric value
          ADD     r0, AUX_R4,     #OFFSET
```

4.4.4 Declaring an Entry Point

In the Keil tools, the ENTRY directive declares an entry point to a program. The syntax is

ENTRY

Your program must have at least one ENTRY point for a program; otherwise, a warning is generated at link time. If you have a project with multiple source files, not every source file will have an ENTRY directive, and any single source file should only have one ENTRY directive. The assembler will generate an error if more than one ENTRY exists in a single source file.

EXAMPLE 4.5

```
AREA ARMex, CODE, READONLY
ENTRY       ; Entry point for the application
```

4.4.5 Allocating Memory and Specifying Contents

When writing programs that contain tables or data that must be configured before the program begins, it is necessary to specify exactly what memory looks like. Strings, floating-point constants, and even addresses can be stored in memory as data using various directives.

4.4.5.1 Keil Tools

One of the more common directives, DCB, actually defines the initial runtime contents of memory. The syntax is

{*label*} DCB *expr*{,*expr*}...

where *expr* is either a numeric expression that evaluates to an integer in the range –128 to 255, or a quoted string, where the characters of the string are stored consecutively in memory. Since the DCB directive affects memory at the byte level, you should use an ALIGN directive afterward if any instructions follow to ensure that the instruction is aligned correctly in memory.

EXAMPLE 4.6

Unlike strings in C, ARM assembler strings are not null-terminated. You can construct a null-terminated string using DCB as follows:

```
C_string DCB "C_string",0
```

If this string started at address 0x4000 in memory, it would look like

Address		ASCII equivalent
0x4000	43	C
0x4001	5F	_
0x4002	73	s
0x4003	74	t
0x4004	72	r
0x4005	69	i
0x4006	6E	n
0x4007	67	g
0x4008	00	

Compare this to the way to that the Code Composer Studio assembler did the same thing using the .cstring directive in Section 4.2.

In addition to the directive for allocating memory at the resolution of bytes, there are directives for reserving and defining halfwords and words, with and without alignment. The DCW directive allocates one or more halfwords of memory, aligned on two-byte boundaries (DCWU does the same thing, only without the memory alignment). The syntax for these directives is

{*label*} DCW{U} *expr*{,*expr*}...

where *expr* is a numeric expression that evaluates to an integer in the range –32768 to 65535.

Another frequently used directive, DCD, allocates one or more words of memory, aligned on four-byte boundaries (DCDU does the same thing, only without the memory alignment). The syntax for these directives is

{*label*} DCD{U} *expr*{,*expr*}

where *expr* is either a numeric expression or a program-relative expression. DCD inserts up to 3 bytes of padding before the first defined word, if necessary, to achieve a 4-byte alignment. If alignment isn't required, then use the DCDU directive.

EXAMPLE 4.7

```
coeff DCW   0xFE37, 0x8ECC   ; defines 2 halfwords
data1 DCD   1,5,20           ; defines 3 words containing
                             ; decimal values 1, 5, and 20
data2 DCD   mem06 + 4        ; defines 1 word containing 4 +
                             ; the address of the label mem06

      AREA  MyData, DATA, READWRITE
      DCB   255              ; now misaligned...
data3 DCDU  1,5,20           ; defines 3 words containing
                             ; 1, 5, and 20 not word aligned
```

4.4.5.2 Code Composer Studio

There are similar directives in CCS for initializing memory, each directive specifying the width of the values being used. For placing one or more values into consecutive bytes of the current section, you can use either the .byte or .char directive. The syntax is

{*label*} .byte $value_1${,...,$value_n$}

where *value* can either be a string in quotes or some other expression that gets evaluated assuming the data is 8-bit signed data.

EXAMPLE 4.8

If you wanted to place a few constants and some short strings in memory, you could say

```
LAB1   .byte   10, -1, "abc", 'a'
```

and in memory the values would appear as

```
0A FF 61 62 63 61
```

For halfword values, there are .half and .short directives which will always align the data to halfword boundaries in the section. For word length values, there are .int, .long, and .word directives, which also align the data to word boundaries in the section. There is even a .float directive (for single-precision floating-point values) and a .double directive (for double-precision floating-point values)!

4.4.6 Aligning Data or Code to Appropriate Boundaries

Sometimes you must ensure that your data and code are aligned to appropriate boundaries. This is typically required in circumstances where it's necessary or optimal to have your data aligned a particular way. For example, the ARM940T processor has a cache with 16-byte cache lines, and to maximize the efficiency of the cache, you might try to align your data or function entries along 16-byte boundaries. For those processors where you can load and store double words (64 bits), such as the ARM1020E or ARM1136EJ-S, the data must be on an 8-byte boundary. A label on a line by itself can be arbitrarily aligned, so you might use ALIGN 4 before the label to align your ARM code, or ALIGN 2 to align Thumb code.

4.4.6.1 Keil Tools

The ALIGN directive aligns the current location to a specified boundary by padding with zeros. The syntax is

```
ALIGN {expr{,offset}}
```

where *expr* is a numeric expression evaluating to any power of two from 2^0 to 2^{31}, and *offset* can be any numeric expression. The current location is aligned to the next address of the form

offset + n * *expr*

If *expr* is not specified, ALIGN sets the current location to the next word (four byte) boundary.

EXAMPLE 4.9

```
            AREA OffsetExample, CODE
            DCB 1               ; This example places the two
            ALIGN 4,3           ; bytes in the first and fourth
            DCB 1               ; bytes of the same word

            AREA Example, CODE, READONLY
start       LDR r6, =label1
            ; code
            MOV pc,lr
```

```
label1        DCB 1          ; pc now misaligned
              ALIGN          ; ensures that subroutine1 addresses
subroutine1   MOV r5, #0x5   ; the following instruction
```

4.4.6.2 Code Composer Studio

The .align directive can be used to align the section Program Counter to a particular boundary within the current section. The syntax is

.align {*size in bytes*}

If you do not specify a size, the default is one byte. Otherwise, a size of 2 aligns code or data to a halfword boundary, a size of 4 aligns to a word boundary, etc.

4.4.7 Reserving a Block of Memory

You may wish to reserve a block of memory for variables, tables, or storing data during routines. The SPACE and .space directives reserve a zeroed block of memory.

4.4.7.1 Keil Tools

The syntax is

{*label*} SPACE *expr*

where *expr* evaluates to the number of zeroed bytes to reserve. You may also want to use the ALIGN directive after using a SPACE directive, to align any code that follows.

EXAMPLE 4.10

```
      AREA MyData, DATA, READWRITE
data1 SPACE 255  ; defines 255 bytes of zeroed storage
```

4.4.7.2 Code Composer Studio

There are actually two directives that reserve memory—the .space and .bes directives. When a label is used with the .space directive, it points to the first byte reserved in memory, while the .bes points to the last byte reserved. The syntax for the two is

{*label*} .space *size (in bytes)*

{*label*} .bes *size (in bytes)*

EXAMPLE 4.11

```
RES_1:   .space 100   ; RES_1 points to the first byte
RES_2:   .bes   30    ; RES_2 points to the last byte
```

As an aside, there is also a .bss directive for reserving uninitialized space—consult Chapter 4 of *ARM Assembly Language Tools User's Guide* for all the details.

4.4.8 Assigning Literal Pool Origins

Literal pools are areas of data that the ARM assembler creates for you at the end of every code section, specifically for constants that cannot be created with rotation schemes or that do not fit into an instruction's supported formats. Chapter 6 discusses literal pools at length, but you should at least see the uses for the LTORG directive here. Situations arise where you might have to give the assembler a bit of help in placing literal pools, since they are placed at the end of code sections, and these ends rely on the AREA directives at the beginning of sections that follow (or the end of your code).

EXAMPLE 4.12

Consider the code below. An LDR pseudo-instruction is used to move the constant 0x55555555 into register r1, which ultimately gets converted into a real LDR instruction with a PC-relative offset. This offset must be calculated by the assembler, but the offset has limits (4 kilobytes). Imagine then that we reserve 4200 bytes of memory just at the end of our code—the literal pool would go *after* the big, empty block of memory, but this is too far away. An LTORG directive is required to force the assembler to put the literal pool after the MOV instruction instead, allowing an offset to be calculated that is within the 4 kilobyte range. In larger programs, you may find yourself making several literal pools, so place them after unconditional branches or subroutine return instructions. This prevents the processor from executing the constants as instructions.

```
       AREA Example, CODE, READONLY
start  BL      func1
func1                              ; function body
                                   ; code
       LDR     r1, = 0x55555555    ; => LDR R1, [pc, #offset to lit
                                   ; pool 1]
       ; code
       MOV     pc,lr               ; end function
       LTORG                       ; lit. pool 1 contains literal
                                   ; 0x55555555
data   SPACE 4200                  ; clears 4200 bytes of memory,
                                   ; starting at current location
       END                         ; default literal pool is empty
```

Note that the Keil tools permit the use of the LDR pseudo-instruction, but Code Composer Studio does not, so there is no equivalent of the LTORG directive in the CCS assembler.

4.4.9 Ending a Source File

This is the easiest of the directives—END simply tells the assembler you're at the end of a source file. The syntax for the Keil tools is

```
END
```

and for Code Composer Studio, it's

```
.end
```

When you terminate your source file, place the directive on a line by itself.

4.5 MACROS

Macro definitions allow a programmer to build definitions of functions or operations once, and then call this operation by name throughout the code, saving some writing time. In fact, macros can be part of a process known as conditional assembly, wherein parts of the source file may or may not be assembled based on certain variables, such as the architecture version (or a variable that you specify yourself). While this topic is not discussed here, you can find all the specifics about conditional assembly, along with the directives involved, in the Directives Reference section of the *RealView Assembler User's Guide* or the Macro Description chapter of the *ARM Assembly Language Tools User's Guide* from TI.

The use of macros is neither recommended nor discouraged, as there are advantages and disadvantages to using them. You can generally shorten your source code by using them, but when the macros are expanded, they may chew up memory space because of their frequent use. Macros can sometimes be quite large. Using macros does allow you to change your code more quickly, since you usually only have to edit one block, rather than multiple instances of the same type of code. You can also define a new operation in your code by writing it as a macro and then calling it whenever it is needed. Just be sure to document the new operation thoroughly, as someone unfamiliar with your code may one day have to read it!

Note that macros are not the same thing as a subroutine call, since the macro definitions are substituted at assembly time, replacing the macro call with the actual assembly code. It is sometimes actually easier to follow the logic of source code if repeated sections are replaced with a macro, but they are not required in writing assembly. Let's examine macros using only the Keil tools—the concept translates easily to Code Composer Studio.

Two directives are used to define a macro: MACRO and MEND. The syntax is

```
        MACRO
{$label} macroname{$cond} {$parameter{,$parameter}...}
        ; code
        MEND
```

where *$label* is a parameter that is substituted with a symbol given when the macro is invoked. The symbol is usually a label. The macro name must not begin with an instruction or directive name. The parameter *$cond* is a special parameter designed to contain a condition code; however, values other than valid condition codes are permitted. The term *$parameter* is substituted when the macro is invoked.

Within the macro body, parameters such as *$label*, *$parameter*, or *$cond* can be used in the same way as other variables. They are given new values each time the macro is invoked. Parameters must begin with $ to distinguish them from ordinary symbols. Any number of parameters can be used. The *$label* field is optional, and the macro itself defines the locations of any labels.

EXAMPLE 4.13

Suppose you have a sequence of instructions that appears multiple times in your code—in this case, two ADD instructions followed by a multiplication. You could define a small macro as follows:

```
    MACRO
    ; macro definition:
    ;
    ; vara = 8 * (varb + varc + 6)

$Label_1 AddMul $vara, $varb, $varc

$Label_1
    ADD $vara, $varb, $varc         ; add two terms
    ADD $vara, $vara, #6            ; add 6 to the sum
    LSL $vara, $vara, #3            ; multiply by 8
    MEND
```

In your source code file, you can then instantiate the macro as many times as you like. You might call the sequence as

```
        ; invoke the macro
CSet1   AddMul r0, r1, r2
        ; the rest of your code
```

and the assembler makes the necessary substitutions, so that the assembly listing actually reads as

```
    ; invoke the macro
CSet1
    ADD  r0, r1, r2
    ADD  r0, r0, #6
    LSL  r0, r0, #3
    ; the rest of your code
```

4.6 MISCELLANEOUS ASSEMBLER FEATURES

While your first program will not likely contain many of these, advanced programmers typically throw variables, literals, and complex expressions into their code to save time in writing assembly. Consult the *RealView Assembler User's Guide* or *ARM Assembly Language Tools User's Guide* for the complete set of rules and allowable expressions, but we can adopt a few of the most common operations for our own use throughout the book.

4.6.1 Assembler Operators

Primitive operations can be performed on data before it is used in an instruction. Note that these operators apply to the data—they are *not* part of an instruction.

Operators can be used on a single value (unary operators) or two values (binary operators). Unary operators are not that common; however, binary operators prove to be quite handy for shuffling bits across a register or creating masks. Some of the most useful binary operators are

	Keil Tools	Code Composer Studio
A modulo B	A:MOD:B	A % B
Rotate A left by B bits	A:ROL:B	
Rotate A right by B bits	A:ROR:B	
Shift A left by B bits	A:SHL:B or A << B	A << B
Shift A right by B bits	A:SHR:B or A >> B	A >> B
Add A to B	A + B	A + B
Subtract B from A	A − B	A − B
Bitwise AND of A and B	A:AND:B	A & B
Bitwise Exclusive OR of A and B	A:EOR:B	A ^ B
Bitwise OR of A and B	A:OR:B	A | B

These types of operators creep into macros especially, and should you find yourself writing conditional assembly files, for whatever reason, you may decide to use these types of operators to control the creation of the source code.

EXAMPLE 4.14

To set a particular bit in a register (say if it were a bit to enable/disable the caches, a branch predictor, interrupts, etc.) you might have the control register copied to a general-purpose register first. Then the bit of interest would be modified using an OR operation, and the control register would be stored back. The OR instruction might look like

```
ORR r1, r1, #1:SHL:3    ; set CCREG[3]
```

Here, a 1 is shifted left three bits. Assuming you like to call register r1 CCREG, you have now set bit 3. The advantage in writing it this way is that you are more likely to understand that you wanted a one in a particular bit location, rather than simply using a logical operation with a value such as 0x8.

You can even use these operators in the creation of constants, for example,

```
DCD (0x8321:SHL:4):OR:2
```

which could move this two-byte field to the left by four bits, and then set bit 1 of the resulting constant with the use of the OR operator. This might be easier to read, since you may need a two-byte value shifted, and reading the original before the shift may help in understanding what the code does. It is not necessary to do this, but again, it provides some insight into the code's behavior.

To create very specific bit patterns quickly, you can string together many operators in the same field, such as

```
MOV  r0, #((1:SHL:14):OR:(1:SHL:12))
```

which may look a little odd, but in effect we are putting the constant 0x5000 into register r0 by taking two individual bits, shifting them to the left, and then ORing the two patterns (convince yourself of this). It would look very similar in the Code Composer Studio tools as

```
MOV  r0, #((1 <<14)  |  (1 <<12))
```

You may wonder why we're creating such a strange configuration and not something simpler, such as

```
MOV  r0, #0x5000
```

which is clearly easier to enter. Again, it depends on the context of the program. The programmer may need to load a configuration register, which often has very specific bit fields for functions, and the former notation will remind the reader that you are enabling two distinct bits in that register.

4.6.2 Math Functions in CCS

There are a number of built-in functions within Code Composer Studio that make math operations a bit easier. Some of the many included functions are

$$cos(*expr*)	Returns the cosine of *expr* as a floating-point value
$$sin(*expr*)	Returns the sine of *expr* as a floating-point value
$$log(*expr*)	Returns the natural logarithm of *expr*, where *expr* > 0
$$max(*expr1*, *expr2*)	Returns the maximum of two values
$$sqrt(*expr*)	Returns the square root of *expr*, where *expr* >= 0, as a floating-point value

You may never use these in your code; however, for algorithmic development, they often prove useful for quick tests and checks of your own routines.

EXAMPLE 4.15

You can build a list of trigonometric values very quickly in a data section by saying something like

```
.float        $$cos(0.434)
.float        $$cos(0.348)
.float        $$sin(0.943)
.float        $$tan(0.342)
```

4.7 EXERCISES

1. What is wrong with the following program?

```
            AREA         ARMex2, CODE, READONLY
            ENTRY
start
MOV         r0, #6
ADD         r1, r2, #2
            END
```

2. What is another way of writing the following line of code?

```
MOV PC, LR
```

3. Use a Keil directive to assign register r6 to the name `bouncer`.

4. Use a Code Composer Studio directive to assign register r2 to the name `FIR_index`.

5. Fill in the missing Keil directive below:

```
SRAM_BASE          0x2000
                   MOV  r12, #SRAM_BASE
                   STR  r6, [r12]
```

6. What is the purpose of a macro?

7. Create a mask (bit pattern) in memory using the DCD directive (Keil) and the SHL and OR operators for the following cases. Repeat the exercise using the .word directive (CCS) and the << and | operators. Remember that bit 31 is the most significant bit of a word and bit 0 is the least significant bit.
   a. The upper two bytes of the word are 0xFFEE and the least significant bit is set.
   b. Bits 17 and 16 are set, and the least significant byte of the word is 0x8F.
   c. Bits 15 and 13 are set (hint: do this with two SHL directives).
   d. Bits 31 and 23 are set.

8. Give the Keil directive that assigns the address 0x800C to the symbol INTEREST.

9. What constant would be created if the following operators are used with a DCD directive? For example,

```
MASK DCD 0x5F:ROL:3
```

   a. 0x5F:SHR:2
   b. 0x5F:AND:0xFC
   c. 0x5F:EOR:0xFF
   d. 0x5F:SHL:12

10. What constant would be created if the following operators are used with a .word directive? For example,

```
MASK .word 0x9B<<3
```

    a. 0x9B>>2
    b. 0x9B & 0xFC
    c. 0x9B ^ 0xFF
    d. 0x9B<<12

11. What instruction puts the ASCII representation of the character "R" in register r11?

12. Give the Keil directive to reserve a block of zeroed memory, holding 40 words and labeled `coeffs`.

13. Give the CCS directive to reserve a block of zeroed memory, holding 40 words and labeled `coeffs`.

14. Explain the difference between Keil's EQU, DCD, and RN directives. Which, if any, would be used for the following cases?
    a. Assigning the Abort mode's bit pattern (0x17) to a new label called Mode_ABT.
    b. Storing sequential byte-sized numbers in memory to be used for copying to another location in memory.
    c. Storing the contents of register r12 to memory address 0x40000004.
    d. Associating a particular microcontroller's predefined memory-mapped register address with a name from the chip's documentation, for example, VIC0_VA7R.

5 Loads, Stores, and Addressing

5.1 INTRODUCTION

Processor architects spend a great deal of time analyzing typical routines on simulation models of a processor, often to find performance bottlenecks. Dynamic instruction usage gives a good indication of the types of operations that are performed the most while code is running. This differs from static usage that only describes the frequency of an instruction in the code itself. It turns out that while typical code is running, about half of the instructions deal with data movement, including data movement between registers and memory. Therefore, loading and storing data efficiently is critical to optimizing processor performance. As with all RISC processors, dedicated instructions are required for loading data from memory and storing data to memory. This chapter looks at those basic load and store instructions, their addressing modes, and their uses.

5.2 MEMORY

Earlier we said that one of the major components of any computing system is memory, a place to store our data and programs. Memory can be conceptually viewed as contiguous storage elements that hold data, each element holding a fixed number of bits and having an address. The typical analogy for memory is a very long string of mailboxes, where data (your letter) is stored in a box with a specific number on it. While there are some digital signal processors that use memory widths of 16 bits, the system that is nearly universally adopted these days has the width of each element as 8 bits, or a byte long. Therefore, we always refer to memory as being so many megabytes* (abbreviated MB, representing 2^{20} or approximately 10^6 bytes), gigabytes (abbreviated GB, representing 2^{30} or approximately 10^9 bytes), or even terabytes (abbreviated TB, representing 2^{40} or approximately 10^{12} bytes).

Younger programmers really should see what an 80 MB hard drive used to look like as late as the 1980s—imagine a washing machine with large, magnetic plates in the center that spun at high speeds. With the advances in magnetic materials and silicon memories, today's programmers have 4 TB hard drives on their desks and think

* The term megabyte is used loosely these days, as 1 kilobyte is defined as 2^{10} or 1024 bytes. A megabyte is 2^{20} or 1,048,576 bytes, but it is abbreviated as 1 million bytes. The distinction is rarely important.

nothing of it! Visit museums or universities with collections of older computers, if only to appreciate how radically storage technology has changed in less than one lifetime.

In large computing systems, such as workstations and mainframes, the memory to which the processor speaks directly is a fixed size, such as 4 GB, but the machine is capable of swapping out areas of memory, or pages, to larger storage devices, such as hard drives, that can hold as much as a terabyte or more. The method that is used to do this lies outside the scope of this book, but most textbooks on computer architecture cover it pretty well. Embedded systems typically need far less storage, so it's not uncommon to see a complete design using 2 MB of memory or less. In an embedded system, one can also ask how much memory is actually needed, since we may only have a simple task to perform with very little data. If our processor is used in an application that takes remote sensor data and does nothing but transmit it to a receiver, what could we possibly need memory for, other than storing a small program or buffering small amounts of data? Often, it turns out, embedded processors spend a lot of time twiddling their metaphorical thumbs, idly waiting for something to do. If a processor such as one in our remote sensor does decide to shut down or go into a quiescent state, it may have to save off the contents of its registers, including control registers, floating-point registers, and status registers. Energy management software may decide to power down certain parts of a chip when idle, and a loss of power may mean a loss of data. It may even have to store the contents of other on-chip memories such as a cache or tightly coupled memory (TCM).

Memory comes in different flavors and may reside at different addresses. For example, not all memory has to be readable and writable—some may be readable only, such as ROM (Read-Only Memory) or EEPROM (Electrically Erasable Programmable ROM)—but the data is accessed the same way for all types of memory. Embedded systems often use less expensive memories, e.g., 8-bit memory over faster, more expensive 32-bit memory, and it is left to the hardware designers to build a memory system for the application at hand. Programmers then write code for the system knowing something about the hardware up front. In fact, maps are often made of the memory system so that programmers know exactly how to access the various memory types in the system. Examining Figure 1.4 again, you'll notice that the address bus on the ARM7TDMI consists of 32 bits, meaning that you could address bytes in memory from address 0 to 2^{32}–1, or 4,294,967,295 (0xFFFFFFFF), which is considered to be 4 GB of memory space. If you look at the memory map of a Cortex-M4-based microcontroller, such as the Tiva TM4C123GH6ZRB shown in Table 5.1, you'll note that the entire address space is defined, but certain address ranges do not exist, such as addresses between 0x44000000 and 0xDFFFFFFF. You can also see that this part has different types of memories on the die—flash ROM memory and SRAM—and an interface to talk to external memory off-chip, such as DRAM. Not all addresses are used, and much of the memory map contains areas dedicated to specific functions, some of which we'll examine further in later chapters. While the memory layout is defined by an SoC's implementation, it is not part of the processor core.

TABLE 5.1
Memory Map of the Tiva TM4C123GH6ZRB

Start	End	Description	For Details, See Page...[a]
		Memory	
0x0000.0000	0x0003.FFFF	On-chip flash	553
0x0004.0000	0x00FF.FFFF	Reserved	—
0x0100.0000	0x1FFF.FFFF	Reserved for ROM	538
0x2000.0000	0x2000.7FFF	Bit-banded on-chip SRAM	537
0x2000.8000	0x21FF.FFFF	Reserved	—
0x2200.0000	0x220F.FFFF	Bit-band alias of bit-banded on-chip SRAM starting at 0x2000.0000	537
0x2210.0000	0x3FFF.FFFF	Reserved	—
		Peripherals	
0x4000.0000	0x4000.0FFF	Watchdog timer 0	798
0x4000.1000	0x4000.1FFF	Watchdog timer 1	798
0x4000.2000	0x4000.3FFF	Reserved	—
0x4000.4000	0x4000.4FFF	GPIO Port A	675
0x4000.5000	0x4000.5FFF	GPIO Port B	675
0x4000.6000	0x4000.6FFF	GPIO Port C	675
0x4000.7000	0x4000.7FFF	GPIO Port D	675
0x4000.8000	0x4000.8FFF	SSI 0	994
0x4000.9000	0x4000.9FFF	SSI 1	994
0x4000.A000	0x4000.AFFF	SSI 2	994
0x4000.B000	0x4000.BFFF	SSI 3	994
0x4000.C000	0x4000.CFFF	UART 0	931
0x4000.D000	0x4000.DFFF	UART 1	931
0x4000.E000	0x4000.EFFF	UART 2	931
0x4000.F000	0x4000.FFFF	UART 3	931
0x4001.0000	0x4001.0FFF	UART 4	931
0x4001.1000	0x4001.1FFF	UART 5	931
0x4001.2000	0x4001.2FFF	UART 6	931
0x4001.3000	0x4001.3FFF	UART 7	931
0x4001.4000	0x4001.FFFF	Reserved	—
		Peripherals	
0x4002.0000	0x4002.0FFF	I^2C 0	1044
0x4002.1000	0x4002.1FFF	I^2C 1	1044
0x4002.2000	0x4002.2FFF	I^2C 2	1044
0x4002.3000	0x4002.3FFF	I^2C 3	1044
0x4002.4000	0x4002.4FFF	GPIO Port E	675
0x4002.5000	0x4002.5FFF	GPIO Port F	675
0x4002.6000	0x4002.6FFF	GPIO Port G	675
0x4002.7000	0x4002.7FFF	GPIO Port H	675
0x4002.8000	0x4002.8FFF	PWM 0	1270

(continued)

TABLE 5.1 (continued)
Memory Map of the Tiva TM4C123GH6ZRB

Start	End	Description	For Details, See Page...[a]
0x4002.9000	0x4002.9FFF	PWM 1	1270
0x4002.A000	0x4002.BFFF	Reserved	—
0x4002.C000	0x4002.CFFF	QEI 0	1341
0x4002.D000	0x4002.DFFF	QEI 1	1341
0x4002.E000	0x4002.FFFF	Reserved	—
0x4003.0000	0x4003.0FFF	16/32-bit Timer 0	747
0x4003.1000	0x4003.1FFF	16/32-bit Timer 1	747
0x4003.2000	0x4003.2FFF	16/32-bit Timer 2	747
0x4003.3000	0x4003.3FFF	16/32-bit Timer 3	747
0x4003.4000	0x4003.4FFF	16/32-bit Timer 4	747
0x4003.5000	0x4003.5FFF	16/32-bit Timer 5	747
0x4003.6000	0x4003.6FFF	32/64-bit Timer 0	747
0x4003.7000	0x4003.7FFF	32/64-bit Timer 1	747
0x4003.8000	0x4003.8FFF	ADC 0	841
0x4003.9000	0x4003.9FFF	ADC 1	841
0x4003.A000	0x4003.BFFF	Reserved	—
0x4003.C000	0x4003.CFFF	Analog Comparators	1240
0x4003.D000	0x4003.DFFF	GPIO Port J	675
0x4003.E000	0x4003.FFFF	Reserved	—
0x4004.0000	0x4004.0FFF	CAN 0 Controller	1094
0x4004.1000	0x4004.1FFF	CAN 1 Controller	1094
0x4004.2000	0x4004.BFFF	Reserved	—
0x4004.C000	0x4004.CFFF	32/64-bit Timer 2	747
0x4004.D000	0x4004.DFFF	32/64-bit Timer 3	747
0x4004.E000	0x4004.EFFF	32/64-bit Timer 4	747
0x4004.F000	0x4004.FFFF	32/64-bit Timer 5	747
0x4005.0000	0x4005.0FFF	USB	1146
0x4005.1000	0x4005.7FFF	Reserved	—
0x4005.8000	0x4005.8FFF	GPIO Port A (AHB aperture)	675
0x4005.9000	0x4005.9FFF	GPIO Port B (AHB aperture)	675
0x4005.A000	0x4005.AFFF	GPIO Port C (AHB aperture)	675
0x4005.B000	0x4005.BFFF	GPIO Port D (AHB aperture)	675
0x4005.C000	0x4005.CFFF	GPIO Port E (AHB aperture)	675
0x4005.D000	0x4005.DFFF	GPIO Port F (AHB aperture)	675
0x4005.E000	0x4005.EFFF	GPIO Port G (AHB aperture)	675
0x4005.F000	0x4005.FFFF	GPIO Port H (AHB aperture)	675
0x4006.0000	0x4006.0FFF	GPIO Port J (AHB aperture)	675
0x4006.1000	0x4006.1FFF	GPIO Port K (AHB aperture)	675
0x4006.2000	0x4006.2FFF	GPIO Port L (AHB aperture)	675
0x4006.3000	0x4006.3FFF	GPIO Port M (AHB aperture)	675

TABLE 5.1 (continued)
Memory Map of the Tiva TM4C123GH6ZRB

Start	End	Description	For Details, See Page...[a]
0x4006.4000	0x4006.4FFF	GPIO Port N (AHB aperture)	675
0x4006.5000	0x4006.5FFF	GPIO Port P (AHB aperture)	675
0x4006.6000	0x4006.6FFF	GPIO Port Q (AHB aperture)	675
0x4006.7000	0x400A.EFFF	Reserved	—
0x400A.F000	0x400A.FFFF	EEPROM and Key Locker	571
0x400B.0000	0x400B.FFFF	Reserved	—
0x400C.0000	0x400C.0FFF	I^2C 4	1044
0x400C.1000	0x400C.1FFF	I^2C 5	1044
0x400C.2000	0x400F.8FFF	Reserved	—
0x400F.9000	0x400F.9FFF	System Exception Module	497
0x400F.A000	0x400F.BFFF	Reserved	—
0x400F.C000	0x400F.CFFF	Hibernation Module	518
0x400F.D000	0x400F.DFFF	Flash memory control	553
0x400F.E000	0x400F.EFFF	System control	237
0x400F.F000	0x400F.FFFF	μDMA	618
0x4010.0000	0x41FF.FFFF	Reserved	—
0x4200.0000	0x43FF.FFFF	Bit-banded alias of 0x4000.0000 through 0x400F.FFFF	—
0x4400.0000	0xDFFF.FFFF	Reserved	—
		Private Peripheral Bus	
0xE000.0000	0xE000.0FFF	Instrumentation Trace Macrocell (ITM)	70
0xE000.1000	0xE000.1FFF	Data Watchpoint and Trace (DWT)	70
0xE000.2000	0xE000.2FFF	Flash Patch and Breakpoint (FPS)	70
0xE000.3000	0xE000.DFFF	Reserved	—
0xE000.E000	0xE000.EFFF	Cortex-M4F Peripherals (SysTick, NVIC, MPU, FPU and SCB)	134
0xE000.F000	0xE003.FFFF	Reserved	—
0xE004.0000	0xE004.0FFF	Trace Port Interface Unit (TPIU)	71
0xE004.1000	0xE004.1FFF	Embedded Trace Macrocell (ETM)	70
0xE004.2000	0xFFFF.FFFF	Reserved	—

[a] See Tiva TM4C123GH6ZRB Microcontroller Data Sheet.

5.3 LOADS AND STORES: THE INSTRUCTIONS

Now that we have some idea of how memory is described in the system, the next step is to consider getting data out of memory and into a register, and vice versa. Recall that RISC architectures are considered to be load/store architectures, meaning that data in external memory must be brought into the processor using an instruction. Operations that take a value in memory, multiply it by a coefficient, add it to another

TABLE 5.2
Most Often Used Load/Store Instructions

Loads	Stores	Size and Type
LDR	STR	Word (32 bits)
LDRB	STRB	Byte (8 bits)
LDRH	STRH	Halfword (16 bits)
LDRSB		Signed byte
LDRSH		Signed halfword
LDM	STM	Multiple words

register, and then store the result back to memory with only a single instruction do not exist. For hardware designers, this is considered to be a very good thing, since some older architectures had so many options and modes for loading and storing data that it became nearly impossible to build the processors without introducing errors in the logic. Without listing every combination, Table 5.2 describes the most common instructions for dedicated load and store operations in the version 4T and version 7-M instruction sets.

Load instructions take a single value from memory and write it to a general-purpose register. Store instructions read a value from a general-purpose register and store it to memory. Load and store instructions have a single instruction format:

LDR|STR{<size>}{<cond>} <Rd>, <addressing_mode>

where <size> is an optional size such as byte or halfword (word is the default size), <cond> is an optional condition to be discussed in Chapter 8, and <Rd> is the source or destination register. Most registers can be used for both load and store instructions; however, there are register restrictions in the v7-M instructions, and for version 4T instructions, loads to register r15 (the PC) must be used with caution, as this could result in changing the flow of instruction execution. The addressing modes allowed are actually quite flexible, as we'll see in the next section, and they have two things in common: a base register and an (optional) offset. For example, the instruction

```
LDR  r9, [r12, r8, LSL #2]
```

would have a base register of r12 and an offset value created by shifting register r8 left by two bits. We'll get to the details of shift operations in Chapter 7, but for now just recognize LSL as a logical shift left by a certain number of bits. The offset is added to the base register to create the effective address for the load in this case.

It may be helpful at this point to introduce some nomenclature for the address—the term *effective address* is often used to describe the final address created from values in the various registers, with offsets and/or shifts. For example, in the instruction above, if the base register r12 contained the value 0x4000 and we added register r8, the offset, which contained 0x20, to it, we would have an effective address of 0x4080 (remember the offset is shifted). This is the address used to access memory.

A shorthand notation for this is ea<operands>, so if we said ea<r12 + r8*4>, the effective address is the value obtained from summing the contents of register r12 and 4 times the contents of register r8.

Sifting through all of the options for loads and stores, there are basically two main types of addressing modes available with variations, both of which are covered in the next section:

- Pre-indexed addressing
- Post-indexed addressing

If you allow for the fact that a simple load such as

```
LDR    r2, [r3]
```

can be viewed as special case of pre-indexed addressing with a zero offset, then loads and stores for the ARM7TDMI and Cortex-M4 processors take the form of an instruction with one of the two indexing schemes. Referring back to Table 5.2, the first three types of instructions simply transfer a word, halfword, or byte to memory from a register, or from memory to a register. For halfword loads, the data is placed in the least significant halfword (bits [15:0]) of the register with zeros in the upper 16 bits. For halfword stores, the data is taken from the least significant halfword. For byte loads, the data is placed in the least significant byte (bits [7:0]) of the register with zeros in the upper 24 bits. For byte stores, the data is taken from the least significant byte.

EXAMPLE 5.1

Consider the instruction

```
LDRH r11, [r0]; load a halfword into r11
```

Assuming the address in register r0 is 0x8000, before and after the instruction is executed, the data appears as follows:

	Memory	Address
r11 before load	0xEE	0x8000
0x12345678	0xFF	0x8001
r11 after load	0x90	0x8002
0x0000FFEE	0xA7	0x8003

Notice that 0xEE, the least significant byte at address 0x8000, is moved to the least significant byte in register r11, the second least significant byte, 0xFF, is moved to second least significant byte of register r11, etc. We'll have much more to say about this ordering shortly.

Signed halfword and signed byte load instructions deserve a little more explanation. The operation itself is quite easy—a byte or a halfword is read from memory, sign extended to 32 bits, then stored in a register. Here the programmer is specifically branding the data as signed data.

EXAMPLE 5.2

The instruction

```
LDRSH r11, [r0]; load signed halfword into r11
```

would produce the following scenario, again assuming register r0 contains the address 0x8000:

r11	Memory	Address
r11 before load	0xEE	0x8000
0x12345678	0x8C	0x8001
r11 after load	0x90	0x8002
0xFFFF8CEE	0xA7	0x8003

As in Example 5.1, the two bytes from memory are moved into register r11, except the most significant bit of the value at address 0x8001, 0x8C, is set, meaning that in a two's complement representation, this is a negative number. Therefore, the sign bit should be extended, which produces the value 0xFFFF8CEE in register r11.

You may not have noticed the absence of signed stores of halfwords or bytes into memory. After a little thinking, you might come to the conclusion that data stored to memory never needs to be sign extended. Computers simply treat data as a sequence of bit patterns and must be told how to interpret numbers. The value 0xEE could be a small, positive number, or it could be an 8-bit, two's complement representation of the number -18. The LDRSB and LDRSH instructions provide a way for the programmer to tell the machine that we are treating the values read from memory as signed numbers. This subject will be brought up again in Chapter 7 when we deal with fractional notations.

There are some very minor differences in the two broad classes of loads and stores, for both the ARM7TDMI and the Cortex-M4. For example, those instructions transferring words and unsigned bytes have more addressing mode options than instructions transferring halfwords and signed bytes, as shown in Table 5.3 and Table 5.4. These are not critical to understanding the instructions, so we'll proceed to see how they are used first.

TABLE 5.3
Addressing Options for Loads and Stores on the ARM7TDMI

	Imm Offset	Reg Offset	Scaled Reg Offset	Examples
Word	12 bits	Supported	Supported	`LDR r0, [r8, r2, LSL #28]`
Unsigned byte				`LDRB r4, [r8, #0xF1A]`
Halfword	8 bits	Supported	Not supported	`STRH r9, [r10, #0xF4]`
Signed halfword				`LDRSB r9, [r2, r1]`
Signed byte				

TABLE 5.4
Addressing Options for Loads and Stores on the Cortex-M4

	Imm Offset	Reg Offset	Scaled Reg Offset	Examples

Unsigned byte
Signed byte
Halfword
Signed halfword
Word
	Depending on instruction, index can range from –255 to 4095[a]	Supported	Supported	`LDRSB r3, [r6, r7, LSL #2]`
				`LDRSH r10, [r2, #0x42]`
				`STRH r3, [r6, r8]`

[a] Due to the way the instructions are encoded, there are actually *different* instructions for LDRSB r3, [r4, #0] and LDRSB r3, [r4, #-0]! Consult the v7-M ARM for other dubious behavior.

EXAMPLE 5.3

Storing data to memory requires only an address. If the value 0xFEEDBABE is held in register r3, and we wanted to store it to address 0x8000, a simple STR instruction would suffice.

```
STR r3, [r8]; store data to 0x8000
```

The registers and memory would appear as:

	Memory	Address
r8 before store	0xBE	0x8000
0x00008000	0xBA	0x8001
r8 after store	0xED	0x8002
0x00008000	0xFE	0x8003

However, we can perform a store operation and also increment our address automatically for further stores by using a post-increment addressing mode:

```
STR r3, [r8], #4; store data to 0x8000
```

The registers and memory would appear as:

	Memory	Address
r8 before store	0xBE	0x8000
0x00008000	0xBA	0x8001
r8 after store	0xED	0x8002
0x00008004	0xFE	0x8003

Other examples of single-operand loads and stores are below. We'll study the two types of addressing and their uses in the next sections.

```
LDR  r5, [r3]              ; load r5 with data from ea<r3>
STRB r0, [r9]              ; store data in r0 to ea<r9>
STR  r3, [r0, r5, LSL #3]  ; store data in r3 to ea<r0+(r5<<3)>
LDR  r1, [r0, #4]!         ; load r1 from ea<r0+4>,r0=r0+4
STRB r7, [r6, #-1]!        ; store byte to ea<r6-1>,r6=r6-1
LDR  r3, [r9], #4          ; load r3 from ea<r9>,r9=r9+4
STR  r2, [r5], #8          ; store word to ea<r5>,r5=r5+8
```

Load Multiple instructions load a subset (or possibly all) of the general-purpose registers from memory. Store Multiple instructions store a subset (or possibly all) of the general-purpose registers to memory. Because Load and Store Multiple instructions are used more for stack operations, we'll come back to these in Chapter 13, where we discuss parameter passing and stacks in detail. Additionally, the Cortex-M4 can load and store two words using a single instruction, but for now, we'll concentrate on the basic loads and stores.

5.4 OPERAND ADDRESSING

We said that the addressing mode for load and store instructions could be one of two types: pre-indexed addressing or post-indexed addressing, with or without offsets. For the most part, these are just variations on a theme, so once you see how one works, the others are very similar. We'll begin by examining pre-indexed addressing first.

5.4.1 Pre-Indexed Addressing

The pre-indexed form of a load or store instruction is

LDR|STR{<size>}{<cond>} <Rd>, [<Rn>, <offset>]{!}

In pre-indexed addressing, the address of the data transfer is calculated by adding an offset to the value in the base register, Rn. The optional "!" specifies writing the effective address back into Rn at the end of the instruction. Without it, Rn contains its original value after the instruction executes. Figure 5.1 shows the instruction

```
STR r0, [r1, #12]
```

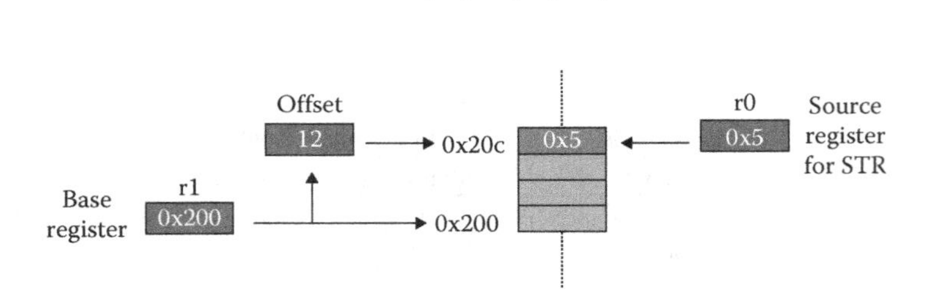

FIGURE 5.1 Pre-indexed store operation.

where register r0 contains 0x5. The store is done by using the value in register r1, 0x200 in this example, as a base address. The offset 12 is added to this address *before* the data is stored to memory, so the effective address is 0x20C. An important point here is the base register r1 is not modified after this operation. If the value needs to be updated automatically, then the "!" can be added to the instruction, becoming

```
STR r0, [r1, #12]!
```

Referring back to Table 5.3, when performing word and unsigned byte accesses on an ARM7TDMI, the offset can be a register shifted by any 5-bit constant, or it can be an unshifted 12-bit constant. For halfword, signed halfword, and signed byte accesses, the offset can be an unsigned 8-bit immediate value or an unshifted register. Offset addressing can use the barrel shifter, which we'll see in Chapters 6 and 7, to provide logical and arithmetic shifts of constants. For example, you can use a rotation (ROR) and logical shift to the left (LSL) on values in registers before using them. In addition, you can either add or subtract the offset from the base register. As you are writing code, limitations on immediate values and constant sizes will be flagged by the assembler, and if an error occurs, just find another way to calculate your offsets and effective addresses.

Further examples of pre-indexed addressing modes for the ARM7TDMI are as follows:

```
STR r3, [r0, r5, LSL #3]      ; store r3 to ea<r0+(r5<<3)> (r0 unchanged)
LDR r6, [r0, r1, ROR #6]!     ; load r6 from ea<r0+(r1>>6)> (r0 updated)
LDR r0, [r1, #-8]             ; load r0 from ea<r1-8>
LDR r0, [r1, -r2, LSL #2]     ; load r0 from ea<r1+(-r2<<2)>
LDRSH r5, [r9]                ; load signed halfword from ea<r9>
LDRSB r3, [r8, #3]            ; load signed byte from ea<r8+3>
LDRSB r4, [r10, #0xc1]        ; load signed byte from ea<r10+193>
```

Referring back to Table 5.4, the Cortex-M4 has slightly more restrictive usage. For example, you cannot use a negated register as an offset, nor can you perform any type of shift on a register other than a logical shift left (LSL), and even then, the shift count must be no greater than 3. Otherwise, the instructions look very similar. Valid examples are

```
LDRSB  r0, [r5, r3, LSL #1]
STR    r8, [r0, r2]
LDR    r12, [r7, #-4]
```

5.4.2 Post-Indexed Addressing

The post-indexed form of a load or store instruction is:

LDR|STR{<size>}{<cond>} <Rd>, [<Rn>], <offset>

In post-indexed addressing, the effective address of the data transfer is calculated from the unmodified value in the base register, Rn. The offset is then added to the

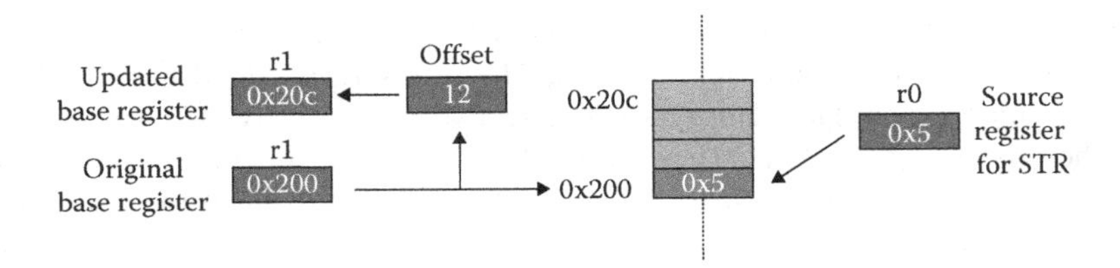

FIGURE 5.2 Post-indexed store operation.

value in Rn, and the sum is written back to Rn. This type of incrementing is useful in stepping through tables or lists, since the base address is automatically updated for you.

Figure 5.2 shows the instruction

```
STR r0, [r1], #12
```

where register r0 contains the value 0x5. In this case, register r1 contains the base address of 0x200, which is used as the effective address. The offset of 12 is added to the base address register *after* the store operation is complete. Also notice the absence of the "!" option in the mnemonic, since post-indexed addressing always modifies the base register.

As for pre-indexed addressing, the same rules shown in Table 5.3 for ARM7TDMI addressing modes and in Table 5.4 for Cortex-M4 addressing modes apply to post-indexed addressing, too. Examples of post-indexed addressing for both cores include

```
STR   r7, [r0], #24   ; store r7 to ea<r0>, then r0 = r0+24
LDRH  r3, [r9], #2    ; load halfword to r3 from ea<r9>, then r9 = r9+2
STRH  r2, [r5], #8    ; store halfword from r2 to ea<r5>, then r5 = r5+8
```

The ARM7TDMI has a bit more flexibility, in that you can even perform rotations on the offset value, such as

```
LDR r2, [r0], r4, ASR #4; load r2 to ea<r0>, add r4/16 after
```

EXAMPLE 5.4

Consider a simple ARM7TDMI program that moves a string of characters from one memory location to another.

```
SRAM_BASE   EQU     0x04000000      ; start of SRAM for STR910FM32
            AREA    StrCopy, CODE
            ENTRY                   ; mark the first instruction
Main        ADR     r1, srcstr      ; pointer to the first string
            LDR     r0, =SRAM_BASE  ; pointer to the second string
strcopy
            LDRB    r2, [r1], #1    ; load byte, update address
            STRB    r2, [r0], #1    ; store byte, update address
            CMP     r2, #0          ; check for zero terminator
            BNE     strcopy         ; keep going if not
```

```
stop        B       stop            ; terminate the program
srcstr      DCB     "This is my (source) string", 0
            END
```

The first line of code equates the starting address of SRAM with a constant so that we can just refer to it by name, instead of typing the 32-bit number each time we need it. In addition to the two assembler directives that follow, the program includes two pseudo-instructions, ADR and a special construct of LDR, which we will see in Chapter 6. We can use ADR to load the address of our source string into register r1. Next, the address of our destination is moved into register r0. A loop is then set up that loads a byte from the source string into register r2, increments the address by one byte, then stores the data into a new address, again incrementing the destination address by one. Since the string is null-terminated, the loop continues until it detects the final zero at the end of the string. The BNE instruction uses the result of the comparison against zero and branches back to the label `strcopy` only if register r2 is not equal to zero. The source string is declared at the end of the code using the DCB directive, with the zero at the end to create a null-terminated string. If you run the example code on an STR910FM32 microcontroller, you will find that the source string has been moved to SRAM starting at address 0x04000000 when the program is finished.

If you follow the suggestions outlined in Appendix A, you can run this exact same code on a Cortex-M4 part, such as the Tiva TM4C123GH6ZRB, accounting for one small difference. On the TI microcontroller, the SRAM region begins at address 0x20000000 rather than 0x04000000. Referring back to the memory map diagram shown in Table 5.1, this region of memory is labeled as bit-banded on-chip SRAM, but for this example, you can safely ignore the idea of a bit-banded region and use it as a simple scratchpad memory. We'll cover bit-banding in Section 5.6.

5.5 ENDIANNESS

The term "endianness" actually comes from a paper written by Danny Cohen (1981) entitled "On Holy Wars and a Plea for Peace." The raging debate over the ordering of bits and bytes in memory was compared to Jonathan Swift's satirical novel *Gulliver's Travels*, where in the book rival kingdoms warred over which end of an egg was to be broken first, the little end or the big end. Some people find the whole topic more like something out of *Alice's Adventures in Wonderland*, where Alice, upon being told by a caterpillar that one side of a perfectly round mushroom would make her grow taller while the other side would make her grow shorter, asks "And now which is which?" While the issue remains a concern for software engineers, ARM actually supports both formats, known as little-endian and big-endian, through software and/or hardware mechanisms.

To illustrate the problem, suppose we had a register that contained the 32-bit value 0x0A0B0C0D, and this value needed to be stored to memory addresses 0x400 to 0x403. Little-endian configurations would dictate that the least significant byte in the register would be stored to the lowest address, and the most significant byte in the register would be stored to the highest address, as shown in Figure 5.3. While it was only briefly mentioned earlier, Examples 5.1, 5.2, and 5.3 are all assumed to be little-endian (have a look at them again).

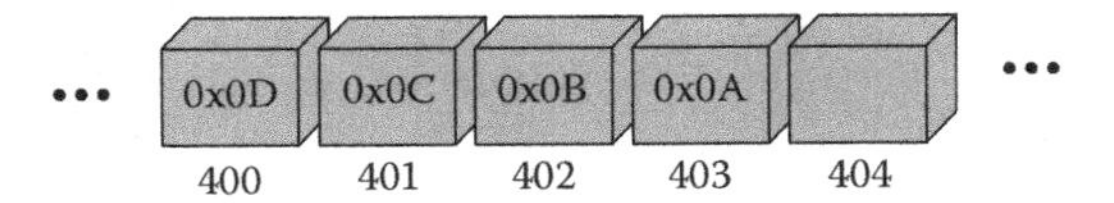

FIGURE 5.3 Little-endian memory configuration.

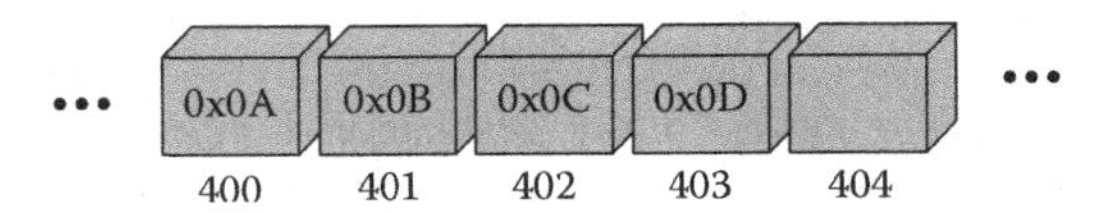

FIGURE 5.4 Big-endian memory configuration.

There is really no reason that the bytes couldn't be stored the other way around, namely having the lowest byte in the register stored at the highest address and the highest byte stored at the lowest address, as shown in Figure 5.4. This is known as *word-invariant* big-endian addressing in the ARM literature. Using an ARM7TDMI, if you are always reading and writing word-length values, the issue really doesn't arise at all. You only see a problem when halfwords and bytes are being transferred, since there is a difference in the data that is returned. As an example, suppose you transferred the value 0xBABEFACE to address 0x400 in a little-endian configuration. If you were to load a halfword into register r3 from address 0x402, the register would contain 0x0000BABE when the instruction completed. If it were a big-endian configuration, the value in register r3 would be 0x0000FACE.

ARM has no preference for which you use, and it will ultimately be up to the hardware designers to determine how the memory system is configured. The default format is little-endian, but this can be changed on the ARM7TDMI by using the BIGEND pin. Nearly all microcontrollers based on the Cortex-M4 are configured as little-endian, but more detailed information on *byte-invariant* big-endian formatting should be reviewed in the *Architectural Reference Manual* (ARM 2007c) and (Yiu 2014), in light of the fact that word-invariant big-endian format has been deprecated in the newest ARM processors. Many large companies have used a particular format for historical reasons, but there are some applications that benefit from one orientation over another, e.g., reading network traffic is simpler when using a big-endian configuration. All of the coding examples in the book assume a little-endian memory configuration.

For programmers who may have seen memory ordered in a big-endian configuration, or for those who are unfamiliar with endianness, a glimpse at memory might be a little confusing. For example, in Figure 5.5, which shows the Keil development tools, the instruction

```
MOV r0, #0x83
```

can be seen in both the disassembly and memory windows. However, the bit pattern for the instruction is 0xE3A00083, but it appears to be backwards starting at 0x1E4 in the memory window, only because the lowest byte (0x83) has been stored at the lowest address. This is actually quite correct—the disassembly window has taken some liberties here in reordering the data for easier viewing. Code Composer Studio does

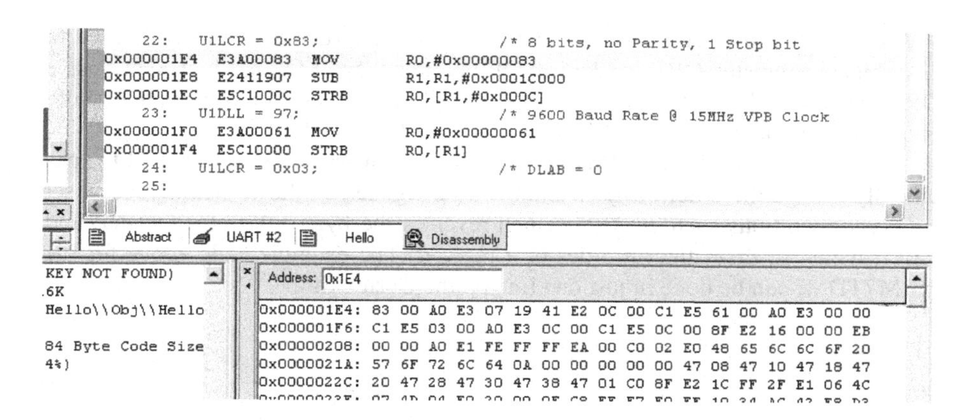

FIGURE 5.5 Little-endian addressing of an instruction.

something similar, so check your tools with a simple test case if you are uncertain. While big-endian addressing might be a little easier to read in a memory window such as this, little-endian addressing can also be easy to read with some practice, and some tools even allow data to be formatted by selecting your preferences.

5.5.1 Changing Endianness

Should it be necessary to swap the endianness of a particular register or a large number or words, the following code can be used for the ARM7TDMI. This method is best for single words.

```
; On entry: r0 holds the word to be swapped
; On exit : r0 holds the swapped word, r1 is destroyed
byteswap                            ; r0=A, B, C, D
        EOR r1, r0, r0, ROR #16     ; r1=A^C,B^D,C^A,D^B
        BIC r1, r1, #0xFF0000       ; r1=A^C, 0, C^A,D^B
        MOV r0, r0, ROR #8          ; r0=D, A, B, C
        EOR r0, r0, r1, LSR #8      ; r0=D, C, B, A
```

The following method is best for swapping the endianness of a large number of words:

```
; On entry: r0 holds the word to be swapped
; On exit : r0 holds the swapped word,
;         : r1, r2 and r3 are destroyed
byteswap                          ; three instruction initialization
      MOV r2, #0xFF               ; r2=0xFF
      ORR r2, r2, #0xFF0000       ; r2=0x00FF00FF
      MOV r3, r2, LSL #8          ; r3=0xFF00FF00
; repeat the following code for each word to swap
                                  ; r0=A B C D
      AND r1, r2, r0, ROR #24     ; r1=0 C 0 A
      AND r0, r3, r0, ROR #8      ; r0=D 0 B 0
      ORR r0, r0, r1              ; r0=D C B A
```

We haven't come across the BIC, ORR, or EOR instructions yet. BIC is used to clear bits in a register, ORR is a logical OR operation, and EOR is a logical exclusive OR operation. All will be covered in more detail in Chapter 7, or you can read more about them in the *Architectural Reference Manual* (ARM 2007c).

After the release of the ARM10 processor, new instructions were added to specifically change the order of bytes and bits in a register, so the v7-M instruction set supports operations such as REV, which reverses the byte order of a register, and RBIT, which reverses the bit order of a register. The example code above for the ARM7TDMI can be done in just one line on the Cortex-M4:

```
byteswap                 ; r0 = A B C D
        REV r1, r0       ; r1 = D C B A
```

5.5.2 Defining Memory Areas

The algorithm has been defined, the microcontroller has been identified, the features are laid out for you, and now it's time to code. When you write your first routines, it will probably be necessary to initialize some memory areas and define variables, and while this is seen again in Chapter 12, it's probably worth elaborating a bit more here. There are some easy ways to set up tables and constants in your program, and the methods you use depend on how readable you want the code to be. For example, if a table of coefficients is needed, and each coefficient is represented in 8 bits, then you might declare an area of memory as

```
table DCB 0xFE, 0xF9, 0x12, 0x34
      DCB 0x11, 0x22, 0x33, 0x44
```

if you are reading each value with a LDRB instruction. Assuming that the table was started in memory at address 0x4000 (the compilation tools would normally determine the starting address, but it's possible to do it yourself), the memory would look like

Address	Data Value
0x4000	0xFE
0x4001	0xF9
0x4002	0x12
0x4003	0x34
0x4004	0x11
0x4005	0x22
0x4006	0x33
0x4007	0x44

If all of the data used will be word-length values, then you'd probably declare an area in memory as

```
table DCD 0xFEF91234
      DCD 0x11223344
```

but notice that its memory listing in a little-endian system would look like

Address	Data Value
0x4000	0x34
0x4001	0x12
0x4002	0xF9
0x4003	0xFE
0x4004	0x44
0x4005	0x33
0x4006	0x22
0x4007	0x11

In other words, the directives used and the endianness of the system will determine how the data is ordered in memory, so be careful. Since you normally don't switch endianness while the processor is running, once a configuration is chosen, just be aware of the way the data is stored.

5.6 BIT-BANDED MEMORY

With the introduction of the Cortex-M3 and M4 processors, ARM gave programmers the ability to address single bits more efficiently. Imagine that some code wants to access only one particular bit in a memory location, say bit 2 of a 32-bit value held at address 0x40040000. Microcontrollers often use memory-mapped registers in place of registers in the core, especially in industrial microcontrollers where you have ten or twenty peripherals, each with its own set of unique registers. Let's further say that a peripheral such as a Controller Area Network (CAN) controller on the Tiva TM4C123GH6ZRB, which starts at memory address 0x40040000, has individual control bits that are set or cleared to enable different modes, read status information, or transmit data. For example, bit 7 of the CAN Control Register puts the CAN controller in test mode. If we wish to set this bit and only this bit, you could use a read-modify-write operation such as:

```
LDR     r3, =0x40040000    ; location of CAN Control Register
LDR     r2, [r3]           ; read the memory-mapped register contents
ORR     r2, #0x80          ; set bit 7
STR     r2, [r3]           ; write the entire register contents back
```

This seems horribly wasteful from a code size and execution time perspective to set just one bit in a memory-mapped register. Imagine then if every bit in a register had its own address—rather than loading an entire register, modifying one bit, then writing it back, an individual bit could be set by just writing to its address. Examining Table 5.1 again, you can see that there are two bit-banded regions of memory: addresses from 0x22000000 to 0x220FFFFF are used specifically for bit-banding the 32KB region from 0x20000000 to 0x20007FFF; and addresses from 0x42000000 to 0x43FFFFFF are used specifically for bit-banding the 1MB region from 0x40000000 to 0x400FFFFF. Figure 5.6 shows the mapping between

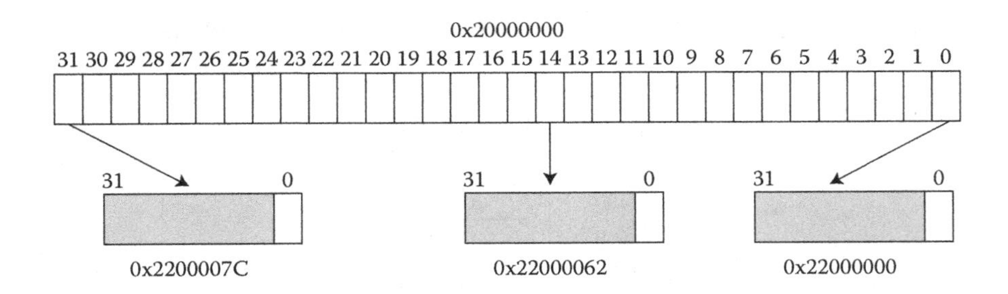

FIGURE 5.6 Mapping bit-banded regions.

the regions. Going back to the earlier CAN example, we could set bit 7 using just a single store operation:

```
LDR     r3, =0x4280001C
MOV     r4, #1
STR     r4, [r3]          ; set bit 7 of the CAN Control Register
```

The address 0x4280001C is derived from

$$\begin{aligned} \text{bit-band alias} &= \text{bit-band base} + (\text{byte offset} \times 32) + (\text{bit number} \times 4) \\ &= 0\text{x}42000000 + (0\text{x}40000 \times 0\text{x}20) + (7 \times 4) \\ &= 0\text{x}42000000 + 0\text{x}800000 + 0\text{x}1\text{C} \end{aligned}$$

As another example, if bit 1 at address 0x40038000 (the ADC 0 peripheral) is to be modified, the bit-band alias is calculated as:

$$0\text{x}42000000 + (0\text{x}38000 \times 0\text{x}20) + (1 \times 4) = 0\text{x}42700004$$

What immediately becomes obvious is that you would need a considerable number of addresses to make a one-to-one mapping of addresses to individual bits. In fact, if you do the math, to have each bit in a 32KB section of memory given its own address, with each address falling on a word boundary, i.e., ending in either 0, 4, 8, or C, you would need

$$32{,}768 \text{ bytes} \times 8 \text{ bits/byte} \times 4 \text{ bytes/bit} = 1\text{MB}$$

The trade-off then becomes an issue of how much address space can be sacrificed to support this feature, but given that microcontrollers never use all 4GB of their address space, and that large swaths of the memory map currently go unused, this is possible. Perhaps in ten years, it might not be.

5.7 MEMORY CONSIDERATIONS

In a typical microcontroller, there are often blocks of volatile memory (SRAM or some other type of RAM) available for you to use, along with different kinds of non-volatile memory (flash or ROM) where your code would live. Simulators such as Keil's RealView Microcontroller Development Kit model those different blocks of memory for you, so you don't necessarily stop to think about how code was loaded into flash or how some variables ended up in SRAM. As a programmer, you write your code, press a few buttons, and voilà—things just work. Describing what happens behind the scenes and all the options associated with emulation and debugging could easy fill another book, but let's at least see how blocks of memory are configured as we declare sections of code.

Consider a directive used in a program to reserve some space for a stack (a stack is a section of memory used during exception processing and subroutines which we'll see in Chapters 13, 14 and 15, but for now we are just telling the processor to reserve a section of RAM for us). Our directive might look like

```
        AREA STACK, NOINIT, READWRITE, ALIGN=3
StackMem
        SPACE  Stack
```

If we are programming something like a microcontroller, then we also have our program that needs to be stored in flash memory, so that when the processor is reset, code *already* exists in memory to be executed. The start of our program might look like

```
        AREA    RESET, CODE, READONLY
        THUMB
;***************************************************************
;
; The vector table.
;
;***************************************************************
        DCD     StackMem+Stack          ;Top of Stack
        DCD     Reset_Handler           ; Reset Handler
        DCD     NmiSR                   ; NMI Handler
        DCD     FaultISR                ; Hard Fault Handler
                .
                .
                .
```

At this point, something is missing—how does a development tool know that there is a block of RAM on our microcontroller for things like stacks, and how does it know the starting address of that block? When you first start your simulation, you likely pick a part from a list of available microcontrollers (if you use the Keil tools), and the map of the memory system is already configured in the tool for you. When you assemble your program, the tools will generate a map file such as the one in Figure 5.7 (Keil) or Figure 5.8 (CCS) which shows where code and variables are actually stored. The linker then uses this information when building an executable to ensure the various sections (in the object files created by the assembler) are placed in the appropriate memories, where sections are built with the AREA directives we have been using. In Figure 5.7, you can see that the section that we called RESET, which is our program, would be stored to ROM starting at address 0x0. Any read-only sections are also stored to this ROM region. Read/write and zero-initialized data would be stored to RAM starting at address 0x04000000, which

```
; *************************************************************
; ***** Scatter-Loading Description File generated by uVision ****
; *************************************************************
LR_IROM1 0x00000000 0x00040000 { ;load region size_region
  ER_IROM1 0x00000000 0x00040000 { ;load address=execution address
    *.o (RESET, +First)
    *(InRoot$$Sections)
    .ANY (+RO)
  }
  RW_IRAM1 0x04000000 0x00010000 { ;RW data
    .ANY (+RW +ZI)
  }
}
```

FIGURE 5.7 Keil memory map tile.

```
/******************************************************************************
 *
 * Default Linker Command file for the Texas Instruments TM4C123GH6PM
 *
 * This is derived from revision 11167 of the TivaWare Library.
 *
 *****************************************************************************/

--retain=g_pfnVectors

MEMORY
{
  FLASH (RX) : origin=0x00000000, length=0x00040000
  SRAM (RWX) : origin=0x20000000, length=0x00008000
}
/* The following command line options are set as part of the CCS project.  */
/* If you are building using the command line, or for some reason want to  */
/* define them here, you can uncomment and modify these lines as needed.   */
/* If you are using CCS for building, it is probably better to make any    */
/* modifications in your CCS project and leave this file alone.            */
/*                                                                         */
/* --heap_size=0                                                           */
/* --stack_size=256                                                        */
/* --library=rtsv7M4_T_le_eabi.lib                                         */

/* Section allocation in memory */

SECTIONS
{
 .intvecs:     > 0x00000000
 .text   :     > FLASH
 .const  :     > FLASH
 .cinit  :     > FLASH
 .pinit  :     > FLASH
 .init_array : > FLASH
 .myCode :     > FLASH

 .vtable :     > 0x20000000
 .data   :     > SRAM
 .bss    :     > SRAM
 .sysmem :     > SRAM
 .stack  :     > SRAM
}
```

FIGURE 5.8 Code Composer Studio linker command file.

is where the SRAM block is located on an STR910FM32 microcontroller, in this example. You can also create your own custom scatter-loading file to feed into the linker, and those details can be found in *RealView Compilation Tools Developer Guide* (ARM 2007a).

Other techniques, like those used in the gnu tools, can be used to assign variables to certain regions of memory. For example, in C, it is possible to tell the linker to place a variable at a specific location in memory. If you were writing code, you might say something like:

```
#include <stdio.h>

extern int cube(int n1);
int gCubed __attribute__((at(0x9000))); // Place at 0x9000

int main()
{

    gCubed = cube(3);
    printf("Your number cubed is: %d\n", gCubed);

}
```

Your global variable called gCubed would be placed at the absolute address 0x9000. In most instances, it is still far easier to control variables and data using directives.

5.8 EXERCISES

1. Describe the contents of register r13 after the following instructions complete, assuming that memory contains the values shown below. Register r0 contains 0x24, and the memory system is little-endian.

Address	Contents
0x24	0x06
0x25	0xFC
0x26	0x03
0x27	0xFF

   a. `LDRSB r13, [r0]`
   b. `LDRSH r13, [r0]`
   c. `LDR r13, [r0]`
   d. `LDRB r13, [r0]`

2. Indicate whether the following instructions use pre- or post-indexed addressing modes:
   a. `STR r6, [r4, #4]`
   b. `LDR r3, [r12], #6`

c. `LDRB r4, [r3, r2]!`
d. `LDRSH r12, [r6]`

3. Calculate the effective address of the following instructions if register r3 = 0x4000 and register r4 = 0x20:
   a. `STRH r9, [r3, r4]`
   b. `LDRB r8, [r3, r4, LSL #3]`
   c. `LDR r7, [r3], r4`
   d. `STRB r6, [r3], r4, ASR #2`

4. What's wrong with the following instruction running on an ARM7TDMI?

```
LDRSB r1,[r6],r3,LSL#4
```

5. Write a program for either the ARM7TDMI or the Cortex-M4 that sums word-length values in memory, storing the result in register r3. Include the following table of values to sum in your code:

```
TABLE  DCD  0xFEBBAAAA, 0x12340000, 0x88881111
       DCD  0x00000013, 0x80808080, 0xFFFF0000
```

6. Assume an array contains 30 words of data. A compiler associates variables *x* and *y* with registers r0 and r1, respectively. Assume the starting address of the array is contained in register r2. Translate the C statement below into assembly instructions:

   x = array[7] + y;

7. Using the same initial conditions as Exercise 6, translate the following C statement into assembly instructions:

   array[10] = array[8] + y;

8. Consider a C procedure that initializes an array of bytes to all zeros, given as

```
init_Indices (int a[], int s) {
        int i;
        for (i=0; i<s; i++)
        a[i]=0; }
```

   Write the assembly language for this initialization routine. Assume $s > 0$ and is held in register r2. Register r1 contains the starting address of the array, and the variable *i* is held in register r3. While loops are not covered until Chapter 8, you can build a simple for loop using the following construction:

```
        MOV r3, #0        ; clear i
loop    instruction
        instruction
        ADD r3, r3, #1    ; increment i
```

```
CMP r3, r2          ; compare i to s
BNE loop            ; branch to loop if not equal
```

9. Suppose that registers belonging to a particular peripheral on a microcontroller have a starting address of 0xE000C000. Individual registers within the peripheral are addressed as offsets from the starting address. If a register called LSR0 is 0x14 bytes away from the starting address, write the assembly and Keil directives that will load a byte of data into register r6, where the data is located in the LSR0 register. Use pre-indexed addressing.

10. Assume register r3 contains 0x8000. What would the register contain after executing the following instructions?
    a. `STR   r6, [r3, #12]`
    b. `STRB  r7, [r3], #4`
    c. `LDRH  r5, [r3], #8`
    d. `LDR   r12, [r3, #12]!`

11. Assuming you have a little-endian memory system connected to the ARM7TDMI, what would register r4 contain after executing the following instructions? Register r6 holds the value 0xBEEFFACE and register r3 holds 0x8000.

```
STR   r6, [r3]
LDRB  r4, [r3]
```

    What if you had a big-endian memory system?

6 Constants and Literal Pools

6.1 INTRODUCTION

One of the best things about learning assembly language is that you deal directly with hardware, and as a result, learn about computer architecture in a very direct way. It's not absolutely necessary to know how data is transferred along busses, or how instructions make it from an instruction queue into the execution stage of a pipeline, but it is interesting to note why certain instructions are necessary in an instruction set and how certain instructions can be used in more than one way. Instructions for moving data, such as MOV, MVN, MOVW, MOVT, and LDR, will be introduced in this chapter, specifically for loading constants into a register, and while floating-point constants will be covered in Chapter 9, we'll also see an example or two of how those values are loaded. The reason we focus so heavily on constants now is because they are a very common requirement. Examining the ARM rotation scheme here also gives us insight into fast arithmetic—a look ahead to Chapter 7. The good news is that a shortcut exists to load constants, and programmers make good use of them. However, for completeness, we will examine what the processor and the assembler are doing to generate these numbers.

6.2 THE ARM ROTATION SCHEME

As mentioned in Chapter 1, an original design goal of early RISC processors was to have fixed-length instructions. In the case of ARM processors, the ARM and many of the Thumb-2 instructions are 32 bits long (16-bit Thumb instructions will be discussed later on). This brings us to the apparent contradiction of fitting a 32-bit constant into an instruction that is only 32 bits long. To see how this is done, let's begin by examining the binary encoding of an ARM MOV instruction, as shown in Figure 6.1.

You can see the fields associated with the class of instruction (bits [27:25], which indicate that this is a data processing instruction), the instruction itself (bits [24:21], which would indicate a MOV instruction), and the least significant 12 bits. These last bits have quite a few options, and give the instruction great flexibility to either use registers, registers with shifts or rotates, or immediate values as operands. We will look at the case where the operand is an immediate data value, as show in Figure 6.2. Notice that the least significant byte (8 bits) can be any number between 0 and 255, and bits [11:8] of the instruction now specify a rotate value. The value is multiplied by 2, then used to rotate the 8-bit value to the right by that many bits, as shown in

31 28	27 26	25	24 21	20	19 16	15 12	11 0
cond	0 0	1	opcode	S	Rn	Rd	shifter_operand

FIGURE 6.1 MOV instruction.

31 28	27 26	25	24 21	20	19 16	15 12	11 8	7 0
cond	0 0	1	opcode	S	Rn	Rd	rotate_imm	8_bit_immediate

FIGURE 6.2 MOV instruction with an immediate operand.

Figure 6.3. This means that if our bit pattern were 0xE3A004FF, for example, the machine code actually translates to the mnemonic

```
MOV     r0, #0xFF, 8
```

since the least-significant 12 bits of the instruction are 0x4FF, giving us a rotation factor of 8, or 4 doubled, and a byte constant of 0xFF.

Figure 6.4 shows a simplified diagram of the ARM7 datapath logic, including the barrel shifter and main adder. While its use for logical and arithmetic shifts is covered in detail in Chapter 7, the barrel shifter is also used in the creation of constants. Barrel shifters are really little more than circuits designed specifically to shift or rotate data, and they can be built using very fast logic. ARM's rotation scheme moves bits to the right using the inline barrel shifter, wrapping the least significant bit around to the most significant bit at the top.

With 12 bits available in an instruction and dedicated hardware for performing shifts, ARM7TDMI processors can generate *classes* of numbers instead of every number between 0 and $2^{32} - 1$. Analysis of typical code has shown that about half of all constants lie in the range between –15 and 15, and about ninety percent of them lie in the range between –511 and 511. You generally also need large, but simple constants, e.g., 0x4000, for masks and specifying base addresses in memory. So while not every constant is possible with *this* scheme, as we will see shortly, it is still possible to put any 32-bit number in a register.

Let's examine some of the classes of numbers that can be generated using this rotation scheme. Table 6.1 shows examples of numbers you can easily generate with

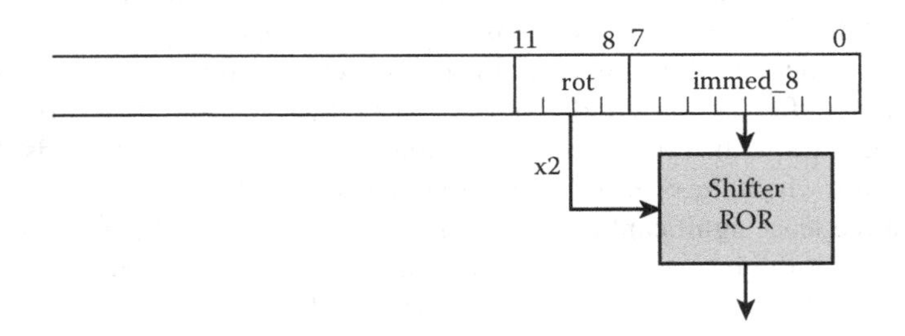

FIGURE 6.3 Byte rotated by an even number of bits.

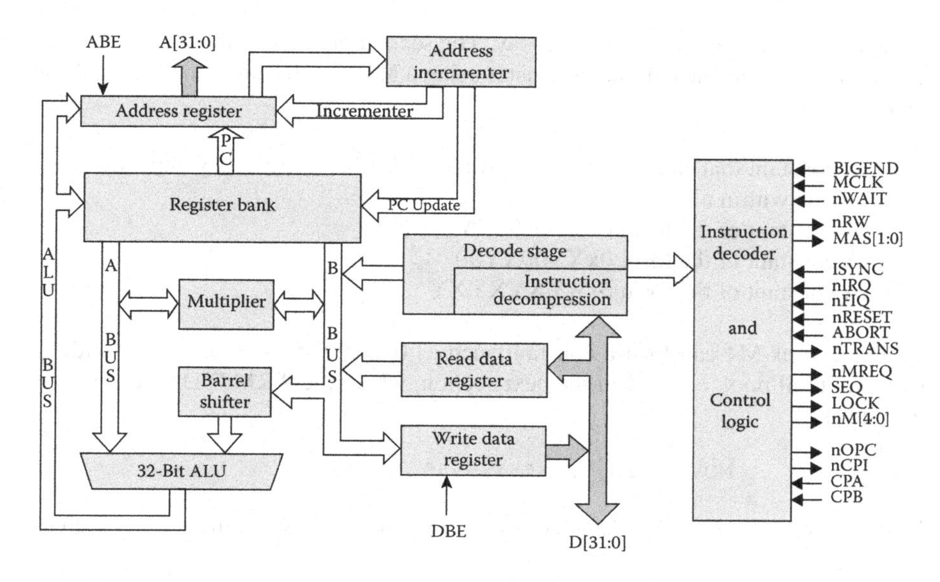

FIGURE 6.4 ARM7 internal datapaths.

a MOV using an ARM7TDMI. You can, therefore, load constants directly into registers or use them in data operations using instructions such as

```
MOV    r0, #0xFF                 ; r0 = 255
MOV    r0, #0x1, 30              ; r0 = 1020
MOV    r0, #0x1, 26              ; r0 = 4096
ADD    r0, r2, #0xFF000000       ; r0 = r2 + 0xFF000000
SUB    r2, r3, #0x8000           ; r2 = r3 − 0x8000
RSB    r8, r9, #0x8000           ; r8 = 0x8000 - r9
```

The Cortex-M4 can generate similar classes of numbers, using similar Thumb-2 instructions; however, the format of the MOV instruction is different, so rotational

TABLE 6.1
Examples of Creating Constants with Rotation

Rotate	Binary	Decimal	Step	Hexadecimal
No rotate	000000000000000000000000xxxxxxxx	0-255	1	0-0xFF
Right, 30 bits	0000000000000000000000xxxxxxxx00	0-1020	4	0-0x3FC
Right, 28 bits	00000000000000000000xxxxxxxx0000	0-4080	16	0-0xFF0
Right, 26 bits	000000000000000000xxxxxxxx000000	0-16320	64	0-0x3FC0
…	…	…	…	…
Right, 8 bits	xxxxxxxx000000000000000000000000	0-255x2^{24}	2^{24}	0-0xFF000000
Right, 6 bits	xxxxxx0000000000000000000000000xx	—	—	—
Right, 4 bits	xxxx0000000000000000000000000xxxx	—	—	—
Right, 2 bits	xx000000000000000000000000000000	—	—	—

values are not specified in the same way. The second operand is more flexible, so if you wish to load a constant into a register using a MOV instruction, the constant can take the form of

- A constant that can be created by shifting an 8-bit value left by any number of bits within a word
- A constant of the form 0x00XY00XY
- A constant of the form 0xXY00XY00
- A constant of the form 0xXYXYXYXY

The Cortex-M4 can load a constant such as 0x55555555 into a register without using a literal pool, covered in the next section, which the ARM7TDMI cannot do, written as

```
MOV    r3, #0x55555555
```

Data operations permit the use of constants, so you could use an instruction such as

```
ADD r3, r4, #0xFF000000
```

that use the rotational scheme. If you using a MOV instruction to perform a shift operation, then the preferred method is to use ASR, LSL, LSR, ROR, or RRX instructions, which are covered in the next chapter.

EXAMPLE 6.1

Calculate the rotation necessary to generate the constant 4080 using the byte rotation scheme.

SOLUTION

Since 4080 is 111111110000_2, the byte 11111111_2 or 0xFF can be rotated to the left by four bits. However, the rotation scheme rotates a byte to the *right*; therefore, a rotation factor of 28 is needed, since rotating to the left *n* bits is equivalent to rotating to the right by (32-*n*) bits. The ARM instruction would be

```
MOV    r0, #0xFF, 28; r0 = 4080
```

EXAMPLE 6.2

A common method used to access peripherals on a microcontroller (ignoring bit-banding for the moment) is to specify a base address and an offset, meaning that the peripheral starts at some particular value in memory, say 0x22000000, and then the various registers belonging to that peripheral are specified as an offset to be added to the base address. The reasoning behind this scheme relies on the addressing modes available to the processor. For example, on the Tiva TM4C123GH6ZRB microcontroller, the system control base starts at address 0x400FE000. This region contains registers for configuring the main clocks, turning the PLL on and off, and enabling various other peripherals. Let's further suppose that we're interested in setting just one bit in a register called RCGCGPIO,

Encoding T2 ARMv7-M

```
MOV{S}<c> .W <Rd>, #<const>
```

15 14 13 12 11	10	9	8 7 6 5	4	3 2 1 0	15	14 13 12	11 10 9 8	7 6 5 4 3 2 1 0
1 1 1 1 0	i	0	0 0 1 0	S	1 1 1 1	0	imm3	Rd	imm8

FIGURE 6.5 MOV operation using a 32-bit Thumb instruction.

which is located at an offset of 0x608 and turns on the clock to GPIO block F. This can be done with a single store instruction such as

```
STR     r1, [r0, r2]
```

where the base address 0x400FE000 would be held in register r0, and our offset of 0x608 would be held in register r2. The most direct way to load the offset value of 0x608 into register r2 is just to say

```
MOV     r2, #0x608
```

It turns out that this value can be created from a byte (0xC1) shifted three bits to the left, so if you were to assemble this instruction for a Cortex-M4, the 32-bit Thumb-2 instruction that is generated would be 0xF44F62C1. From Figure 6.5 below you can see that the rotational value 0xC1 occupies the lowest byte of the instruction.

The MVN (move negative) instruction, which moves a one's complement of the operand into a register, can also be used to generate classes of numbers, such as

```
MVN     r0, #0          ; r0 = 0xFFFFFFFF
MVN     r3, #0xEE       ; r3 = 0xFFFFFF11
```

for the ARM7TDMI and Cortex-M4, and

```
MVN     r0, #0xFF, 8    ; r0 = 0x00FFFFFF
```

for the ARM7TDMI.

These rotation schemes are fine, but as a programmer, you might find this entire process a bit tiring if you have to enter dozens of constants for a data-intensive algorithm. This brings us back to our shortcut, and to numbers that cannot be built using the various methods above.

6.3 LOADING CONSTANTS INTO REGISTERS

We covered the topic of memory in detail in the last chapter, and we saw that there are specific instructions for loading data from memory into a register—the LDR instruction. You can create the address required by this instruction in a number of different ways, and so far we've examined addresses loaded directly into a register. Now the idea of an address created from the Program Counter is introduced, where register r15 (the PC) is used with a displacement value to create an address. And we're also going to bend the LDR instruction a bit to create a pseudo-instruction that the assembler understands.

First, the shortcut: When writing assembly, you should use the following pseudo-instruction to load constants into registers, as this is by far the easiest, safest, and most maintainable way, assuming that your assembler supports it:

```
LDR    <Rd>, =<numeric constant>
```

or for floating-point numbers

```
VLDR.F32    <Sd>, =<numeric constant>
VLDR.F64    <Dd>, =<numeric constant>
```

so you could say something like

```
LDR         r8, =0x20000040; start of my stack
```

or

```
VLDR.F32    s7, =3.14159165; pi
```

It may seem unusual to use a pseudo-instruction, but there's a valid reason to do so. For most programmers, constants are declared at the start of sections of code, and it may be necessary to change values as code is written, modified, and maintained by other programmers. Suppose that a section of code begins as

```
SRAM_BASE    EQU     0x04000000
             AREA    EXAMPLE, CODE, READONLY
;
; initialization section
;
             ENTRY
             MOV     r0, #SRAM_BASE
             MOV     r1, #0xFF000000
             .
             .
             .
```

If the value of SRAM_BASE ever changed to a value that couldn't be generated using the byte rotation scheme, the code will generate an error. If the code were written using

```
LDR    r0, =SRAM_BASE
```

instead, the code will always assemble no matter what value SRAM_BASE takes. This immediately raises the question of how the assembler handles those "unusual" constants.

When the assembler sees the LDR pseudo-instruction, it will try to use either a MOV or MVN instruction to perform the given load before going further. Recall that we can generate classes of numbers, but not every number, using the rotation schemes mentioned earlier. For those numbers that cannot be created, a *literal pool*, or a block of constants, is created to hold them in memory, usually very near the instructions that asked for the data, along with a load instruction that fetches the constant from memory. By default, a literal pool is placed at every END directive, so a load instruction would look just beyond the last instruction in a block of code for your number. However, the addressing mode that is used to do this, called a PC-relative address, only has a range of 4 kilobytes (since the offset is only 12 bits), which means that a very large block of code can cause a problem if we don't correct for it. In fact, even a short block of code can potentially cause problems. Suppose we have the following ARM7TDMI code in memory:

```
        AREA Example, CODE
        ENTRY                         ; mark first instruction
        BL      func1                 ; call first subroutine
        BL      func2                 ; call second subroutine
stop    B       stop                  ; terminate the program
func1   LDR     r0, =42               ; => MOV r0, #42
        LDR     r1, =0x12345678       ; => LDR r1, [PC, #N]
                                      ; where N=offset to literal pool 1
        LDR     r2, =0xFFFFFFFF       ; => MVN r2, #0
        BX      lr                    ; return from subroutine
        LTORG                         ; literal pool 1 has 0x12345678
func2   LDR     r3, =0x12345678       ; => LDR r3, [PC, #N]
                                      ; N=offset back to literal pool 1
        ;LDR    r4, =0x87654321       ; if this is uncommented, it fails.
                                      ; Literal pool 2 is out of reach!
        BX      lr                    ; return from subroutine
BigTable
        SPACE 4200                    ; clears 4200 bytes of memory,
                                      ; starting here
        END                           ; literal pool 2 empty
```

This contrived program first calls two very short subroutines via the branch and link (BL) instruction. The next instruction is merely to terminate the program, so for now we can ignore it. Notice that the first subroutine, labeled func1, loads the number 42 into register r0, which is quite easy to do with a byte rotation scheme. In fact, there is no rotation needed, since 0x2A fits within a byte. So the assembler generates a MOV instruction to load this value. The next value, 0x12345678, is too "odd" to create using a rotation scheme; therefore, the assembler is forced to generate a literal pool, which you might think would start after the 4200 bytes of space we've reserved at the end of the program. However, the load instruction cannot reach this far, and if we do nothing to correct for this, the assembler will generate an error. The second load instruction in the subroutine, the one setting all the bits in register r2, can be performed with a MVN instruction. The final instruction in the subroutine transfers the value from the Link Register (r14) back into the Program Counter (register r15), thereby forcing the processor to return to the instruction following the first BL instruction. Don't worry about subroutines just yet, as there is an entire chapter covering their operation.

By inserting an LTORG directive just at the end of our first subroutine, we have forced the assembler to build its literal pool between the two subroutines in memory, as shown in Figure 6.6, which shows the memory addresses, the instructions, and the actual mnemonics generated by the assembler. You'll also notice that the LDR instruction at address 0x10 in our example appears as

```
LDR    r1, [PC,#0x0004]
```

which needs some explanation as well. As we saw in Chapter 5, this particular type of load instruction tells the processor to use the Program Counter (which always contains the address of *the instruction being fetched* from memory) modify that number (in this case add the number 8 to it) and then use this as an address. When we used the LTORG directive and told the assembler to put our literal pool between the subroutines in memory, we fixed the placement of our constants, and the assembler can then calculate how far those constants lie from the address in the Program Counter. The important thing to note in all of this is where the Program Counter is when the LDR instruction is in the pipeline's execute stage. Again, referring to Figure 6.6, you can see that if the LDR instruction is in the execute stage of the ARM7TDMI's pipeline, the MVN is in the decode stage, and the BX instruction is in the fetch stage. Therefore, the difference between the address 0x18 (what's in the PC) and where we need to be to get our constant, which is 0x1C, is 4, which is the offset used to modify the PC in the LDR instruction. The good news is that you don't ever have to calculate these offsets yourself—the assembler does that for you.

There are two more constants in the second subroutine, only one of which actually gets turned into an instruction, since we commented out the second load instruction. You will notice that in Figure 6.6, the instruction at address 0x20 is another PC-relative address, but this time the offset is negative. It turns out that the instructions can share the data already in a literal pool. Since the assembler just generated this constant for the first subroutine, and it just happens to be very near our instruction (within 4 kilobytes), you can just subtract 12 from the value of the Program Counter when the LDR instruction is in the execute stage of the pipeline. (For those

Address		Instruction			
0x00000000	EB000001	BL	0x0000000C		
0x00000004	EB000005	BL	0x00000020		
0x00000008	EAFFFFFE	B	0x00000008		
0x0000000C	E3A0002A	MOV	RO,#0x0000002A		
0x00000010	E59F1004	LDR	R1,[PC,#0x0004]		EXECUTE
0x00000014	E3E02000	MVN	R2,#0x00000000		DECODE
0x00000018	E12FFF1E	BX	R14	← PC	FETCH
0x0000001C	12345678			← PC + 4	
0x00000020	E51F300C	LDR	R3,[PC,#-0x000C]		
0x00000024	E12FFF1E	BX	R14		

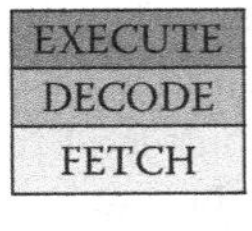

FIGURE 6.6 Disassembly of ARM7TDMI program.

readers really paying attention: the Program Counter seems to have fetched the next instruction from beyond our little program—is this a problem or not?) The second load instruction has been commented out to prevent an assembler error. As we've put a table of 4200 bytes just at the end of our program, the nearest literal pool is now more than 4 kilobytes away, and the assembler cannot build an instruction to reach that value in memory. To fix this, another LTORG directive would need to be added just before the table begins.

If you tried to run this same code on a Cortex-M4, you would notice several things. First, the assembler would generate code using a combination of 16-bit and 32-bit instructions, so the disassembly would look very different. More importantly, you would get an error when you tried to assemble the program, since the second subroutine, func2, tries to create the constant 0x12345678 in a second literal pool, but it would be beyond the 4 kilobyte limit due to that large table we created. It cannot initially use the value already created in the first literal pool like the ARM7TDMI did because the assembler creates the shorter (16-bit) version of the LDR instruction. Looking at Figure 6.7, you can see the offset allowed in the shorter instruction is only 8 bits, which is scaled by 4 for word accesses, and it cannot be negative. So now that the Program Counter has progressed beyond the first literal pool in memory, a PC-relative load instruction that cannot subtract values from the Program Counter to create an address will not work. In effect, we cannot see backwards. To correct this, a very simple modification of the instruction consists of adding a ".W" (for wide) extension to the LDR mnemonic, which forces the assembler to use a 32-bit Thumb-2 instruction, giving the instruction more options for creating addresses. The code below will now run without any issues.

```
          BL      func1              ; call first subroutine
          BL      func2              ; call second subroutine
stop      B       stop               ; terminate the program
func1     LDR     r0, =42            ; => MOV r0, #42
          LDR     r1, =0x12345678    ; => LDR r1, [PC, #N]
                                     ; where N=offset to literal pool 1
          LDR     r2, =0xFFFFFFFF    ; => MVN r2, #0
          BX      lr                 ; return from subroutine
          LTORG                      ; literal pool 1 has 0x12345678
```

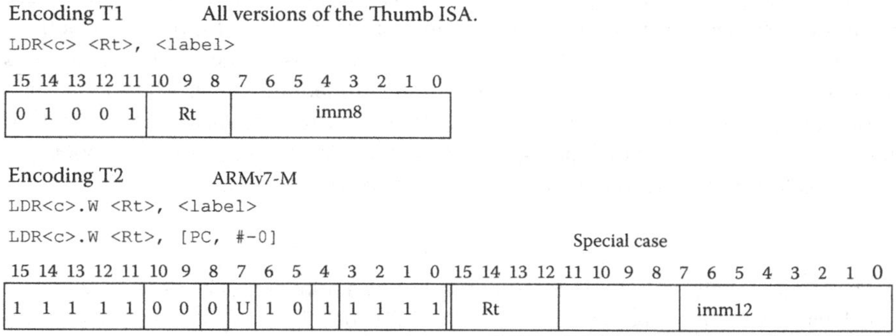

Encoding T1 All versions of the Thumb ISA.

LDR<c> <Rt>, <label>

15	14	13	12	11	10	9	8	7	6	5	4	3	2	1	0
0	1	0	0	1	Rt			imm8							

Encoding T2 ARMv7-M

LDR<c>.W <Rt>, <label>

LDR<c>.W <Rt>, [PC, #-0] Special case

15	14	13	12	11	10	9	8	7	6	5	4	3	2	1	0
1	1	1	1	1	0	0	0	U	1	0	1	1	1	1	1

15	14	13	12	11	10	9	8	7	6	5	4	3	2	1	0
Rt								imm12							

FIGURE 6.7 LDR instruction in Thumb and Thumb-2.

```
func2       LDR.W   r3, =0x12345678   ; => LDR r3, [PC, #N]
                                      ; N=offset back to literal pool 1
            ;LDR    r4, =0x98765432   ; if this is uncommented, it fails.
                                      ; Literal pool 2 is out of reach!
            BX      lr                ; return from subroutine
BigTable
            SPACE 4200                ; clears 4200 bytes of memory,
                                      ; starting here
```

So to summarize:

Use LDR <Rd > , =< numeric constant> to put a constant into an integer register.

Use VLDR <Sd > , =< numeric constant> to put a constant into a floating-point register. We'll see this again in Section 9.9.

Literal pools are generated at the end of each section of code.

The assembler will check if the constant is available in a literal pool already, and if so, it will attempt to address the existing constant.

On the Cortex-M4, if an error is generated indicating a constant is out of range, check the width of the LDR instruction.

The assembler will attempt to place the constant in the next literal pool if it is not already available. If the next literal pool is out of range, the assembler will generate an error and you will need to fix it, probably with an LTORG or adjusting the width of the instruction used.

If you do use an LTORG, place the directive after the failed LDR pseudo-instruction and within ±4 kilobytes. You must place literal pools where the processor will not attempt to execute the data as an instruction, so put the literal pools after unconditional branch instructions or at the end of a subroutine.

6.4 LOADING CONSTANTS WITH MOVW, MOVT

Earlier we saw that there are several ways of moving constants into registers for both the ARM7TDMI and the Cortex-M4, and depending on the type of data you have, the assembler will try and optimize the code by using the smallest instruction available, in the case of the Cortex-M4, or use the least amount of memory by avoiding literal pools, in the case of both the ARM7TDMI and the Cortex-M4. There are two more types of move instructions available on the Cortex-M4; both instructions take 16 bits of data and place them in a register. MOVW is the same operation as MOV, only the operand is restricted to a 16-bit immediate value. MOVT places a 16-bit value in the top halfword of a register, so the pair of instructions can load any 32-bit constant into a destination register, should your assembler not support the LDR pseudo-instruction, e.g., Code Composer Studio.

EXAMPLE 6.3

The number 0xBEEFFACE cannot be created using a rotational scheme, nor does it fall into any of the formats, such as 0xXY00XY00, that allow a single MOV instruction to load this value into a register. You can, however, use the combination of MOVT and MOVW to create a 32-bit constant in register r3:

```
        MOVW            r3, #0xFACE
        MOVT            r3, #0xBEEF
```

6.5 LOADING ADDRESSES INTO REGISTERS

At some point, you will need to load the address of a label or symbol into a register. Usually you do this to give yourself a starting point of a table, a list, or maybe a set of coefficients that are needed in a digital filter. For example, consider the ARM7TDMI code fragment below.

```
SRAM_BASE EQU     0x04000000
          AREA    FILTER, CODE

dest      RN0     ; destination pointer
image     RN1     ; image data pointer
coeff     RN2     ; coefficient table pointer
pointer   RN3     ; temporary pointer
          ENTRY
          CODE32
Main
          ; initialization area
          LDR     dest, =#SRAM_BASE    ; move memory base into dest
          MOV     pointer, dest        ; current pointer is destination
          ADR     image, image_data    ; load image data pointer
          ADR     coeff, cosines       ; load coefficient pointer
          BL      filter               ; execute one pass of filter
          .
          .
          .
          ALIGN
image_data
          DCW     0x0001,0x0002,0x0003,0x0004
          DCW     0x0005,0x0006,0x0007,0x0008
          .
          .
          .
cosines
          DCW     0x3ec5,0x3537,0x238e,0x0c7c
          DCW     0xf384,0xdc72,0xcac9,0xc13b
          .
          .
          .
          END
```

While the majority of the program is still to be written, you can see that if we were to set up an algorithm, say an FIR filter, where you had some data stored in memory and some coefficients stored in memory, you would want to set pointers to the start of each set. This way, a register would hold a starting address. To access a particular data value, you would simply use that register with an offset of some kind.

We have seen the directives EQU and RN already in Chapter 4, but now we actually start using them. The first line equates the label SRAM_BASE to a number, so that when we use it in the code, we don't have to keep typing that long address, similar to the #DEFINE statement in C. The RN directives give names to our registers r0, r1, r2, and r3, so that we can refer to them by their function rather than by their number. You don't have to do this, but often it's helpful to know a register's use while programming. The first two instructions load a known address (called an absolute address, since it doesn't move if you relocate your code in memory) into registers r0 and r3. The third and fourth instructions are the pseudo-instruction ADR, which is particularly useful at loading addresses into a register. Why do it this way? Suppose that this section of code was to be used along with other blocks. You wouldn't necessarily know exactly where your data starts once the two sections are assembled, so it's easier to let the assembler calculate the addresses for you. As an example, if image_data actually started at address 0x8000 in memory, then this address gets moved into register r1, which we've renamed. However, if we change the code, move the image data, or add another block of code that we write later, then this address will change. By using ADR, we don't have to worry about the address.

EXAMPLE 6.4

Let's examine another example, this time to see how the ADR pseudo-instruction actually gets converted into real ARM instructions. Again, the code in this example doesn't actually do anything except set up pointers, but it will serve to illustrate how ADR behaves.

```
      AREA    adrlabel,CODE,READONLY
      ENTRY                          ; mark first instruction to execute

Start BL      func                   ; branch to subroutine
stop  B       stop                   ; terminate
      LTORG                          ; create a literal pool
func  ADR     r0, Start              ; => SUB r0, PC, #offset to Start
      ADR     r1, DataArea           ; => ADD r1, PC, #offset to DataArea
      ;ADR    r2, DataArea + 4300    ; This would fail because the offset
                                     ; cannot be expressed by operand2 of ADD
      ADRL    r2, DataArea + 4300    ; => ADD r2, PC, #offset1
                                     ; ADD r2, r2, #offset2
      BX      lr                     ; return
DataArea
      SPACE   8000                   ; starting at the current location,
                                     ; clears an 8000-byte area of memory to 0
      END
```

You will note that the program calls a subroutine called func, using a branch and link operation (BL). The next instruction is for ending the program, so we really only need to examine what happens after the LTORG directive. The subroutine begins with a label, `func`, and an ADR pseudo-instruction to load the starting address of our main program into register r0. The assembler actually creates either an ADD or SUB instruction with the Program Counter to do this. Similar to the LDR pseudo-instruction we saw previously, by knowing the value of the Program Counter at the time when this ADD or SUB reaches the execute stage of the pipeline, we can simply take that value and modify it to generate an address. The catch is that the offset must be a particular type of number. For ARM instructions, that number must be one that can be created using a byte value rotated by an even number of bits, exactly as we saw in Section 6.2 (if rejected by the assembler, it will generate an error message to indicate that an offset cannot be represented by 0–255 and a rotation). For 32-bit Thumb instructions, that number must be within ±4095 bytes of a byte, half-word, or word-aligned address. If you notice the second ADR in this example, the distance between the instruction and the label `DataArea` is small enough that the assembler will use a simple ADD instruction to create the constant.

The third ADR tries to create an address where the label is on the other side of an 8000-byte block of memory. This doesn't work, but there is another pseudo-instruction: ADRL. Using two operations instead of one, the ADRL will calculate an offset that is within a range based on the addition of two values now, both created by the byte rotation scheme mentioned above (for ARM instructions). There is a fixed range for 32-bit Thumb instructions of ±1MB. You should note that if you invoke an ADRL pseudo-instruction in your code, it will generate two operations even if it could be done using only one, so be careful in loops that are sensitive to cycle counts. One other important point worth mentioning is that the label used with ADR or ADRL *must be within the same code section*. If a label is out of range in the same section, the assembler faults the reference. As an aside, if a label is out of range in other code sections, the linker faults the reference.

There is yet another way of loading addresses into registers, and it is exactly the same as the LDR pseudo-instruction we saw earlier for loading constants. The syntax is

```
LDR   <Rd>, =label
```

In this instance, the assembler will convert the pseudo-instruction into a load instruction, where the load reads the address from a literal pool that it creates. As with the case of loading constants, you must ensure that a literal pool is within range of the instruction. This pseudo-instruction differs from ADR and ADRL in that labels *outside* of a section can be referenced, and the linker will resolve the reference at link time.

EXAMPLE 6.5

The example below shows a few of the ways the LDR pseudo-instruction can be used, including using labels with their own offsets.

```
      AREA    LDRlabel, CODE, READONLY
      ENTRY                       ; Mark first instruction to execute
start
      BL      func1               ; branch to first subroutine
      BL      func2               ; branch to second subroutine
stop  B       stop                ; terminate

func1
      LDR     r0, =start          ;=> LDR R0, [PC, #offset into Literal Pool 1]
      LDR     r1, =Darea + 12     ;=> LDR R1, [PC, #offset into Lit. Pool 1]
      LDR     r2, =Darea + 6000   ;=> LDR R2, [PC, #offset into Lit. Pool 1]
      BX      lr                  ; return

      LTORG
func2
      LDR     r3, =Darea + 6000   ; => LDR R3, [PC, #offset into Lit. Pool 1]
                                  ; (sharing with previous literal)
      ; LDR   r4, =Darea + 6004   ; if uncommented produces an error
```

```
                                    ; as literal pool 2 is out of range
        BX      lr                  ; return
Darea
        SPACE   8000                ; starting at the current location, clears
                                    ; an 8000-byte area of memory to zero
        END                         ; literal pool 2 is out of range of the LDR
                                    ; instructions above
```

You can see the first three LDR statements in the subroutine `func1` would actually be PC-relative loads from a literal pool that would exist in memory at the LTORG statement. Additionally, the first load statement in the second subroutine could use the same literal pool to create a PC-relative offset. As the SPACE directive has cleared an 8000-byte block of memory, the second load instruction cannot reach the second literal pool, since it must be within 4 kilobytes.

So to summarize:

Use the pseudo-instruction

ADR <Rd>, label

to put an address into a register whenever possible. The address is created by adding or subtracting an offset to/from the PC, where the offset is calculated by the assembler.

If the above case fails, use the ADRL pseudo-instruction, which will calculate an offset using two separate ADD or SUB operations. Note that if you invoke an ADRL pseudo-instruction in your code, it will generate two operations even if it could be done using only one.

Use the pseudo-instruction

LDR <Rd>, =label

if you plan to reference labels in other sections of code, or you know that a literal table will exist and you don't mind the extra cycles used to fetch the literal from memory. Use the same caution with literal pools that you would for the construct

LDR <Rd>, =constant

Consult the *Assembler User's Guide* (ARM 2008a) for more details on the use of ADR, ADRL and LDR for loading addresses.

6.6 EXERCISES

1. What constant would be loaded into register r7 by the following instructions?
   a. `MOV r7, #0x8C, 4`
   b. `MOV r7, #0x42, 30`
   c. `MVN r7, #2`
   d. `MVN r7, #0x8C, 4`

2. Using the byte rotation scheme described for the ARM7TDMI, calculate the instruction and rotation needed to load the following constants into register r2:
   a. 0xA400
   b. 0x7D8
   c. 0x17400
   d. 0x1980

3. Tell whether or not the following constants can be loaded into an ARM7TDMI register without creating a literal pool and using only a single instruction:
   a. 0x12340000
   b. 0x77777777
   c. 0xFFFFFFFF
   d. 0xFFFFFFFE

4. Tell whether or not the following constants can be loaded into a Cortex-M4 register without creating a literal pool and using only a single instruction:
   a. 0xEE00EE00
   b. 0x09A00000
   c. 0x33333373
   d. 0xFFFFFFFE

5. What is the best way to put a numeric constant into a register, assuming your assembler supports the method?

6. Where is the best place to put literal pool data?

7. Suppose you had the following code:

```
            AREA SAMPLE, CODE,READONLY
            ENTRY
start
            MOV   r12, #SRAM_BASE
            ADD   r0, r1, r2
            MOV   r0, #0x18
            BL    routine1
            .
            .
            .
routine1
            STM   sp!, {r0-r3,lr}
            .
            .
            .
            END
```

   Describe two ways to load the label `routine1` into register r3, noting any restrictions that apply.

8. Describe the difference between ADR and ADRL.

9. Give the instruction(s) to perform the following operations for both the ARM7TDMI and the Cortex-M4:
   a. Add 0xEC00 to register r6, placing the sum in register r4.
   b. Subtract 0xFF000000 from register r12, placing the result in register r7.
   c. Add the value 0x123456AB to register r7, placing the sum in register r12.
   d. Place a two's complement representation of –1 into register r3.

10. Suppose you had the following code and you are using the Keil tools:

```
          .
          .
          .

          BL func1                ; call first subroutine
          BL func2                ; call second subroutine
stop      B stop                  ; terminate the program

func1     MOV r2, #0
          LDR r1, =0xBABEFACE
          LDR r2, =0xFFFFFFFC
          MOV pc, lr              ; return from subroutine

          LTORG                   ; literal pool 1 has 0xBABEFACE
func2     LDR r3, =0xBABEFACE
          LDR r4, =0x66666666
          MOV pc, lr              ; return from subroutine
BigTable
          SPACE 3700              ; clears 3700 bytes of memory,
                                  ; starting here
          .
          .
          .
```

    On an ARM7TDMI, will loading 0x66666666 into register r4 cause an error? Why or why not? What about on a Cortex-M4?

11. The ARM branch instruction—B—provides a ±32 MB branching range. If the Program Counter is currently 0x8000 and you need to jump to address 0xFF000000, how do you think you might do this?

12. Assuming that the floating-point hardware is already enabled (CP10 and CP11), write the instructions to load the floating-point register s3 with a quiet NaN, or 0x7FC00000. (Hint: use Program 5 from Chapter 3 as a guide.) You can write it using either the Keil tools or the CCS tools.

7 Integer Logic and Arithmetic

7.1 INTRODUCTION

This is a long chapter, but for good reasons. Arithmetic operations are probably one of the more common types of instructions you will use, especially if the software being written involves manipulating large amounts of data, such as an incoming audio signal. Graphics algorithms, speech processing algorithms, digital controllers, and audio processing all involve a great deal of arithmetic work, so it's important to understand the types of data that you have and how to perform the operation needed in the shortest amount of time and/or space. We begin with a discussion of flags, examine the basic arithmetic instructions for both the ARM7TDMI and the Cortex-M4, quickly examine some of the new DSP extensions in the Cortex-M4, and then wrap up with an overview of fractional arithmetic. Once you gain a good understanding of the concepts behind integer arithmetic, you should then be able to tackle floating-point arithmetic in Chapters 9, 10, and 11, and even go on to more advanced functions, such as trigonometric functions, exponentials, and square roots.

7.2 FLAGS AND THEIR USE

Recall from Chapter 2 that the Program Status Register holds the current state of the machine: the flags, the mode, the interrupt bits, and the Thumb bit for the ARM7TDMI, and the flags, the exception number, the Interrupt-Continuable Instruction (ICI) bits, and the Thumb bit for the Cortex-M4, shown in Figure 7.1. There are four bits, N, Z, C, and V, in the uppermost nibble that help determine whether or not an instruction will be conditionally executed. The flags are set and cleared based on one of four things:

- Instructions that are specifically used for setting and clearing flags, such TST or CMP
- Instructions that are told to set the flags by appending an "S" to the mnemonic. For example, EORS would perform an exclusive OR operation and set the flags afterward, since the S bit is set in the instruction. We can do this with all of the ALU instructions, so we control whether or not to update the flags
- A direct write to the Program Status Register, where you explicitly set or clear flags
- A 16-bit Thumb ALU instruction, which will be covered both here and in Chapter 17

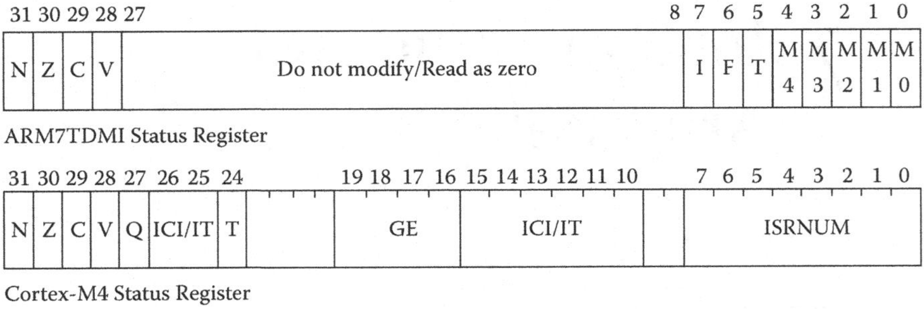

31	30	29	28	27 … 8	7	6	5	4	3	2	1	0
N	Z	C	V	Do not modify/Read as zero	I	F	T	M 4	M 3	M 2	M 1	M 0

ARM7TDMI Status Register

31	30	29	28	27	26 25	24		19 18 17 16	15 14 13 12 11 10		7 6 5 4 3 2 1 0
N	Z	C	V	Q	ICI/IT	T		GE	ICI/IT		ISRNUM

Cortex-M4 Status Register

FIGURE 7.1 Status registers.

The Q flag on the Cortex-M4 indicates a value has saturated and different rules govern its behavior, so it is discussed separately in Section 7.4.4. In the next sections, we'll examine each flag individually—some are quite easy and some require a little thought.

7.2.1 The N Flag

This flag is useful when checking for a negative result. What does this mean, negative? This definition sits in the context of a two's complement number system, and as we saw in the last few chapters, a two's complement number is considered to be negative if the most significant bit is set. Be careful, though, as you could easily have two perfectly good positive numbers add together to produce a value with the uppermost bit set.

EXAMPLE 7.1

Adding –1 to –2 is easy enough, and the result has the most-significant bit set, as expected. In two's complement notation, this would be represented as

```
  FFFFFFFF
+ FFFFFFFE
----------
  FFFFFFFD
```

If we were to code this on the ARM7TDMI as

```
MOV   r3, #-1
MOV   r4, #-2
ADDS  r3, r4, r3
```

we would expect to see the N bit set in the CPSR, as shown in Figure 7.2, which it is, as the most significant bit of register r3 was set as a result of the addition.

EXAMPLE 7.2

If we add the values below, the addends are positive in two's complement notation, but the sum is negative, i.e.,

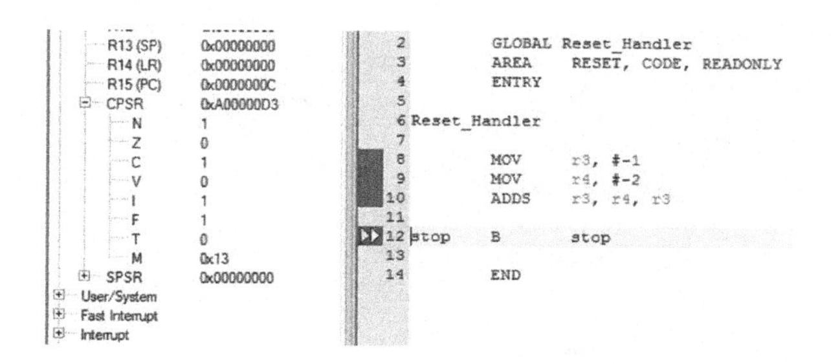

FIGURE 7.2 Status flags in the CPSR.

```
  7B000000
+ 30000000
----------
  AB000000
```

which means that something might be wrong. First, notice that since the most-significant bit is now set, this forces the N bit to be set if our ADD instruction actually sets the flags (remember, it doesn't have to). Second, if you aren't working with two's complement numbers, then perhaps we don't really care what the value of the N bit is. Finally, in a two's complement representation, notice that we originally meant to add two positive numbers together to get a bigger positive sum, but the result indicates that this positive sum cannot be represented in 32 bits, so the result effectively overflowed the precision we had available. So perhaps we need one more flag to work with signed values.

7.2.2 The V Flag

When performing an operation like addition or subtraction, if we calculate the V flag as an exclusive OR of the carry bit going into the most significant bit of the result with the carry bit coming out of the most significant bit, then the V flag accurately indicates a signed overflow. Overflow occurs if the result of an add, subtract, or compare is greater than or equal to 2^{31}, or less than -2^{31}.

EXAMPLE 7.3

Two signed values, assumed to be in two's complement representations, are added to produce the sum

```
  A1234567
+ B0000000
----------
 151234567
```

which does not fit into 32 bits. More importantly, since the numbers are considered to be in a two's complement format, then we overflowed, since we added two fairly large, negative numbers together, and the most significant bit of the 32-bit result is clear (notice the 5 in the most significant byte of the result).

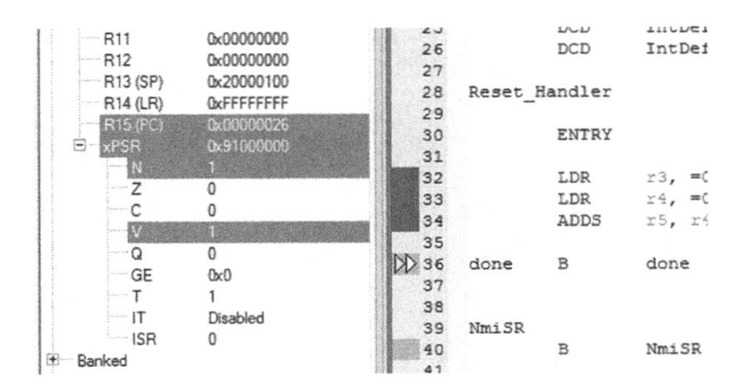

FIGURE 7.3 Status flags indicating an overflow.

Let's examine Example 7.2 again. When we added 0x7B000000 to 0x30000000, the result did, in fact, fit into 32 bits. However, the result would be interpreted as a negative number when we started off adding two positive numbers, so is this an overflow case? The answer is yes. Both the N and the V bits would be set in the xPSR, as shown in Figure 7.3, if you were to run the following code on the Cortex-M4:

```
LDR   r3, =0x7B000000
LDR   r4, =0x30000000
ADDS  r5, r4, r3
```

Notice that the 'S' extension is added to the ADD mnemonic, indicating that we want the flags updated as a result of the addition.

7.2.3 The Z Flag

This is one of the easiest to understand, as the only thing the Z flag tells us is that the result of an operation produces zero, meaning all 32 bits must be zero. This might be the result of a counter expiring, or a routine might need to examine an operand before performing some other kind of arithmetic routine, such as division.

EXAMPLE 7.4

In Chapter 16, we'll create a short program to change the color of the LED on the Tiva Launchpad, part of which is shown below.

```
      MOVT  r7, #0xF4   ; set counter to 0xF40000
spin
      SUBS  r7, r7, #1 ; just twiddling our thumbs....
      BNE   spin
```

In order to hold the LED at a particular color for a second or two, a short loop sets a register to a fixed value then subtracts one, setting the flags in the process, until the register equals zero. The Z flag is used to determine when the counter hits zero, where the BNE (branch if not equal to zero) instruction uses the value of the Z flag. If it is clear, then the program jumps back to the SUBS instruction and repeats. Otherwise, the loop is exhausted and the program continues doing something else.

7.2.4 THE C FLAG

The Carry flag is set if the result of an addition is greater than or equal to 2^{32}, if the result of a subtraction is positive, or as the result of an inline barrel shifter operation in a move or logical instruction. Carry is a useful flag, allowing us to build operations with greater precision should we need it, e.g., creating routines to add 64-bit numbers, which we will see in a moment. If we were to add the two values shown in the code below, the C bit will be set in the status register, since the sum is greater than 2^{32}.

```
LDR   r3, =0x7B000000
LDR   r7, =0xF0000000
ADDS  r4, r7, r3  ; value exceeds 32 bits, generates C out
```

Like older processors, such as the MC68000 and its predecessors, the carry flag is *inverted* after a subtraction operation, making the carry bit more like a borrow bit, primarily due to the way subtraction is implemented in hardware. For example, these instructions will set the carry bit to a one, since the operation produces no carry out and the bit is inverted:

```
LDR   r0, =0xC0000000
LDR   r2, =0x80000000
SUBS  r4, r0, r2  ; r4 = r0 - r2 (watch the order!)
```

Let's further suppose that we really want to subtract two 64-bit numbers:

0x7000BEEFC0000000
– 0x3000BABE80000000

We know the answer should be 0x4000043140000000, and to get this, we use the following code:

```
LDR   r0, =0xC0000000   ; lower 32-bits
LDR   r1, =0x7000BEEF   ; upper 32-bits
LDR   r2, =0x80000000   ; lower 32-bits
LDR   r3, =0x3000BABE   ; upper 32-bits
SUBS  r4, r0, r2        ; set C bit for next subtraction
SBC   r5, r1, r3        ; upper 32 bits use the carry flag
```

The first subtraction operation will set the status flags for us. We saw earlier that the C flag is set, since there is no carry out for the first operation, e.g., 0xC minus 0x8 produces no carry. The second subtraction is a subtract with carry (SBC) operation, using the carry bit to perform a normal subtraction. If the carry bit had been clear, the SBC instruction would have subtracted one more from its result.

7.3 COMPARISON INSTRUCTIONS

Apart from using the S bit with instructions to set flags, there are also four instructions that do nothing *except* set the condition codes or test for a particular bit in a register. They are:

CMP—Compare. CMP subtracts a register or an immediate value from a register value and updates the condition codes. You can use CMP to quickly check the contents of a register for a particular value, such as at the beginning or end of a loop.

CMN—Compare negative. CMN adds a register or an immediate value to another register and updates the condition codes. CMN can also quickly check register contents. This instruction is actually the inverse of CMP, and the assembler will replace a CMP instruction when appropriate. For example, if you typed

```
CMP r0, #-20
```

the assembler will instead generate

```
CMN r0, #0x14
```

TST—Test. TST logically ANDs an arithmetic value with a register value and updates the condition codes without affecting the V flag. You can use TST to determine if many bits of a register are all clear or if at least one bit of a register is set.

TEQ—Test equivalence. TEQ logically exclusive ORs an arithmetic value with a register value and updates the condition codes without affecting the V flag. You can use TEQ to determine if two values are the same.

The syntax for these instructions is

instruction{<cond>} <Rn>, <operand2>

where {<cond>} is one of the optional conditions covered in Chapter 8, and <operand2> can be a register with an optional shift, or an immediate value. Typical instructions might look like

```
CMP     r8, #0          ; r8 = =0?
BEQ     routine         ; yes, then go to my routine

TST     r4, r3          ; r3 = 0xC0000000 to test bits 31, 30

TEQ     r9, r4, LSL #3
```

Recall that the condition code flags are kept in the Program Status Register, along with other state information, such as the mode (for the ARM7TDMI) or the current exception number (for the Cortex-M4). For both processors, you can use the MRS (Move PSR to general-purpose register) instruction to read the flags, and the MSR (Move general-purpose register to PSR) to write the flags.

The ARM7TDMI has both Current and Saved Program Status Registers, so the MRS instruction will read the flags in the CPSR and any of the SPSRs. For example, the two instructions

```
MRS r0, CPSR
MRS r1, SPSR
```

will load the contents of the CPSR and SPSR into registers r0 and r1, respectively. From there, you can examine any flags that you like. The restrictions here are that you cannot use register r15 as the destination register, and you must not attempt to access an SPSR in User mode, since the register does not exist. The *ARM Architectural Reference Manual* (ARM 2007c) defines the results of this operation as UNPREDICTABLE.

The Cortex-M4 has only one status register, but it can be referenced in three different views—APSR, IPSR, or EPSR—or all at once as PSR. The flags are held only in the APSR, so you could read or change the values using

```
MRS     r3, APSR        ; read flag information into r3
MSR     APSR, r2        ; write to just the flags
MSR     PSR, r7         ; write all status information to r7
```

As we begin to write more complex programs, individual flags will become less important, and you will more than likely use the condition codes along with a branch (B) instruction or another instruction to create loops and conditional assembly routines without actually having to read the flags, for example, the BEQ instruction above. This topic is covered in much more detail in Chapter 8.

7.4 DATA PROCESSING OPERATIONS

You would expect any microprocessor, even the simplest, to include the fundamental operations such as add, subtract, and shift, and from these you could build more advanced operations such as divide, multiply, and square root. The ARM microprocessors are designed to be used in embedded applications, which are very sensitive to power dissipation and die size. Ideally, the processor would provide a wide range of data processing instructions without making the gate count of the part too high or make the area requirements too large. By combining a barrel shifter, a 32-bit ALU, and a hardware multiplier, the ARM7TDMI provides a rich instruction set while saving power. With significant advances in CMOS processes, more transistors can be used in VLSI designs the size of the ARM7TDMI or smaller, allowing for even more arithmetic functionality to be added to the Cortex-M4, such as a hardware divider, saturated math and DSP operations, and even a floating-point unit!

An example of a data processing instruction, ADD, might look like

ADDS{<cond>} r0, r1, <operand2>

where “S” indicates that the status flags should be updated and {<cond>} is one of the optional conditions covered in Chapter 8, e.g., EQ, LT, GT, or PL. The second operand, <operand2>, can be an immediate value, a register, or a register with a shift or rotate associated with it. The last option turns out to be quite handy, as we'll see below. As an aside, this syntax has been updated for Unified Assembly Language (UAL)—the older style of mnemonic would have been written as ADD{<cond>}S,* and we will see how even shifts and rotations have changed their format.

7.4.1 Boolean Operations

Both the ARM7TDMI and the Cortex-M4 support Boolean logic operations using two register operands, shown in Table 7.1. Although we saw MOV instructions in previous chapters, the MOVN instruction can also be used to logically invert all bits in a register, since it takes the one's complement negation of an operand. A very fast way to load the two's complement representation of –1 into a register is to logically invert zero, since the 32-bit value 0xFFFFFFFF is –1 in a two's complement notation, written as

```
MOVN r5, #0    ; r5 = -1 in two's complement
```

Examples of the remaining operations include

```
AND r1, r2, r3 ; r1 = r2 AND r3
ORR r1, r2, r3 ; r1 = r2 OR r3
EOR r1, r2, r3 ; r1 = r2 exclusive OR r3
BIC r1, r2, r3 ; r1 = r2 AND NOT r3
```

The first three instructions are fairly straightforward—AND, OR, and exclusive OR are basic logic functions. The fourth instruction is the Bit Clear operation, which can be used to clear selected bits in a register. For each bit in the second operand, a 1 clears the corresponding bit in the first operand (a register), and a 0 leaves it unchanged. According to the data processing instruction format, we can also use an immediate value for the second operand. For example,

TABLE 7.1
Boolean Operations

ARM7TDMI Instruction	Cortex-M4 Instruction	Comment
AND	AND	Logically ANDs two operands
ORR	ORR	Logically ORs two operands
	ORN	OR of operand 1 with NOT operand 2
EOR	EOR	Exclusive OR of two operands
MOVN	MVN	Move negative—logically NOTs all bits
BIC	BIC	Bit Clear—clears selected bits in a register

* Since older ARM7TDMI code is quite common, it is very likely you will see both formats of instruction.

```
BIC r2, r3, #0xFF000000
```

clears the upper byte of register r3 and moves the result to register r2.

The Cortex-M4 has one additional Boolean operation called ORN, for OR Not, which logically ORs the first operand with the one's complement of the second operand.

7.4.2 Shifts and Rotates

Figure 7.4 shows part of the internal data path of the ARM7TDMI, where the data for an instruction come down two busses leading to the main ALU. Only one of those busses goes through the barrel shifter, which is a dedicated hardware block of logic to rotate or shift data left or right. Because of this asymmetry, we can rotate or shift only one of the operands in the instruction, but in general, this is enough functionality. With the addition of a few instructions, we can overcome any limitations that are introduced with this type of design. In fact, the very idea to have a barrel shifter sit between the register bank (an architecture term that describes the physical registers r0 through r15) and the main ALU allows for 32-bit constants to be used in ALU and MOV instructions, despite having only 32 bits for the instruction itself. We saw this in Chapter 6 with literals and constants.

The types of shifts and rotates that the ARM processors can perform are shown in Figure 7.5. There are two types of logical shifts, where the data is treated as unsigned, an arithmetic shift where the data is treated as signed, and two types of rotates. The absence of rotate left can be explained by the fact that a rotate left by m bits is the same as a rotate to the right by ($32-m$) bits (except for the effect on the carry bit), and can, therefore, be done using the same instruction. Another instruction that may appear to have gone missing is ASL, or an arithmetic shift left. With a little thought, it becomes apparent that you would never need such an instruction, since arithmetic shifts need to preserve the sign bit, and shifting signed data to the left will do so as long as the number doesn't overflow. As an example, the number –1 in two's complement notation is 0xFFFFFFFF, and shifting it left results in 0xFFFFFFFE, which is –2 and correct. A 32-bit number such as 0x8000ABCD will overflow if shifted left, resulting in 0x0001579A, which is now a positive number.

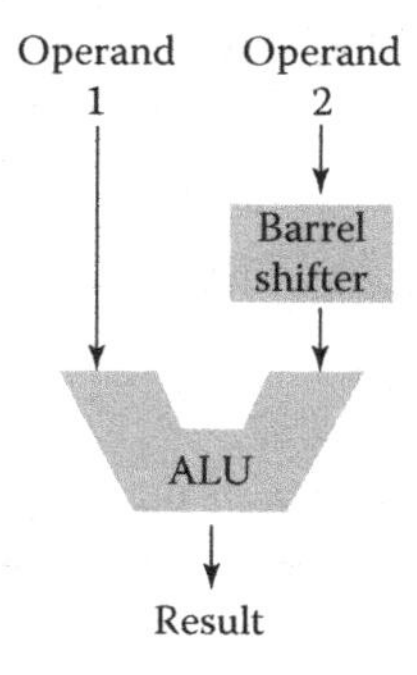

FIGURE 7.4 The ARM7TDMI barrel shifter.

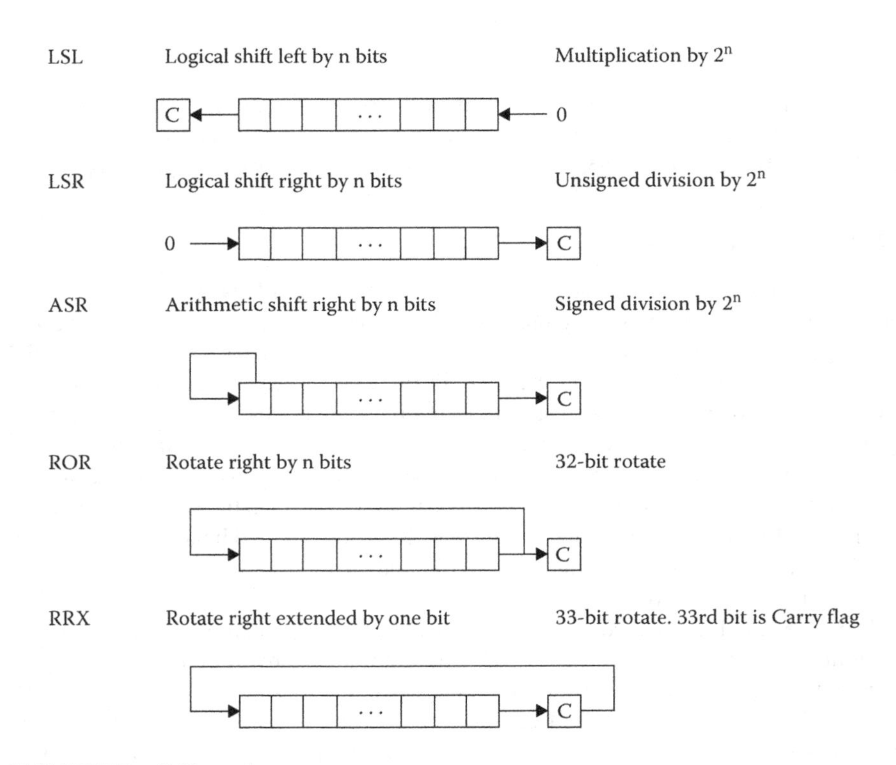

FIGURE 7.5 Shifts and rotates.

If you want to shift or rotate data without performing another operation such as an add, then a MOV instruction works well. Recall from Chapter 6 that MOV can transfer data from a register to a register, so by adding an optional shift operation, the instruction gets slightly more elaborate. When you follow the UAL conventions, an instruction that just shifts data to the left should use the LSL mnemonic; an instruction that just shifts data to the right should use the LSR mnemonic. The assembler will choose the best instruction to use under the guidelines of the assembly directives that are given. For example, code that is written as

```
LSL  r3, r4, #1
```

for an ARM7TDMI will be replaced with

```
MOV  r3, r4, LSL #1
```

assuming that you have not explicitly told the assembler you want Thumb instructions. On a Cortex-M4, the assembler will replace this with the 32-bit Thumb-2 instruction

```
LSL  r3, r4, #1
```

because the original mnemonic does not set the flags, and neither will the 32-bit version. Thumb arithmetic instructions set flags automatically, so if you said

```
LSLS  r3, r3, #1
```

you would get the 16-bit Thumb instruction

```
LSL  r3, #1
```

More will be said about Thumb in Chapter 17.

EXAMPLE 7.5

The following instructions show how simple shifts and rotates are written.

```
LSL    r4, r6, #4     ; r4 = r6 << 4 bits
LSL    r4, r6, r3     ; r4 = r6 << # specified in r3
ROR    r4, r6, #12    ; r4 = r6 rotated right 12 bits
                      ; r4 = r6 rotated left 20 bits
```

All shift operations take one clock cycle to execute, except register-specified shifts, which take an extra cycle as there are only two read ports on the register bank, and an extra read is required. When performing shifts, the shift count can be either an unsigned 5-bit value, i.e., 0 to 31, as in the first example, or the bottom byte in a register, as in the second example.

EXAMPLE 7.6

The shift and logical operations can also be used to move data from one byte to another. Suppose we need to move the uppermost byte from register r2 and put it at the bottom of register r3. The contents of register r3 are shifted left by 8 bits first. Two instructions could be used to do this:

```
LSR  r0, r2, #24          ; extract top byte from R2 into R0
ORR  r3, r0, r3, LSL #8   ; shift up r3 and insert r0
```

EXAMPLE 7.7 HAMMING CODES

In the 1940s, a mathematician named Richard Hamming developed and formally defined ways of not only detecting errors in bit streams but correcting them as well. For example, if you were going to transmit 8 bits of data from a computer across a channel (and here, a channel could be seen as something like a piece of wire, an optical link, or maybe even a wireless interface) to a receiver, you would hope that the value you sent matches the value received exactly. If there are errors, it's critical to know this. More interestingly, if there is a way to correct the bit error, this byte of information would not need to be resent. The field of error correcting codes has grown substantially since then, and more modern coding schemes such as Reed-Solomon code, Binary Golay code, and BCH code can be found in Roth (2006). While the theory behind them is rather complicated, simple Hamming codes can be built easily, so we'll examine an algorithm to detect up to two bit errors in an 8-bit value. This algorithm can also correct a single bit error.

Consider the idea of adding a bit, called a checksum, to a value that indicates the *parity* of the bits in that value. For example, if you had the 7-bit number

1010111

and we counted the number of ones in the value, 5 in this case, adding a 1 at the beginning of the value would make the parity *even*, since the number of ones (including the parity bit) is an even number. Our new value would be

11010111

If the data were transmitted this way, the receiver could detect an error in the byte sent if one of the data bits changes, since the parity would suddenly become *odd*. Note that if two of the bits changed, then we could not detect an error, since the parity remains even.

One type of Hamming code can be constructed by using four checksum bits placed in strategic locations. If a 12-bit value is constructed using 8 bits of data and four checksum bits as shown below, then we can use the checksum bits to detect up to two errors in the data and even correct a single bit error.

Original 8-bit value

d_7	d_6	d_5	d_4	d_3	d_2	d_1	d_0

Modified 8-bit value

11	10	9	8	7	6	5	4	3	2	1	0
d_7	d_6	d_5	d_4	c_3	d_3	d_2	d_1	c_2	d_0	c_1	c_0

The checksum bits c_3, c_2, c_1, and c_0 are computed as follows:

Checksum bit c_0 should produce even parity for bits 0, 2, 4, 6, 8, and 10. In other words, we're checking a bit, skipping a bit, checking a bit, etc.

Checksum bit c_1 should produce even parity for bits 1, 2, 5, 6, 9, and 10. In other words, we're checking two bits, skipping two bits, checking two bits, etc.

Checksum bit c_2 should produce even parity for bits 3, 4, 5, 6, and 11. Now we're checking four bits, skipping four bits, etc.

Checksum bit c_3 should produce even parity for bits 7, 8, 9, 10, and 11.

As an example, suppose we wanted to generate checksums for the binary value 10101100. The first checksum bit c_0 would be 1, since this would produce even parity for the bits 0, 0, 1, 0, and 0. Using the same method, the remaining checksum bits would show

$$c_1 = 1$$

$$c_2 = 1$$

$$c_3 = 0$$

resulting in the 12-bit value 101001101011.

The code on the next page shows the assembly code to build a 12-bit Hamming code, making efficient use of the barrel shifter during logical operations.

```
    AREA HAMMING, CODE

    ENTRY

; Registers used:
; R0 - temp
; R1 - used to hold address of data
; R2 - holds value to be transmitted
; R4 - temp
main
    MOV r2, #0                  ; clear out transmitting reg
    ADR r1, arraya              ; start of constants
    LDRB r0, [r1]

    ;
    ; calculate c0 using bits   76543210
    ;                            * ** **
    ; even parity, so result of XORs is the value of c0
    ;
    MOV r4, r0                  ; make a copy
    EOR r4, r4, r0, ROR #1      ; 1 XOR 0
    EOR r4, r4, r0, ROR #3      ; 3 XOR 1 XOR 0
    EOR r4, r4, r0, ROR #4      ; 4 XOR 3 XOR 1 XOR 0
    EOR r4, r4, r0, ROR #6      ; 6 XOR 4 XOR 3 XOR 1 XOR 0
    AND r2, r4, #1              ; create c0 -> R2
    ;
    ; calculate c1 using bits   76543210
    ;                            ** ** *

    MOV r4, r0
    EOR r4, r4, r0, ROR #2      ; 2 XOR 0
    EOR r4, r4, r0, ROR #3      ; 3 XOR 2 XOR 0
    EOR r4, r4, r0, ROR #5      ; 5 XOR 3 XOR 2 XOR 0
    EOR r4, r4, r0, ROR #6      ; 6 XOR 5 XOR 3 XOR 2 XOR 0
    AND r4, r4, #1              ; isolate bit
    ORR r2, r2, r4, LSL #1      ; 7 6 5 4 3 2 c1 c0
    ;
    ; calculate c2 using bits   76543210
    ;                           *   ***
    ROR r4, r0, #1              ; get bit 1
    EOR r4, r4, r0, ROR #2      ; 2 XOR 1
    EOR r4, r4, r0, ROR #3      ; 3 XOR 2 XOR 1
    EOR r4, r4, r0, ROR #7      ; 7 XOR 3 XOR 2 XOR 1
    AND r4, r4, #1              ; isolate bit
    ORR r2, r2, r4, ROR #29     ; 7 6 5 4 c2 2 c1 c0
    ;
    ; calculate c3 using bits   76543210
    ;                           ****
    ROR r4, r0, #4              ; get bit 4
    EOR r4, r4, r0, ROR #5      ; 5 XOR 4
```

```
        EOR r4, r4, r0, ROR #6       ; 6 XOR 5 XOR 4
        EOR r4, r4, r0, ROR #7       ; 7 XOR 6 XOR 5 XOR 4
        AND r4, r4, #1
        ;
        ; build the final 12-bit result
        ;
        ORR r2, r2, r4, ROR #25      ; rotate left 7 bits
        AND r4, r0, #1               ; get bit 0 from original
        ORR r2, r2, r4, LSL #2       ; add bit 0 into final
        BIC r4, r0, #0xF1            ; get bits 3,2,1
        ORR r2, r2, r4, LSL #3       ; add bits 3,2,1 to final
        BIC r4, r0, #0x0F            ; get upper nibble
        ORR r2, r2, r4, LSL #4       ; r2 now contains 12 bits
                                     ; with checksums
done    B       done
        ALIGN

arraya
        DCB 0xB5
        DCB 0xAA
        DCB 0x55
        DCB 0xAA

        END
```

Our starting 8-bit value is in memory location arraya and is loaded into register r0. A fast way to generate even parity is to use the result of exclusive OR operations as the checksum bit, e.g., if you take an odd number of bits in a pattern and exclusive OR them together, the result will be 1; therefore, the checksum should be a 1 to make an even number. The first checksum is generated by using an EOR instruction on the original data many times, ignoring all bits except bit 0. Note that the first logical instruction

```
EOR r4, r4, r0, ROR #1
```

takes the original data and exclusive ORs bit 0 with bit 1 of the copied data, all in a single instruction. Subsequent EOR instructions take the copied data and rotate the necessary bits down to bit 0. Ultimately, we're only interested in bit 0, so we logically AND the final result with 1 to clear out all the bits except the least significant bit, since ANDing a value with 0 produces 0, and ANDing a value with 1 just gives back the original value.

The other checksums are calculated in much the same way, always shifting the necessary bits down to bit 0 before EORing them with intermediate results. The final 12-bit value is constructed from the original 8-bit value and the four checksums using logical functions. Notice that rotates to the left are done using ROR instructions, since a rotate to the left by *n* bits is the same as a rotate to the right by (32-*n*) bits, and there is no ROL instruction. The final value is kept in register r2. The first 8-bit value read, 0xB5, should generate the 12-bit Hamming value 0xBA6 in register r2.

To detect an error in the transmitted value, the four checksum bits c3, c2, c1, and c0 are examined. If it turns out that one of the checksum bits is incorrect (this can be verified by looking at the data in the 12-bit value), then it is the checksum bit itself that is incorrect. If there are two checksum bits that are incorrect, say c_n and c_m, then the bit position of the incorrect bit, j, can be found by

$$j = (2^n + 2^m) - 1$$

For example, if checksum bits c_3 and c_2 are incorrect, then the error lies with bit 11. Since this is the only error, it can be corrected.

7.4.3 Addition/Subtraction

The arithmetic instructions in the ARM and Thumb-2 instruction sets include operations that perform addition, subtraction, and reverse subtraction, all with and without carry. Examples include:

```
ADD     r1, r2, r3     ; r1 = r2 + r3
ADC     r1, r2, r3     ; r1 = r2 + r3 + C
SUB     r1, r2, r3     ; r1 = r2 - r3
SUBC    r1, r2, r3     ; r1 = r2 - r3 + C - 1
RSB     r1, r2, r3     ; r1 = r3 - r2
RSC     r1, r2, r3     ; r1 = r3 - r2 + C - 1
```

From the discussion on flags we noted that the Carry flag could be used to indicate that an operation produced a carry bit in the most significant bit of the result. The ADC, SUBC, and RSC instructions make use of this flag by adding the Carry flag into the operation. Suppose we wish to perform a 64-bit addition. Since the registers are only 32 bits wide, we would need to store the two addends in two registers each, and the sum would have to be stored in two registers.

EXAMPLE 7.8

The following two instructions add a 64-bit integer contained in registers r2 and r3 to another 64-bit integer contained in registers r0 and r1, and place the result in registers r4 and r5:

```
ADDS    r4,r0,r2       ; adding the least significant words
ADC     r5,r1,r3       ; adding the most significant words
```

You can see in Figure 7.6 that the carry out from the lower 32-bit sum is added into the upper 32-bit sum to produce the final 64-bit result.

EXAMPLE 7.9

The second operand can make use of the barrel shifter when performing adds and subtracts, a topic we'll explore more shortly, such as

```
SUB     r0, r0, r2, LSL #2     ; r0 = r0 - (r2 <<2)
ADD     r1, r1, r3, LSR #3     ; r1 = r1 + (r3 >>3)
```

There are two very unusual, but useful, instructions listed above: RSB and RSC, which are reverse subtracts. The reverse subtract instruction comes about from

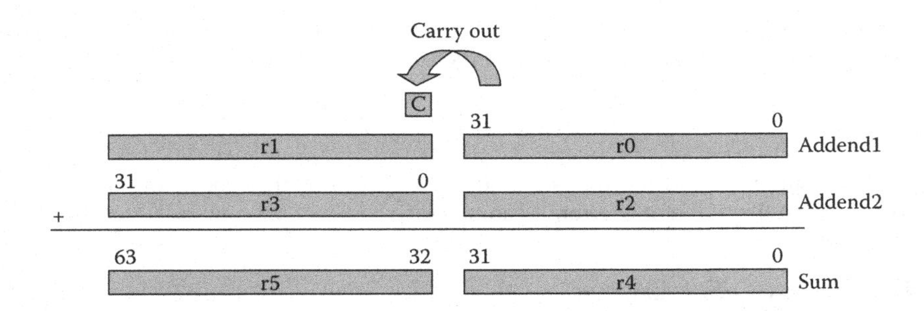

FIGURE 7.6 64-bit addition.

having a barrel shifter on only one of the busses going from the register bank to the main ALU, as shown earlier in Figure 7.4. Consider the case where we want to perform the following operation:

```
SUB r0, r2, r3, LSL #2 ; r0=r2 - r3*4
```

We could do this quite easily with this single instruction. However, suppose we want modify (shift) register r2 before the subtraction instead of register r3. Since subtraction is not a commutative operation, i.e.,

$$x - y \neq y - x, \quad y,x \neq 0$$

register r2 must somehow be made to appear on the bus that contains the barrel shifter. This is done using the reverse subtract operation, where the instruction would be written as

```
RSB r0, r3, r2, LSL #2 ; r0=r2*4 - r3
```

This same instruction can be used to great effect, since the second operand can also be a constant, so you could conceivably subtract a register value *from* a constant, instead of the other way around.

EXAMPLE 7.10

Write an ARM7TDMI assembly program to perform the function of absolute value. Register r0 will contain the initial value, and r1 will contain the absolute value. The pseudo-instruction would look like

```
ABS r1, r0
```

Try to use only two instructions (not counting instructions to terminate the program or any directives).

Solution

Recall that the absolute value function always returns a positive value for the argument, so f(x) = |x| just changes the sign of the argument if the value is negative. We can do this with one more instruction and one instruction to change the sign:

```
          AREA Prog7a, CODE, READONLY
          ENTRY
          MOVS    r1, r0
          RSBLT   r1, r1, #0
done      B       done
          END
```

The program first sets the status flags to see if anything needs to be done. If the argument is zero, the result is zero. If the argument is negative (LT indicates Less Than zero, but we'll cover this in more detail in Chapter 8), the reverse subtract instruction subtracts r1 *from* zero, effectively changing its sign. Notice the conditional execution of the RSB instruction, since a positive value will fail the condition of being less than zero.

7.4.4 Saturated Math Operations

Algorithms for handling speech data, adaptive control algorithms, and routines for filtering are often sensitive to quantization effects when implemented on a microprocessor or microcontroller. A careful analysis of the both the implementation (e.g., direct or indirect, recursive or non-recursive) and the coefficients used in the filter gives programmers a better idea of precautions that must be made in advance. Sometimes it is required that limitations be placed on both the input data and the algorithm's coefficients to prevent overflow conditions or to prevent an algorithm from becoming unstable. In other cases, the software can mitigate any problems by forcing intermediate values to stay within boundaries should they stray. Saturated math is one such approach, especially when dealing with signed data. For example, consider a digital waveform in Figure 7.7, possibly the output of an adaptive predictor, where the values are represented by 16-bit signed integers; in other words, the largest positive value in a register would be 0x00007FFF and the largest negative value would be 0xFFFF8000. If this signal were scaled in some way, it's quite possible that the largest value would overflow, effectively flipping the MSB of a value so that a positive number suddenly becomes negative, and the waveform might appear as in Figure 7.8. Using saturated math instructions, the signal would get clipped, and the waveform might appear as in Figure 7.9, not correcting the values but at least keeping them within limits.

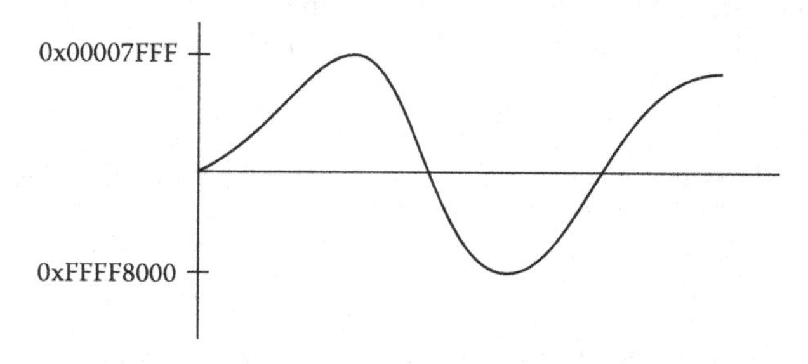

FIGURE 7.7 Signal represented by 16-bit signed integers.

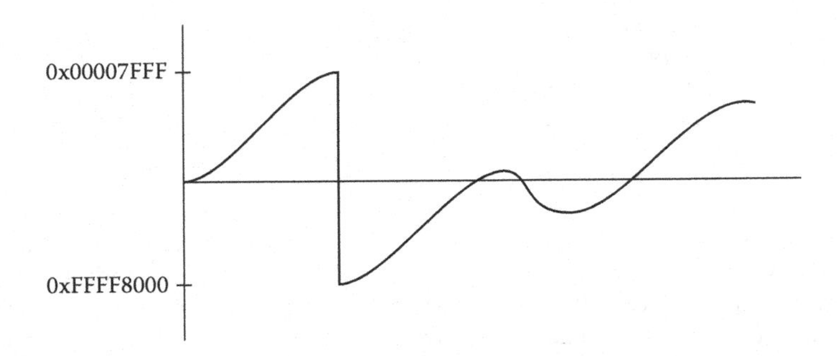

FIGURE 7.8 Digital waveform exceeding bounds.

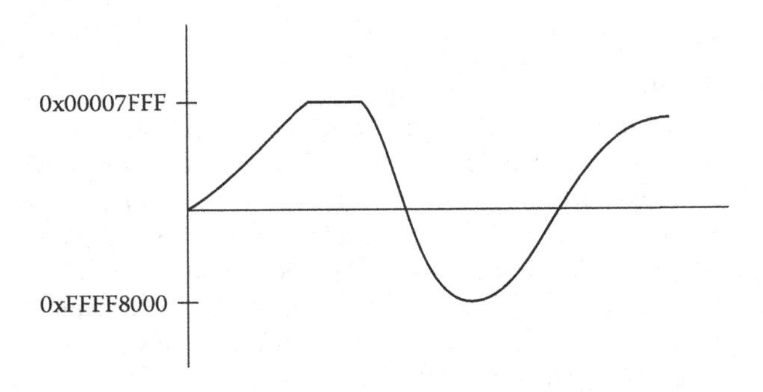

FIGURE 7.9 Digital waveform with saturation.

In version 6 cores and higher, many new instructions were added for manipulating saturated values, such as QADD, QADD8, QADD16, UQADD8, etc., and the Cortex-M4 includes addition, signed multiplication, subtraction, and parallel operations for working with saturated math, one of which we'll examine in Section 7.5. These instructions will return maximum or minimum values based on the results of the operation if those values are exceeded. An additional status bit, the Q bit that we saw in Chapter 2, indicates that saturation has occurred and resides in the APSR shown in Figure 7.1. This bit is considered "sticky" in that once it is set, it must be written to a zero to clear it. In practice, you might use the saturated operation at the end of a loop or once data has been read as an input to an algorithm to ensure that values used in further processing are within acceptable limits.

EXAMPLE 7.11

A 32-bit signed value is to be saturated into a 16-bit signed value. If the value in register r3 is 0x00030000, then the instruction

```
SSAT   r4, #16, r3
```

will place the value 0x00007FFF into register r4, since the input value is above the threshold of 0x7FFF, and the Q bit will be set indicating that the value saturated. If

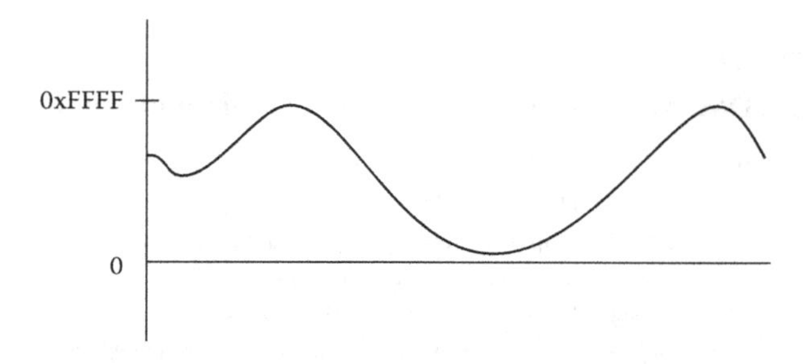

FIGURE 7.10 16-bit unsigned signal.

the value in register r3 is 0xFFFF7FFF, then the returned value in register r4 would be 0xFFFF8000, since the initial value is below the threshold of 0xFFFF8000, the most negative number represented in 16 bits. Again, the Q bit is set.

EXAMPLE 7.12

Unsigned saturation works analogously. Figure 7.10 shows a 16-bit signal where 0x0000 is the smallest value it would have and 0xFFFF is the largest. If register r3 contains the 32-bit signed value 0xFFFF8000, the unsigned saturation instruction

```
USAT    r4, #16, r3
```

would return the value 0x00000000 and the Q bit is set, since the 32-bit signed input is well below the smallest unsigned value of zero. An input value of 0x00030000 would return 0x0000FFFF and the Q bit is set since the input exceeds the maximum unsigned value of 0xFFFF. See the *ARM v7-M Architectural Reference Manual* (ARM 2010a) for more detailed information about the use of SSAT and USAT.

7.4.5 Multiplication

Binary multiplication is provided on nearly every processor these days, but it comes at a cost. As an operation, it's quite common. As a block of digital hardware, it's expensive in that multipliers usually consume quite a bit of area and power relative to the other parts of a microprocessor. Older microcontrollers would often use a shift-add iterative routine to perform multiplication, avoiding building a large multiplier array; however, this tends to be quite slow. Modern designs usually perform multiplications in a single cycle or two, but again, because of power considerations, if there is a way to avoid using the array, an ARM compiler will try to produce code without multiply instructions, as we will see shortly. Microprocessors and/or DSP engines are often selected based on their ability to perform fast multiplication, especially in areas of speech and signal processing, signal analysis, and adaptive control.

Table 7.2 shows all of the supported instructions available in the ARM7TDMI, which are a small subset of those supported on the Cortex-M4. MUL and MLA are multiply and multiply-and-accumulate instructions that produce 32-bit results. MUL multiplies the values in two registers, truncates the result to 32 bits, and stores the

TABLE 7.2
ARM7TDMI Multiply and Multiply-Accumulate Instructions

Instruction	Comment
MUL	32x32 multiply with 32-bit product
MLA	32x32 multiply added to a 32-bit accumulated value
SMULL	Signed 32x32 multiply with 64-bit product
UMULL	Unsigned 32x32 multiply with 64-bit product
SMLAL	Signed 32x32 multiply added to a 64-bit accumulated value
UMLAL	Unsigned 32x32 multiply added to a 64-bit accumulated value

product in a third register. MLA multiplies two registers, truncates the results to 32 bits, adds the value of a third register to the product, and stores the result in a fourth register, for example,

```
MUL     r4, r2, r1          ; r4 = r2 * r1
MULS    r4, r2, r1          ; r4 = r2 * r1, then set the flags
MLA     r7, r8, r9, r3      ; r7 = r8 * r9 + r3
```

Both MUL and MLA can optionally set the N and Z condition code flags. For multiplications that produce only 32 bits of result, there is no distinction between signed and unsigned multiplication. Only the least significant 32 bits of the result are stored in the destination register, and the sign of the operands does not affect this value. There is an additional multiply and subtract instruction (MLS) available on the Cortex-M4 which multiplies two 32-bit values together and then subtracts this product from a third value.

EXAMPLE 7.13

Multiply long instructions produce 64-bit results. They multiply the values of two registers and store the 64-bit result in a third and fourth register. SMULL and UMULL are signed and unsigned multiply long instructions:

```
SMULL r4, r8, r2, r3    ; r4 = bits 31-0 of r2*r3
                        ; r8 = bits 63-32 of r2*r3
UMULL r6, r8, r0, r1    ; {r8,r6} = r0*r1
```

These instructions multiply the values of two registers, add the 64-bit value from a third and fourth register, and store the 64-bit result in the third and fourth registers:

```
SMLAL r4, r8, r2, r3    ; {r8,r4} = r2*r3 + {r8,r4}
UMLAL r5, r8, r0, r1    ; {r8,r5} = r0*r1 + {r8,r5}
```

All four multiply long instructions can optionally set the N and Z condition code flags. If any source operand is negative, the most significant 32 bits of the result are affected.

7.4.6 Multiplication by a Constant

In our discussion of shifts and rotates, we saw that the inline barrel shifter in the ARM7TDMI's datapath can be used in conjunction with other instructions, such as ADD or SUB, in effect getting a multiplication for free. This feature is used to its full advantage when certain multiplications are done using the barrel shifter instead of the multiplier array. Consider the case of multiplying a number by a power of two. This can be written using only an LSL instruction, i.e.,

```
LSL r1, r0, #2 ; r1 = r0*4
```

But what if we wanted to multiply two numbers, one of which is not a power of two, like five? Examine the following instruction:

```
ADD r0, r1, r1, LSL #2 ; r0 = r1 + r1*4
```

This is the same thing as taking a value, shifting it to the left two bits (giving a multiplication by four), and then adding the original value to the product. In other words, multiply the number by five. Why do it this way? Consider the size and power usage of a multiplier array, which is highlighted in Figure 7.11 for the ARM10200 microprocessor. In very low power applications, it's often necessary to play every trick in the book to save power: not clocking logic that is not being used, powering down caches or the entire processor if it is not needed, reducing voltages and frequencies, etc. By using only the 32-bit adder and a barrel shifter, the ARM processors can actually generate multiplications by 2^n, $2^n - 1$, and $2^n + 1$ in a single

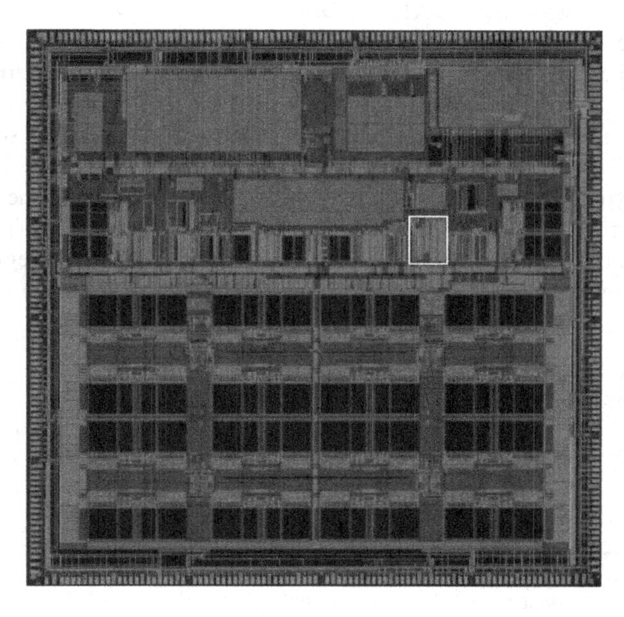

FIGURE 7.11 ARM10200 die photo with multiplier array highlighted.

cycle, without having to use a multiplier array. This also potentially saves some execution time. For example,

```
RSB r0, r2, r2, LSL #3 ; r0 = r2*7
```

will perform a multiplication by 7 by taking register r2, shifting it left by 3 bits, yielding a multiplication by 8, and then subtracting register r2 from the product. Note that the reverse subtract instruction was used here, since an ordinary subtraction will produce the wrong result. By chaining together multiplications, for example, multiplying by 5 and then by 7, larger constants can be created. Examine the following code to see that you can, in fact, create multiplier arguments that are not powers of two:

```
ADD  r0, r1, r1, LSL #1   ; r0 = r1*3
SUB  r0, r0, r1, LSL #4   ; r0 = (r1*3) - (r1*16) = r1* - 13
ADD  r0, r0, r1, LSL #7   ; r0 = (r1* - 13) + (r1*128) = r1*115
```

7.4.7 Division

Binary division is a subject that can get quite complicated very quickly. Historically, ARM cores did not include a binary integer divider in hardware, mostly because division is so infrequently used (and can therefore be done using a software routine), a divider can take up too much area and/or power to consider using on an embedded processor, and there are ways of avoiding division entirely. However, with denser geometries being made available to VLSI designers, it is possible to include division instructions in the newer ARM ISAs without too much overhead, and so we'll examine the divider in the Cortex-M4. This is not to say that good software routines are not still available for processors like the ARM7TDMI. Factors to be considered in choosing a divider routine include the type of data you have (either fractional data or integer data), the speed of the algorithm needed, and the size of the code permitted to perform an algorithm in software. For an excellent treatment of the topic, consider reading Sloss, Symes, and Wright (2004). Runtime libraries include a division routine, so if you happen to be writing in C or C++, generally the compiler will take care of the division algorithm for you. But our focus is assembly, so we'll consider at least one simple case.

The following code, which is a variation of a shift-subtract algorithm, can be used to divide two unsigned, 32-bit values, where the dividend is in register Ra and the divisor is in register Rb, producing a quotient in register Rc and a remainder in register Ra.

```
        AREA Prog7b, CODE, READONLY
Rcnt    RN      0                          ; assign R0 to Rcnt
Ra      RN      1                          ; assign R1 to Ra
Rb      RN      2                          ; assign R2 to Rb
Rc      RN      3                          ; assign R3 to Rc

        ENTRY

        ; Place your dividend in Ra
        ; Place your divisor in Rb
```

```
        MOV     Rcnt, #1                 ; bit to control the
                                         ; division
Div1    CMP     Rb, #0x80000000          ; move Rb until
                                         ; greater than Ra
        CMPCC   Rb, Ra
        LSLCC   Rb, Rb, #1
        LSLCC   Rcnt, Rcnt, #1
        BCC     Div1
        MOV     Rc, #0
Div2    CMP     Ra, Rb                   ; test for possible
                                         ; subtraction
        SUBCS   Ra, Ra, Rb               ; subtract if OK
        ADDCS   Rc, Rc, Rcnt             ; put relevant bit
                                         ; into result
        LSRS    Rcnt, Rcnt, #1           ; shift control bit
        LSRNE   Rb, Rb, #1               ; halve unless
                                         ; finished
        BNE     Div2                     ; divide result in Rc
                                         ; remainder in Ra
done    B       done
        END
```

EXAMPLE 7.14

Let's compare the execution times using both the software routine above and the hardware divider on the Cortex-M4. We will divide 0xFF000000 by 0x98. Since these values can be created using two MOV instructions, load the registers Ra and Rb using:

```
MOV     Ra, #0xFF000000     ; loads register r1
MOV     Rb, #0x98           ; loads register r2
```

Running the code, register r3 contains the value 0x1AD7943 (the quotient) and register r1 contains 0x38 (the remainder). Using the Keil simulation tools, this code takes 450 cycles or 28.125 microseconds to complete on a Tiva TM4C123GH6ZRB microcontroller. Using three lines of code, we can reduce the execution time considerably:

```
MOV     r1,   #0xFF000000
MOV     r2,   #0x98
UDIV    r3,   r1, r2          ; r3 = r1/r2
```

Again, register r3 holds the quotient, but this code takes 0.813 microseconds, or 13 cycles!

The Cortex-M4 gives you the option to handle division by zero one of two ways: a fault exception or placing zero in the destination register. We'll see an example in Chapter 15 showing how to configure the Nested Vectored Interrupt Controller (NVIC) to allow this type of exception.

7.5 DSP EXTENSIONS

The Cortex-M4 is the first M-class ARM core targeted at signal processing applications, and as such, offers more of the DSP instructions defined in the *ARM v7-M*

Architectural Reference Manual (ARM 2010a) to handle cryptography routines, graphics algorithms, speech and video processing, etc. An instruction set summary can be found in the *Cortex-M4 Technical Reference Manual* (ARM 2009), detailing the behavior of each instruction. In this section, we'll look at two examples of DSP operations that go beyond the usual multiply-accumulate calculations found in digital filters.

EXAMPLE 7.15

Some instructions are designed for particular algorithms; some instructions are designed to work on certain types of data, for example on Q31 formatted data (to be explained shortly). Figure 7.12 shows how the instructions SMMLA and SMMLAR work. They both take two 32-bit operands and multiply them together. The most-significant word of the product is then added to a 32-bit accumulated value. If the R bit in the instruction is set (SMMLAR does this), then the value 0x80000000 is added prior to the truncation of the final result. Otherwise, nothing is added before truncation. For example, if we loaded registers r2 and r3 with 32-bit signed operands, and we used register r4 as an accumulated value, then

```
LDR        r3, =0xFFFE6487
LDR        r2, =0x80008F71;
LDR        r4, =0xFFFF0003; accumulator
SMMLAR     r9, r2, r3, r4
```

will produce the final result of 0xFFFFCDBF in register r9 as follows. The 32-bit multiply would produce the 64-bit product 0x0000CDBB9971C897. The accumulator value 0xFFFF0003 is then added producing the intermediate result:

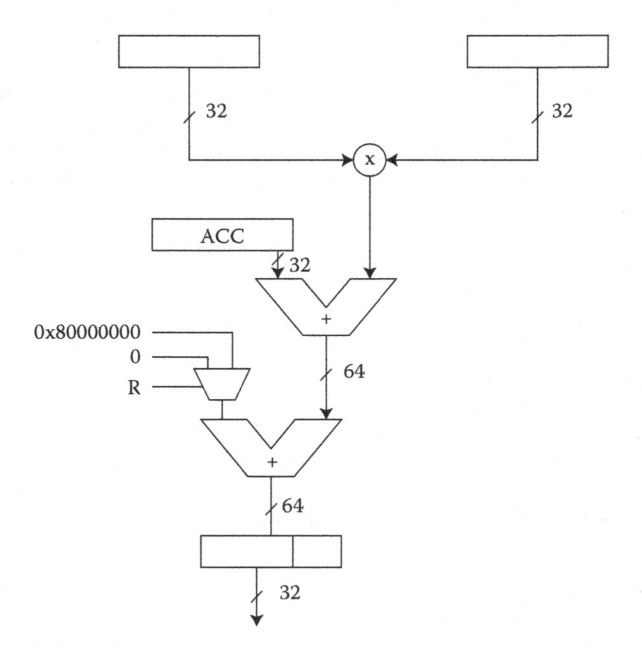

FIGURE 7.12 Operations involved in SMMLA and SMMLAR.

```
  0x0000CDBB9971C897
+ 0xFFFF0003
  0xFFFFCDBE9971C897
```

Because we have selected the rounding version of the instruction, 0x80000000 is added to the sum, giving us

```
  0xFFFFCDBE9971C897
+         0x80000000
  0xFFFFCDBF1971C897
```

At this point, the upper 32 bits are kept, so that register r9 contains 0xFFFFCDBF.

EXAMPLE 7.16

Without context, the USAD8 and USADA8 instructions make little sense, as they're quite specific. These operations calculate the sum of absolute differences and the accumulated sum of absolute differences, respectively. It turns out that these calculations are useful for object recognition, motion estimation, and graphic compression algorithms, such as MPEG and H.264. When comparing two images, one simple block metric that can be found is the L_1 norm of the difference image (or the Manhattan distance between two images), so if you wanted to compare, for example, image blocks of $N \times N$ squares in one image, say m_1, with another $N \times N$ square in image m_2 at a particular point (x,y), then you might compute the accumulated sum of absolute differences according to:

$$acc(x,y) = \sum_{i=0}^{N-1}\sum_{j=0}^{N-1}\left|m_1(x+i,y+j) - m_2(i,j)\right|$$

Four 8-bit pixel values can be read from each of two registers, say registers r3 and r4. The absolute differences are then found between corresponding bytes in registers r3 and r4. These four values are then summed in a destination register with the instruction

```
USAD8{<cond>}    Rd, Rm, Rs          ; sum of absolute differences
```

If you were to calculate an accumulated value, you could then follow at some point in your code with

```
USADA8{<cond>}  {Rd,} Rm, Rs, Rn    ; Rn is accumulated value to include
```

For both of these instructions, {<cond>} refers to an optional condition code that might be used (see Chapter 8).

7.6 BIT MANIPULATION INSTRUCTIONS

There are instructions in the Cortex-M4 that allow the manipulation of individual bits in a register: BFI (Bit Field Insert), UBFX (Unsigned Bit Field Extract), SBFX (Signed Bit Field Extract), BFC (Bit Field Clear), and RBIT (Reverse Bit order). Industrial and automotive applications typically require processing large amounts of general-purpose I/O data, and cycles can be wasted just moving individual bits around from sensors or other interfaces. In older architectures, it was necessary to use several instructions to copy contents, modify bits, and then store them; now with a single instruction, fields can be modified or extracted without changing the entire bit pattern in a register.

BFI, SBFX, and UBFX have the following syntax:

```
BFI{<cond>}      <Rd>, <Rn>, <#lsb>, <#width>
SBFX{<cond>}     <Rd>, <Rn>, <#lsb>, <#width>
UBFX{<cond>}     <Rd>, <Rn>, <#lsb>, <#width>
```

BFC has the following syntax:

```
BFC{<cond>}      <Rd>, <#lsb>, <#width>
```

RBIT is the simplest, and has the syntax:

```
RBIT{<cond>}     <Rd>, <Rn>
```

The parameter #lsb indicates the least significant bit of the bitfield (in other words, where to start in the bit pattern going from right to left) and should be in the range of 0 to 31. The #width parameter indicates the width of the bitfield; this parameter should be in the range of 1 to (32-lsb). This makes the most significant bit position of the field lsb + width-1.

Suppose we start with register r0 equal to 0xABCDDCBA and register r1 equal to 0xFFFFFFFF. If we employ the instruction

```
BFI    r1, r0, #8, #8
```

then registers r0 and r1 would appear as

r0 before	r0 after
0xABCDDCBA	0xABCDDCBA

r1 before	r1 after
0xFFFFFFFF	0xFFFFBAFF

The 8 lower bits of register r0 are inserted into register r1, starting at bit 8. Signed and unsigned bit field extract instructions work in a similar fashion. The instruction

```
UBFX   r1, r0, #12, #8
```

takes an unsigned bit field from a register (inserting zeros), and it leaves register r0 and r1 as follows:

r0 before	r0 after
0xABCDDCBA	0xABCDDCBA

r1 before	r1 after
0xFFFFBAFF	0x000000DD

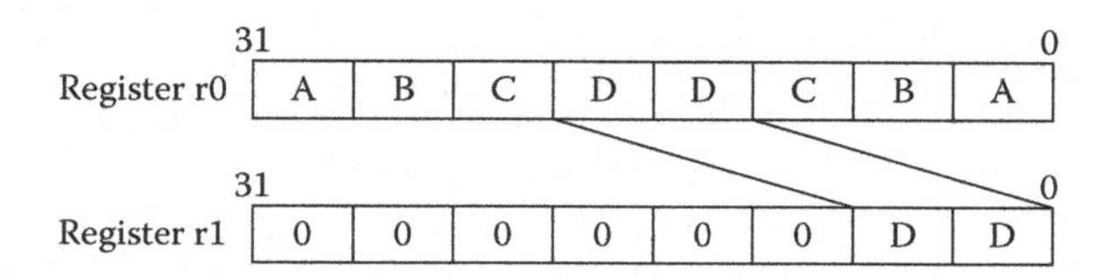

FIGURE 7.13 Unsigned bit field extract instruction.

Figure 7.13 shows the operation. Signed bit field extract sign extends the most significant bit of the field before inserting it into a register. Clearing a bit field can be done using BFC, so if we wished to clear out a nibble from register r1, we would say

```
BFC     r1, #4, #4
```

which leaves register r1 as

r1 before	r1 after
0x000000DD	0x0000000D

Should it become necessary to reverse the bit order of an entire register, you can use RBIT. The instruction

```
RBIT   r1, r1
```

would produce the following results:

r1 before	r1 after
0x0000000D	0xB0000000

7.7 FRACTIONAL NOTATION

When learning assembly, one of the first issues that often arises is how to put values like e and π in a program. This is a question of importance, given that you are likely to come across a value such as $\sqrt{2}$ in practical code, and clearly the ARM processor works only with 32-bit integer values unless you have a floating-point unit available. Or does it? As we've seen in Chapter 1, the processor works with data, raw numbers, bit patterns. These bit patterns are interpreted as different things by the *programmer*, not the processor (unless specifically told to do so). Suppose you have the 32-bit value 0xF320ABCD in register r3. Is this number positive or negative? Under normal circumstances, you simply don't know. You could interpret this number as –215,962,675 in decimal if it were a two's complement representation. You could also interpret this as 4,079,004,621 if it were just an ordinary, unsigned integer. These decisions are largely based on the algorithm and the type of arithmetic operations that are being done on these numbers. Normally, the program expects data in a particular form and uses it accordingly—if an adaptive filter routine is written using only signed numbers, then the programmer takes care to treat the results as signed values.

To take this argument one step further, where is the binary point in the number 0xF320ABCD? In other words, where does the integer portion of this number start

Integer				Fraction			
2^3	2^2	2^1	2^0	2^{-1}	2^{-2}	2^{-3}	2^{-4}
1	0	1	1				
	1	0	1	1			
		1	0	1	1		
			1	0	1	1	
				1	0	1	1

FIGURE 7.14 Binary interpretations of fractional values.

and where does the fractional portion start? Does it even have a fractional portion? Again, the answer lies with the programmer. It is useful at this point to look at small values first and then generalize the ideas to 32-bit numbers.

If the binary number 1011 is considered to be an unsigned number, the base 10 equivalent is $8 + 2 + 1 = 11$. However, you could also assume that a binary point exists just before the last digit, as shown in Figure 7.14, which means that 101.1 would equal $4 + 1 + 0.5 = 5.5$. This representation only gives two possible values for a fractional part—0 or 0.5, which isn't terribly useful. If the binary point is assumed to be one more place to the left, the number becomes 10.11, which is $2 + 0.5 + 0.25 = 2.75$ in base 10. Notice now that the two bits to the right of the binary point provide the four possibilities for a fractional value: 0, 0.25, 0.5, or 0.75. The resolution of the fraction is now 0.25, or in other words, the difference between any two fractional values can be no less than 0.25. Moving the binary point all the way to the left gives us the number 0.1011, which is $0.5 + 0.125 + 0.0625 = 0.6875$ in base ten. Having an entirely fractional value is limiting in some ways, but the resolution of our fractional value is 0.0625, which is something of an improvement.

Recall from Chapter 1 that if a base ten number n is represented as an m-bit two's complement number, with b being an individual bit's value, the value is calculated as

$$n = -b_{m-1}2^{m-1} + \sum_{i=0}^{m-2} b_i 2^i$$

so an 8-bit, two's complement number such as 10110010_2 equals

$$-2^7 + 2^5 + 2^4 + 2^1 = -78.$$

But we just said that the binary point in any representation is entirely up to the programmer. So what if we assumed that this 8-bit number was divided by 2^7? The bit pattern is identical—the interpretation is different. It turns out that the base 10 number n would be calculated as

$$n = -1 + 2^{-2} + 2^{-3} + 2^{-6} = -0.609375.$$

In fact, we could divide any m-bit number by 2^{m-1}, giving us just a fractional value n, such that

$$-1 \leq n \leq (1 - 2^{-(m-1)}).$$

Going back to our question about e and π, if we wanted to represent a number with only some fractional part, then we could scale the number by something less than 2^{m-1}. Suppose we have 16 bits, and we want to use e in a calculation. We know there are at least two bits needed to represent the integer portion of it (since the number is 2.71828...), so we would have at most 13 bits left for the fraction, given that the most-significant bit is the sign bit. In the literature, sometimes this is called Q notation, where our number might be called a Q13 number. Fortunately, the rules for working with Q-notated numbers are straightforward. Numbers that are added or subtracted in this notation should always have their binary points aligned, so $Qn + Qn = Qn$. When two numbers are multiplied, a Qn number times a Qm number will produce a result in a $Q(n + m)$ format.

EXAMPLE 7.17

Convert the transcendental number e into a Q13 format.

SOLUTION

To produce e in Q13 notation, take the value e, multiply it by 2^{13}, and then convert this number to hexadecimal (here, a calculator is often handy). So we have

$$e \times 2^{13} = 22{,}268.1647 = \text{0x56FC}.$$

Note we convert only the integer portion to hex. If this number is interpreted in Q13 notation, we can see that we do indeed have two bits to the left of the imaginary binary point and 13 bits to the right:

```
              sign bit
                 ↓
0x56FC  =  0101011011111100
                   ↑
         imaginary binary point
```

EXAMPLE 7.18

Convert $\sqrt{3}/2$ into a Q15 format.

SOLUTION

We want our value to have 15 bits of fraction and one sign bit, or something that looks like

$$s.f_{14}f_{13}f_{12}f_{11}f_{10}f_9f_8f_7f_6f_5f_4f_3f_2f_1f_0$$

First, we compute the decimal value for $\sqrt{3}/2$, which turns out to be 0.8660254038. Now multiply this value times 2^{15}, giving 28,377.9204. Convert only the integer portion to a hexadecimal value, giving 0x6ED9.

EXAMPLE 7.19

Let's take our illustration one step further to actually have the machine calculate *e* times $\sqrt{2}$. We need to convert this second number into fractional notation, too, so let's use Q13 notation again. If the same rules above are followed, the Q13 representation of $\sqrt{2}$ is 0x2D41. This short block of code below will perform the multiplication.

```
LDR     r3, =0x56FC      ; e in Q13 notation
LDR     r2, =0x2D41      ; sqrt(2) in Q13 notation
MUL     r5, r2, r3       ; product is in Q26 notation
```

If you run the code, you should find the value 0xF6061FC in register r5. Since the product is now in Q26 notation, we must convert the value to decimal and then divide by 2^{26} for the final result, which turns out to be 3.84412378. The actual product is 3.84423102, but you would expect that some precision would be lost immediately from the original values being represented in a limited-precision notation. If there were an infinite number of bits available to perform the operation, then our results would be exact. We haven't said what you would do with this value sitting in the register—yet. That 32-bit number could still represent just a big positive number or something odd like a mask for a configuration register in a cache controller! The processor has no idea what you're doing.

EXAMPLE 7.20

Let's do another example, except this time one of the operands is negative, as this introduces a few more twists in handling these notations. The two values to be multiplied are $\pi/4$ and the value of a digital signal, say −0.3872. The two values should be in Q15 notation, represented by 16 bits. In other words, the representation looks like

$$s.f_{14}f_{13}f_{12}f_{11}f_{10}f_9f_8f_7f_6f_5f_4f_3f_2f_1f_0,$$

or one sign bit, one imaginary binary point, and 15 bits of fractional value. To convert $\pi/4$ into a Q15 representation, we do as before, taking the decimal value and multiplying by 2^{15}, giving

$$\pi/4 \times 2^{15} = 25{,}735.927 = \text{0x6487 (convert only the integer portion to hex)}$$

The other, negative number will require a little more thinking. The easiest way to handle negative numbers is to convert a positive value first, then negate the result. So, to convert a positive Q15 value first, we have

$$|-0.3872| \times 2^{15} = 12{,}687.7696 = \text{0x318F}.$$

To negate 0x318F, you can either do it by hand (not recommended) or use a calculator or computer to perform a two's complement negation. What results is a 16-bit value with the most significant bit set—it had better be, or the value isn't negative. So negating 0x318F produces 0xCE71. (As an aside, some calculators will sign extend this value—just remember that you've chosen only 16 bits to represent the number!) As a sanity check, we can look at this value as

s		f	f	f	f	f	f	f	f	f	f	f	f	f	f	f
1	•	1	0	0	1	1	1	0	0	1	1	1	0	0	0	1
	↑	2^{-1}	2^{-2}	2^{-3}	2^{-4}	2^{-5}	2^{-6}	2^{-7}	2^{-8}	2^{-9}	2^{-10}	2^{-11}	2^{-12}	2^{-13}	2^{-14}	2^{-15}

imaginary binary point

Since we said earlier that this could be viewed as

$$-1 + \sum \text{all fractional bits} =$$

$$-1 + 2^{-1} + 2^{-4} + 2^{-5} + 2^{-6} + \cdots =$$

$$-1 + 0.6128 = -0.3871765,$$

we're getting what we expect. To code the multiplication, we do the same thing as before, only *you must sign extend the negative operands to 32 bits* if you use the ARM7TDMI. Why? The multiplier in the ARM7TDMI will take two 32-bit operands and multiply them together (returning only the lower 32 bits if the MUL instruction is used), so if we put a 16-bit value in one of the registers, the results will not be correct, as this is effectively a positive, two's complement number in 32 bits—to the processor anyway. The first simulation uses the version 4T instruction set, so it's necessary to do things the slightly old-fashioned way. The code would look like the following:

```
LDR     r3, =0x6487             ; pi/4 in Q15 notation
LDR     r2, =0xFFFFCE71         ; -0.3872 in Q15 notation

MUL     r5, r2, r3              ; product is in Q30 notation
LSL     r5, r5, #1              ; shift out extra sign bit
```

The result you find in register r5 is 0xD914032E. To interpret this number, it's easiest to negate it first (again, use a calculator), giving 0x26EBFCD2, since we know the result is a negative number. It's also a Q31 representation, so convert to base ten and then divide by 2^{31}, giving 0.3041. Why the extra shift at the end? Remember that a multiplication by two Q15 numbers will result in a Q30 product; however, there are 32 bits of result, which means that we end up with a superfluous sign bit in the most significant bit. In order to align the binary point again, everything is shifted left one bit. The final result could be taken from the upper half-word (16 bits) of register r5, resulting in another Q15 number. We didn't see this shifting in the first example because the operands were positive (hence there was no sign bit set) and we didn't do any realigning—we just stopped with a positive Q26 number.

If you use the Cortex-M4 processor, which has newer instructions that take 16-bit values and sign extend them for you, then you don't need to add the extra bits to the operands. You would use the instruction SMULBB, which tells the processor that you want to take the bottom half of two registers, multiply them together, and treat all the values as signed two's complement values. The code would be:

```
LDR     r3, =0x6487   ; pi/4 in Q15 notation
LDR     r2, =0xCE71   ; -0.3872 in Q15 notation
```

```
        SMULBB   r5, r2, r3     ; product is in Q30 notation
        LSL      r5, r5, #1     ; shift one bit left
done    B        done
```

Depending on the application and the precision needed in an algorithm, the data may be truncated at some point. So for graphics data where all the values may range only from 0 to 0xFF, once the algorithm produces a result, the entire fractional portion may be truncated anyway before the result is stored. For audio data or variables in a digital controller, you might keep some or all of the fractional precision before sending the result to a digital-to-analog converter (DAC), for example. The application will have a great influence on the way you handle the data. For further reading, see (Hohl and Hinds 2008; Oshana 2006).

7.8 EXERCISES

1. What's wrong with the following ARM instructions? You may want to consult the *ARM Architectural Reference Manual* to see the complete instruction descriptions and limitations.
   a. `ADD        r3, r7, #1023`
   b. `SUB        r11, r12, r3, LSL #32`
   c. `RSCLES     r0, r15, r0, LSL r4`
   d. `EORS       r0, r15, r3, ROR r6`

2. Without using the MUL instruction, give instructions that multiply register r4 by:
   a. 135
   b. 255
   c. 18
   d. 16,384

3. Write a compare routine to compare two 64-bit values, using only two instructions. (Hint: the second instruction is conditionally executed, based on the first comparison.)

4. Write shift routines that allow you to arithmetically shift 64-bit values that are stored in two registers. The routines should shift an operand left or right by one bit.

5. Write the following decimal values in Q15 notation:
   a. 0.3487
   b. −0.1234
   c. −0.1111
   d. 0.7574

6. Write the following signed, two's complement Q8 values in decimal:
   a. 0xFE32
   b. 0x9834

c. 0xE800
d. 0xF000

7. Write the assembly code necessary to detect an error in a 12-bit Hamming code, where your code tests the 4 checksum bits c3, c2, c1, and c0. Place your corrupted data in memory. Assume that only a single error occurs in the data and store your corrected value in register r6.

8. Write a program to calculate $\pi \times 48.9$ in Q10 notation.

9. Show the representation of sin(82°) in Q15 notation.

10. Show the representation of sin(193°) in Q15 notation.

11. Temperature conversion between Celsius and Fahrenheit can be computed using the relationship

$$C = \frac{5}{9}(F - 32)$$

where C and F are in degrees. Write a program that converts a Celsius value in register r0 to degrees Fahrenheit. Convert the fraction into a Q15 representation and use multiplication instead of division in your routine. Load your test value from a memory location called CELS and store the result in memory labeled FAHR. Remember that you will need to specify the starting address of RAM for the microcontroller that you use in simulation. For example, the LPC2132 microcontroller has SRAM starting at address 0x40000000.

12. Write a program for either the ARM7TDMI or the Cortex-M4 that counts the number of ones in a 32-bit value. Store the result in register r3.

13. Using the *ARM Architectural Reference Manual* (or the Keil or CCS tools), give the bit pattern for the following ARM instructions:

a. `RSB     r0, r3, r2, LSL #2`
b. `SMLAL   r3, r8, r2, r4`
c. `ADD     r0, r0, r1, LSL #7`

14. A common task that microcontrollers perform is ASCII-to-binary conversion. If you press a number on a keypad, for example, the processor receives the ASCII representation of that number, not the binary representation. A small routine is necessary to convert the data into binary for use in other arithmetic operations. Looking at the ASCII table in Appendix C, you will notice that the digits 0 through 9 are represented with the ASCII codes 0x30 to 0x39. The digits A through F are coded as 0x41 through 0x46. Since there is a break in the ranges, it's necessary to do the conversion using two checks.

The algorithm to do the conversion is

Mask away the parity bit (bit 7 of the ASCII representation), since we don't care about it.
Subtract a bias away from the ASCII value.
Test to see if the digit is between 0 and 9.
If so, we're done. Otherwise subtract 7 to find the value.

Write this routine in assembly. You may assume that the ASCII representation is a valid character between 0 and F.

15. Write four different instructions that clear register r7 to zero.

16. Suppose register r0 contains the value 0xBBFF0000. Give the Thumb-2 instruction and register value for r1 that would insert the value 0x7777 into the lower half of register r0, so that the final value is 0xBBFF7777.

17. Write ARM instructions that set bits 0, 4, and 12 in register r6 and leave the remaining bits unchanged.

18. Write a program that converts a binary value between 0 and 15 into its ASCII representation. See Exercise 14 for background information.

19. Assume that a signed long multiplication instruction is not available on the ARM7TDMI. Write a program that performs a 32 × 32 multiplication, producing a 64-bit result, using only UMULL and logical operations. Run the program to verify its operation.

20. Write a program to add 128-bit numbers together, placing the result in registers r0, r1, r2, and r3. The first operand should be placed in registers r4, r5, r6, and r7, and the second operand should be in registers r8, r9, r10, and r11.

21. Write a program that takes character data "a" through "z" and returns the character in uppercase.

22. Give three different methods to test the equivalence of two values held in registers r0 and r1.

23. Write assembly code for the ARM7TDMI to perform the following signed division:

$$r1 = r0/16$$

24. Multiply 0xFFFFFFFF (−1 in a two's complement representation) and 0x80000000 (the largest negative number in a 32-bit two's complement representation) on the ARM7TDMI. Use the MUL instruction. What value do you get? Does this number make sense? Why or why not?

25. A Gray code is an ordering of 2^n binary numbers such that only one bit changes from one entry to the next. One example of a 2-bit Gray code is $10\ 11\ 01\ 00_2$. The spaces in this example are for readability. Write ARM assembly to turn a 2-bit Gray code held in register r1 into a 3-bit Gray code in register r2. Note that the 2-bit Gray code occupies only bits [7:0] of register r1, and the 3-bit Gray code occupies only bits [23:0] of register r2. You can ignore the leading zeros. One way to build an n-bit Gray code from an (n – 1)-bit Gray code is to prefix every (n – 1)-bit element of the code with 0. Then create the additional n-bit Gray code elements by taking each (n – 1)-bit Gray code element in reverse order and prefixing it with a one. For example, the 2-bit Gray code above becomes

```
010 011 001 000 100 101 111 110
```

26. Write a program that calculates the area of a circle. Register r0 will contain the radius of the circle in Q3 notation. Represent π in Q10 notation, and store the result in register r3 in Q3 notation.

8 Branches and Loops

8.1 INTRODUCTION

Branches are a necessary evil. Software cannot avoid using them, and hardware engineers treat them as anathema. So much so that computer architects will go to extreme lengths to get rid of them. In fact, researchers spend years and years trying to come up with new strategies to either predict their effects before they arrive or avoid them entirely. A quick read through most computer architecture literature will highlight the elaborate hardware that is included with every modern design: static branch predictors, dynamic branch predictors, two-level adaptive branch predictors, instruction trace caches—the research continues. Certainly, ARM is no stranger to the phenomenon. However, the whole notion of how to remove and predict branches is beyond the scope of this book, so for now, we're going to examine one way around even having to use a branch instruction in assembly code. In our discussion, the use of conditional execution will demonstrate that even though you can't remove them completely, some branches can be avoided or removed. While the concepts underlying the ability to change the flow of a program are identical for both A- and M-class processors, the details are sufficiently different to warrant separate discussions. We'll examine the ARM7TDMI first, then look at some of the new instructions that were made available for the v7-M processors.

8.2 BRANCHING

One way to see the effects of a branch in an instruction stream, and the reason they present obstacles to optimizing code, is to look at a pipeline diagram for the ARM7TDMI, shown in Figure 8.1. The three-stage pipeline can fetch one instruction from memory, decode another instruction, and execute a third instruction, all in the same clock cycle. The analogy for a pipeline is washing dishes—one man washes a plate, one man rinses the plate previously washed, and a third man dries the plate previously rinsed, all at the same time. Once a man is finished, he passes his item to the next person in line; each man stays busy doing his task until all the dishes are done. You can see from the diagram that an ADD, SUB, and MOV instruction presents no problems for a pipelined architecture, since there is nothing present that would cause an instruction to stall or force the processor to wait for it to complete. However, a BL instruction, or any other type of branch, will cause the entire pipeline to be flushed—a branch instruction effectively tells the machine to start fetching new instructions from a different address in memory.

From the diagram, you can see that in cycle 1, the branch (BL) has entered the Execute stage of the pipeline, and two instructions have already been fetched (one

Cycle				1	2	3	4	5
Address	**Operation**							
0x8000	BL	Fetch	Decode	Execute	Linkret	Adjust		
0x8004	X	____	Fetch	Decode				
0x8008	XX	____	____	Fetch				
0x8FEC	ADD	____	____	____	Fetch	Decode	Execute	
0x8FF0	SUB	____	____	____	____	Fetch	Decode	Execute
0x8FF4	MOV	____	____	____	____	____	Fetch	Decode
								Fetch

FIGURE 8.1 ARM7TDMI pipeline diagram.

from address 0x8004 and one from 0x8008). Since the branch says to begin fetching new instructions from address 0x8FEC, those unused instructions must be thrown away. In a three-stage pipeline, the effects are not nearly as deleterious, but consider what would happen in a very deep pipeline, say 24 stages—a branch that is not handled correctly could force the processor to abandon significant amounts of work. It's worth noting at this point what's happening in cycles 2 and 3. Since the branch and link instruction saves a return address for us in the Link Register, the processor takes the Program Counter and moves it into register r14. However, we've already noted that the Program Counter points to the instruction being *fetched* in any given cycle, so at the time that the BL instruction is in the execute stage of the pipeline, the Program Counter is pointing to 0x8008. In cycle 2, this value is moved into the Link Register, but notice that we really need to return to address 0x8004. To correct for this, the processor subtracts four from the value in the Link Register in cycle 3 without introducing any stalls in the pipeline.

8.2.1 Branching (ARM7TDMI)

Any event that modifies the Program Counter (register r15) can be defined as a change of flow, and this can be accomplished by either explicitly modifying the Program Counter by writing to it or using one of the branch instructions. The three types of branch instructions on the ARM7TDMI are:

- B—Branch. This is the simplest form of branch, where condition codes may also be used to decide whether or not to branch to a new address in the code.
- BX—Branch indirect (formerly called Branch and eXchange). In addition to providing a branch using a registered value, this instruction provides a mechanism to switch from 32-bit ARM instructions to 16-bit Thumb instructions. We will cover Thumb in more detail in Chapter 17.
- BL—Branch and Link. Here, the Link Register (r14) is used to hold a return address just after a branch instruction, so that if we want to execute a subroutine and return, the processor merely has to put the value of the Link Register into the Program Counter at the end of the subroutine. We saw a few examples of this already in Chapter 6.

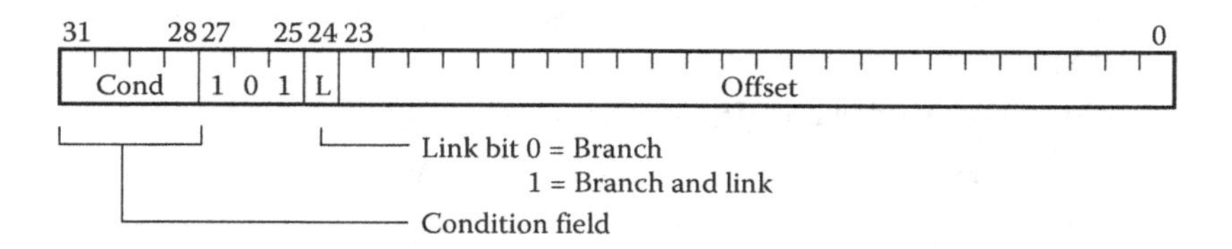

FIGURE 8.2 The B and BL instruction.

The branch instructions in the ARM instruction set, B and BL shown in Figure 8.2, have 24-bit fields for a branch offset. When the processor executes a branch instruction, this offset is added to the Program Counter, and the machine begins fetching instructions from this new address. Since a 32-bit instruction cannot hold a 32-bit address, a couple of questions immediately arise. First, if only 24 bits are available, what's the best way to effectively use them? Rather than just adding this offset to register 15, the 24 bits are shifted left by two bits first, since all ARM instructions must be word-aligned anyway, i.e., the least significant two bits of the address are always zero. This gives a range of ±32 MB using this method and brings up the second question: how do you jump more than 32 MB away from the current address? Remember that register 15, the Program Counter, is just another register, so you can say something like

```
LDR pc, =0xBE000000
```

or

```
MOV pc, #0x04000000
```

which forces an address directly into the Program Counter.

For the most part, this chapter deals with one particular type of branch instruction—B—leaving the discussion of BL and BX for later chapters, but we'll examine both conditional and unconditional branches. The unconditional instruction B alone simply forces the code to jump to some new address. However, it's likely you'll want to condition this decision with more criteria; for example, did a counter just expire or did an earlier subtraction result in a negative number? Table 8.1 shows various combinations of flags in ARM processors that can be used with branches. It is quite possible, then, to say

```
CMP     r0, r1
BLT     Sandwich        ; programmers get hungry...
```

where this means if register r0 is less than register r1, branch to a label called Sandwich. Recall that the job of a comparison instruction is to set the flags in the CPSR and little else, so the branch instruction can immediately use that information to make decisions. We'll certainly see more examples of conditional branches throughout the book, and shortly we'll find that on the ARM7TDMI, almost any instruction can be conditionally executed.

TABLE 8.1
Condition Codes and Their Meaning

Field Mnemonic	Condition Code Flags	Meaning	Code
EQ	Z set	Equal	0000
NE	Z clear	Not equal	0001
CS/HS	C set	Unsigned ≥	0010
CC/LO	C clear	Unsigned <	0011
MI	N set	Negative	0100
PL	N clear	Positive or zero	0101
VS	V set	Overflow	0110
VC	V clear	No overflow	0111
HI	C set and Z clear	Unsigned >	1000
LS	C clear and Z set	Unsigned ≤	1001
GE	N ≥ V	Signed ≥	1010
LT	N ≠ V	Signed <	1011
GT	Z clear, N = V	Signed >	1100
LE	Z set, N ≠ V	Signed ≤	1101
AL	Always	Default	1110

EXAMPLE 8.1

Suppose that you need to compare two signed numbers, where they are assumed to be in two's complement form, with 0xFF000000 in register r0 and 0xFFFFFFFF in register r1. If you wanted to branch to some code only if the first number was less than the second, you might have something like

```
CMP    r0, r1    ; r0<r1?
BLT    algor
```

For this case, the branch would be taken, as register r0 holds a large, negative number and register r1 holds –1. If you assume that the two numbers are unsigned, BCC should be used instead, as register r1 would hold the larger number.

Since any real code will have something more than just arithmetic and control instructions, the whole notion of looping through code needs to be addressed. We need the ability to execute a section of code multiple times, so we're going to start off the discussion of loops by looking at a real problem. Suppose that we had a register containing a binary value that needed to be normalized. In other words, we need to have the leading 1 in the most significant bit, even if we have to shift it to get it there. This does actually come up in numerical algorithms, such as the Newton–Raphson division algorithm, logarithmic routines, and some priority decoders. This problem is so significant that ARM decided to add a new instruction (CLZ, or Count Leading Zeros) to the version 5TE architectures and beyond, just to reduce the cycle count of certain mathematical routines. Since the ARM7TDMI does not have this instruction, it makes a good example to code.

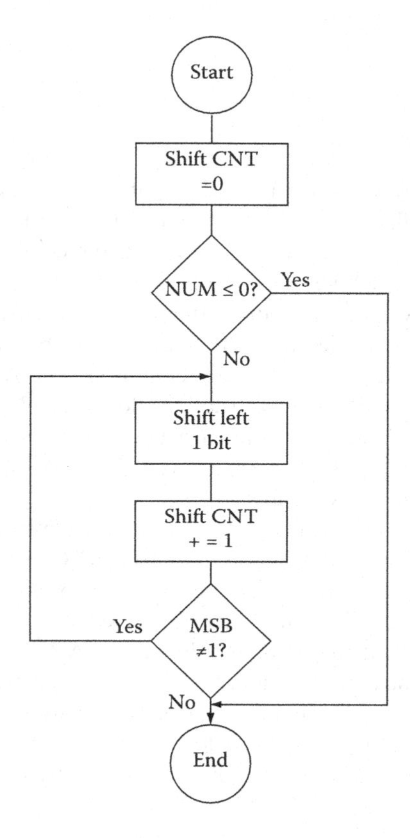

FIGURE 8.3 Flowchart for normalization algorithm.

For our normalization task, the flowchart in Figure 8.3 might help to decide how this algorithm will be implemented. The first thing to test for is whether the argument is either zero or already normalized—if there's nothing to do, then the routine should just stop. Otherwise, we want to shift it by one bit to the left and increment the shift counter, which could be used by another routine to tell how much the original value had to be shifted. The routine should check to see if the most significant bit is now set, as this would be the place to stop. If it's not, the code should go back and repeat the shift/increment/test portion again.

The code for this algorithm might look like the following:

```
       AREA Prog8a, CODE, READONLY
       ENTRY
main
       MOV          r4, #0          ; clear shift count
       CMP          r3, #0          ; is the original value <= 0?
       BLE          finish          ; if yes, we're done
loop   LSLS         r3, r3, #1      ; shift one bit
       ADD          r4, r4, #1      ; increment shift counter
```

```
        BPL           loop
finish
        B             finish
        END
```

The first type of branch we see is the BLE just above the loop statement. The comparison above sets the flags in the CPSR, and if the number is negative (indicated by the most significant bit being set) or zero, then the condition forces the code to branch to the label finish. Otherwise, it continues to the first instruction inside the loop, which shifts the value one place to the left. An important point to note here is that an "S" has been appended to the LSL instruction because we have to tell the machine to set the flags again for the loop condition. The ADD will not have any effect on the flags, since arithmetic instructions do not set flags unless told to do so. The BPL, or Branch if Positive or zero, instruction simply says to check the flags again, and as long as the most significant bit is clear, i.e., the value is still not normalized, to branch back to the label loop.

While this code would not occupy much memory, the issue that is probably not obvious is the cycle time that it would take to execute. Consider the worst case scenario where a 1 is in the least significant bit of the register, forcing 31 shifts to get it to the most significant bit. The LSL instruction takes one cycle to execute, as does the ADD. However, the branch instruction flushes the ARM7's pipeline and causes a change of flow in the instruction stream, and the code repeats all of this 31 times. In total, this adds up to a significant number of cycles. Toward the end of the chapter, we will see two much more efficient ways to do this. For now, we'll continue to examine ways to branch and how different types of loops are written.

8.2.2 Version 7-M Branches

Version 7-M cores have more branch instructions than the ARM7TDMI, but the types of allowable branches have some limitations:

- B—Branch. This is the simplest form of branch, where condition codes may be used to decide whether or not to branch to a new address in the code.
- BX—Branch indirect. A registered value is used as a branch target. If bit[0] of the address is a zero, a usage fault exception will occur. Use this instruction carefully, as the assembler will not be generating offsets or addresses for you, and the value in the register must have bit[0] set.
- BL—Branch with Link. As with the ARM7TDMI, the Link Register will hold a return address after a branch.
- BLX—Branch indirect with Link. This instruction is similar to BL, only the address is held in a register.
- CBZ, CBNZ—Compare and Branch if Zero, Compare and Branch if Nonzero. These two instructions are useful in looping and can reduce the number of instructions.
- IT blocks—IF-THEN blocks. The IT instruction can be used to avoid branching entirely with up to four instructions in a block.

The B, BX, and BL instructions work in the same way as in v4T architectures, namely to change the Program Counter to a new address from which the processor can begin fetching instructions. Both B and BL can be thought to contain immediate addresses, meaning that the address is encoded in the instruction itself. For example, you might say

```
B  myroutine
```

and the linker will calculate the PC-relative offset necessary to jump to myroutine.

The BX and BLX instructions use an address contained in a register; for example,

```
BX  r9
```

will load the value in register r9 into the Program Counter, and fetching begins from this new address.

Unlike v4T branch instructions, which always have a range of –32MB to 32MB, in version 7-M the range varies depending on which branch instruction you use. A 32-bit branch instruction has a range of –16MB to +16MB. A conditional branch used inside of an IT block (discussed shortly) has a range of –16MB to +16MB, while a conditional branch used outside of an IT block has a shorter range of –1MB to +1MB. In some cases, it might be necessary to force the longer instruction to be used to get the maximum range, for example

```
        BEQ.W  label
```

where the .W suffix denotes "wide". For complete details, consult either (Yiu 2014) or the *ARM v7-M Architectural Reference Manual* (ARM 2010a).

With the introduction of Thumb-2 and Unified Assembly Language (UAL), it's worth pointing out here that you're likely to see some mixed use of the BX instruction. BX can be used to change the state of the machine from ARM to Thumb on the ARM7TDMI (covered in detail in Chapter 17), but it can also be used as a simple branch instruction, too, as long as the least significant bit is not set (this would throw us into Thumb state). This instruction takes a register value and loads it into the Program Counter. An example instruction might be

```
BX r4
```

where the value held in register r4 is moved to the PC and then execution begins from this new address. This leaves us with another way to return from subroutines, which are covered in Chapter 13. Rather than using the older instruction

```
MOV pc, lr
```

which transfers the contents of the Link Register into the Program Counter, you should now say

```
BX lr
```

which does the same thing. As you study code samples from other sources, you are likely to see both styles, so just keep this in mind as you read documentation and write your own code.

The Compare and Branch if Nonzero (CBNZ) and Compare and Branch if Zero (CBZ) instructions can be used to avoid changing the condition code flags during loops. As an example, if you assume that the CMP instruction does not change the flags, instead of saying

```
CMP    r2, #0
BEQ    label
```

you would use the single instruction

```
CBZ    r2, label
```

as they are functionally equivalent statements. These two instructions come with a few restrictions worth noting. First, the only registers allowed must be in the range of r0–r7. Second, the branch destination must be within 4–130 bytes following the instruction. Finally, the CBZ and CBNZ instructions cannot be used within an IT block, which brings us to the subject of Section 8.3, looping.

8.3 LOOPING

Nearly all embedded code will have some form of loop construct, especially if an operating system is running or the application requires the processor to periodically check an input or peripheral. We'll examine three easy loop structures—the while loop, the for loop, and the do-while loop, along with code samples that show their construction.

8.3.1 While Loops

Certainly, one of the more common constructs in C or C++, or any high-level language really, is the while loop, and its cousin, the for loop. Since the number of iterations of a while loop is not a constant, these structures tend to be somewhat simple. Suppose we had the following C code:

```
j=100;
while (j!=0) {
//do something
j--;}
```

The while loop can be constructed on an ARM7TDMI as

```
        MOV    r3, #0x64
        B      Test
Loop    .
        .                   ; do something
        .
        SUB    r3, r3, #1   ; j--
Test    ..                  ; evaluate condition j=0?
        BNE    Loop
```

While loops evaluate the loop condition before the loop body. There is only one branch in the loop body itself. The first branch actually throws you into the first iteration of the loop.

The loop can be constructed for the Cortex-M4 using version 7-M instructions as

```
        MOV     r3, #0x64
Loop    CBZ     r3, Exit
        ; do something
        SUB     r3, #1          ; j--
        B       Loop
Exit
```

Here the initial test is done at the start of the loop. The Compare and Branch if Zero (CBZ) instruction will test the counter against zero, and if it is equal to zero, branch outside the loop to Exit. Note that the CBZ instruction will only support forward branches, meaning only to addresses that add to the Program Counter, not those that subtract from it.

8.3.2 For Loops

The other common loop, the for loop, is actually just a variation of the while loop. Suppose you wish to create a for loop to implement a counter of some kind using a control expression to manage an index *j*, which is declared as an integer:

```
for (j=0; j<10; j++) {instructions}
```

The first control expression (`j = 0`) just clears a variable and can execute before the loop begins. The second control expression (`j < 10`) is evaluated on each pass through the loop and determines whether or not to exit. The index increments at the end of each pass to prepare for a branch back to the start of the loop. In assembly, it might be tempting to code this loop as

```
        MOV     r1, #0                  ; j=0
LOOP    CMP     r1, #10                 ; j<10?
        BGE     DONE                    ; if j>=10, finish
        .
        .       ; instructions
        .
        ADD     r1, r1, #1              ; j++
        B       LOOP
DONE    ..
```

A much better way to do this is to count down rather than up. A for loop can be constructed using only one branch at the end, subtracting one from the counter register, and branching back to the top only when the counter value is not equal to zero, like this:

```
        MOV     r1, #10         ; j=10
LOOP
        .
        .                       ; instructions
        .
        SUBS    r1, r1, #1      ; j=j-1
        BNE     LOOP            ; if j=0, finish
DONE            ..
```

This is actually more efficient in that a branch is removed and a comparison against zero comes for free, since we set the condition codes with the SUB instruction and use the B instruction to test whether or not the counter is now zero.

EXAMPLE 8.2

Let's translate the following C code to assembly using an ARM7TDMI-based microcontroller.

```
for (i=0; i<8; i++) {
     a[i]=b[7-i];
       }
```

The index *i* is declared as an integer, and assume the arrays a and b contain only byte-wide data. We also need to have the array a be located in writable memory, so for this example, you will need to select a target device that contains some RAM. Since we'll be using the LPC2132 microcontroller from NXP in Chapter 16, we can select this one as the target device now. It has 16 KB of on-chip RAM, and programming it now only requires that we know the starting address of RAM, which is 0x40000000. The code below implements the above for loop.

```
         AREA Prog8b, CODE, READONLY
SRAM_BASE EQU 0x40000000
         ENTRY
         MOV      r0, #7           ; i
         ADR      r1, arrayb       ; load address of array
         MOV      r2, #SRAM_BASE   ; a[i] starts here
Loop
         RSB      r3, r0, #7       ; index=7-i
         LDRB     r5, [r1, r3]     ; load b[7-i]
         STRB     r5, [r2, r0]     ; store into a[i]
         SUBS     r0, r0, #1       ; i--
         BGE      Loop
done     B        done
         ALIGN
arrayb            DCB 0xA,0x9,0x8,0x7,0x6,0x5,0x4,0x3
         END
```

The code starts by setting the index *i* to 7. The address of array b, which is located in memory just after our program code, is loaded into register r1. The address of array a, which will be located in SRAM on the chip, is placed in register r2. The reverse subtract operation calculates the difference between

7 and *i* to use as a pointer into memory. The data is loaded from memory into register r5 with a load byte instruction (LDRB), and then stored into array a using a store byte instruction (STRB). The counter is decremented, setting the flags for our upcoming comparison. The BGE (Branch if Greater than or Equal to zero) examines the flags, and based on the state of the N and the V flags, branches to the start of the loop. Notice that the data for array b is placed in the code using a DCB statement, which simply places byte-wide (8-bit) constants in the instruction memory. The other important thing to note here is that our loop is created with just one branch statement, since the comparison is built into the branch instruction.

EXAMPLE 8.3

In this next example for the ARM7TDMI, suppose we have six 32-bit integers that need to be summed together, where the integer data is stored in memory. This might be equivalent to a C statement such as

```
sum=0;
for (i=0; i<6; i++) {
     sum += a[i];
}
```

While simple loops don't often require a flowchart, we might sketch one out to help define the steps necessary to write the assembly, as shown in Figure 8.4.

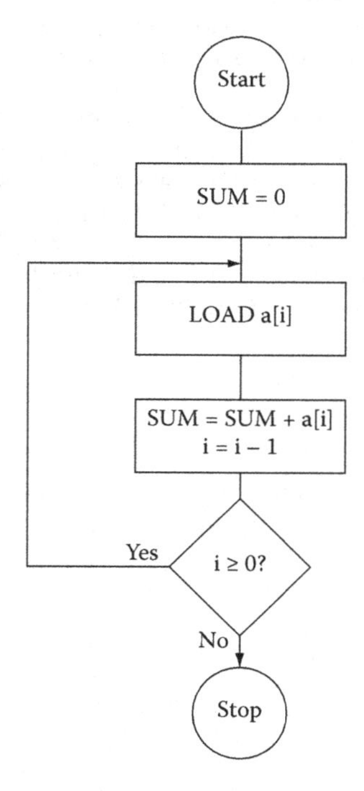

FIGURE 8.4 Flowchart for summing six integers.

The code can be written in just a few lines, with one branch and no CMP instructions by simply counting down:

```
        AREA Prog8c, CODE, READONLY
        ENTRY
        MOV     r0, #0                  ; sum=0
        MOV     r1, #5                  ; # of elements -1
        ADR     r2, arraya              ; load start of array
Loop
        LDR     r3,[r2,r1,LSL #2]       ; load value from memory
        ADD     r0, r3, r0              ; sum += a[i]
        SUBS    r1, r1, #1              ; i=i-1
        BGE     Loop                    ; loop only if i>=0
done    B       done
        ALIGN
arraya DCD      -1,-2,-3,-4,-5,-6
        END
```

The code begins by clearing out the accumulated sum held in register r0. While there are six elements to add, we only load the value 5 into an index register because the values will be loaded from memory using an address with an offset, and therefore, we can use the fact that one of the elements is addressed with an offset of zero. The start of the array in memory is loaded into register r2 using the pseudo-instruction ADR. Notice that the data is declared at the end of the code with a DCD directive and ends up being located just at the end of our code in instruction memory. However, this data could just as easily have been located somewhere else in memory, such as in a peripheral, in SRAM, or elsewhere on the microcontroller.

At the beginning of the loop, register r3 is loaded with one word of data, and the value is then added into the accumulated sum. The counter for the loop is decremented with the SUBS instruction that sets the flags. Recall that we can use the condition codes for a variety of branch types, as well as conditional execution, which we'll see in the next section. The BGE instruction causes the processor to branch back to our label Loop only if the subtraction produced a value that was greater than or equal to zero. Once the counter becomes negative, the loop terminates.

8.3.3 Do-While Loops

Here the loop body is executed before the condition is evaluated. The structure is the same as the while loop but without the initial branch:

```
LOOP
        ...
        ; loop body
        ...
        ; evaluate condition
        BNE LOOP
EXIT
...
```

8.4 CONDITIONAL EXECUTION

As we saw in the beginning of the chapter, branches can potentially cause very large delays in code, so if there were a way to remove a branch entirely, not only would our execution time decrease but our code size would decrease, too. Conditional execution provides this ability, since we can precondition an instruction as it goes through the pipeline—if it's not even necessary to execute the instruction, it passes through without affecting anything. It still takes a clock cycle, and still holds a place in the pipeline, but nothing happens.

8.4.1 v4T Conditional Execution

All version 4T ARM instructions can be conditionally executed based on the four-bit field in the upper nibble of the instruction, shown in Figure 8.5. Fortunately, you can still specify these conditions using the same mnemonics that we use for branches from Table 8.1. Careful readers will have noticed that there are 15 different field mnemonics, such as GT, GE, LT, etc., but there are actually 16 combinations—this is a four-bit field. Figure 8.6 shows an arbitrary Thumb-2 instruction, a 32-bit wide ADD instruction, which allows more flexibility than the 16-bit ADD instruction. Notice that bits 28 through 31, the upper nibble of the instruction, are all ones. In earlier ARM architectures, this encoding was used for the condition Never (NV), which seems a little unusual given that one normally hopes to have instructions used at least *once* in compiled code! By using this encoding to identify some of the new, 32-bit Thumb-2 instructions, the instruction space was given a bit of breathing room, allowing for more operations to be added.

It's at this point that we can also begin to see why Thumb-2 instructions could not be conditionally executed like those in the v4T ISA, primarily for two reasons: 16-bit Thumb instructions have no extra bits for a conditional field, and if another construct is used, something beyond the traditional Thumb instructions will be needed. By adding a new instruction called IT to build small IF-THEN loops, this limitation can be overcome, as we'll see shortly. For now, let's examine how ARM instructions can be conditionally executed using the conditional field.

31 28	27 0
cond	

FIGURE 8.5 Condition code field.

Encoding T3 ARMv7-M

```
ADD{S}<c> .W <Rd>,<Rn>, #<const>
```

15 14 13 12 11	10	9	8 7 6 5	4	3 2 1 0	15	14 13 12	11 10 9 8	7 6 5 4 3 2 1 0
1 1 1 1 0	i	0	1 0 0 0	S	Rn	0	imm3	Rd	imm8

FIGURE 8.6 32-bit wide Thumb instruction.

EXAMPLE 8.4

Suppose you had to test a string for the presence of either a "!" or a "?" character, and you had one-byte character data called char. The test condition might be written in C as

```
if (char=='!' || char=='?')
    found++;
```

Assuming that the character data char was held in register r0 and the variable found was in register r1, you could write the assembly for this as

```
    TEQ    r0,#'!'
    TEQNE  r0,#'?'
    ADDEQ  r1,r1,#1
```

Recall that the TEQ instruction tests equivalence of two things by using the exclusive OR operation, always setting the flags in the CPSR afterward. If the two numbers were in fact the same, the Z flag would be set, so the second TEQ instruction would not be executed. The third instruction would be executed, since the Z flag has not been changed since the first comparison and the condition is still true.

EXAMPLE 8.5

At the risk of being almost overused as an example, the greatest common divisor algorithm is still worth presenting here, as it demonstrates the power of conditional execution. Euclid's algorithm for computing the GCD of two positive integers (a,b) can be written as

```
while (a != b) {
        if (a>b) a=a - b;
        else b=b - a;
        }
```

To illustrate how this works, if you had the numbers 18 and 6, you would always subtract the smaller number from the larger until the two are equal. This gives 12 and 6 on the first pass, and 6 and 6 on the second and final pass.

Assuming that the numbers a and b are held in registers r0 and r1, respectively, the assembly code might look something like this if only the branches are executed conditionally:

```
gcd     CMP    r0,r1        ; a>b?
        BEQ    end          ; if a=b we're done
        BLT    less         ; a<b branches
        SUB    r0,r0,r1     ; a=a-b
        B      gcd          ; loop again
less    SUB    r1,r1,r0     ; b=b-a
        B      gcd
```

The most efficient way to do this is to avoid the branches altogether and conditionally execute the instructions, as

```
gcd  CMP    r0, r1
     SUBGT  r0, r0, r1
     SUBLT  r1, r1, r0
     BNE    gcd
```

Not only does this code execute more quickly, it contains fewer instructions. Note that in the second case, the code compares the two numbers, setting the flags. The two subsequent instructions are mutually exclusive, so there will never be a case where one of the numbers is less than and greater than the other at the same time. The final case where the two numbers are equal after the compare forces both subtraction instructions to be ignored in the pipeline, and the final branch instruction falls through as well because they are, in fact, equal. Not having the extra branches in the code makes a huge difference, since the pipeline does not get flushed repeatedly when the branches are taken.

8.4.2 v7-M Conditional Execution: The IT Block

It was stated in the last section that a new instruction was combined with the older Thumb instruction set to allow small IF-THEN blocks to be built. Like conditional execution, the goal is to remove or avoid branches as much as possible in Thumb-2 code. The IT instruction is used in conjunction with other operations to build blocks using the following syntax:

IT*xyz* condition

where the *x*, *y*, and *z* fields specify either T for THEN (true) or E for ELSE (false). For example, a simple IF-THEN statement such as

```
if (r3 < r8){
        r3 = r3 + r8;
        r4 = 0;}
else
        r3 = 0;
```

might be coded as

```
ITTE   LT
ADDLT  r3, r3, r8
MOVLT  r4, #0
SUBGE  r3, r3, r3
```

Here the ADD and MOV instructions have the same condition specified in the ITTE instruction (LT), and the ELSE instruction reflects the inverse condition (GE). There are up to four instructions in an IF-THEN block and a few simple rules that govern its construction:

- The condition field must be one of the fields listed in Table 8.1, except Always.
- The first statement following the IT instruction must be the true-then-execute case (THEN).
- The number of T's and E's in the IT instruction itself should match the number of THEN and ELSE instructions in the block. If you specify an instruction such as ITTEE, there should be two THEN instructions and two ELSE instructions following the IT instruction.
- Branches to any instruction in the IT block are not permitted, apart from those performed by exception returns.
- Any branches used in an IT block must be the last instruction in the block.
- The ELSE condition must be the inverse of the THEN condition. If you refer to Table 8.1 again, you will notice that these two fields differ only in the LSB of the encoding. In other words, GE, which is 1010, is the inverse of LT, which is 1011.

Note that the IT instruction does not affect the condition code flags. If you use 16-bit instructions in the IT block, other than CMP, CMN, and TST, they do not set the condition code flags either.

EXAMPLE 8.6

In Chapter 16, we will examine a program that changes the color of the LEDs on a Tiva Launchpad. One small section of the code can be stated in C as

```
if (Color == 8)
        Color=2;
else
        Color=Color * 2;
```

This forces the Color variable to take on the values 2, 4, or 8, and then to cycle through those same values over and over. Assuming our variable is held in register r6, the assembly for the Cortex-M4 would look like

```
CMP            r6, #8
ITE            LT
LSLLT          r6, r6, #1      ; LED=LED * 2
MOVGE          r6, #2          ; reset to 2 otherwise
```

The first comparison tests against our upper limit (8) and sets the flags for our conditional instructions coming up. Notice that the IT instruction specifies only one Less Than instruction (LSL) and one Else instruction (MOV). The IT block then begins with a logical shift of the value in register r6 if the value was either two or four. Otherwise, the value is reset to the starting value of two with a simple MOV. The Else instruction is predicated with the inverse condition of LT.

8.5 STRAIGHT-LINE CODING

Now that we've seen how branches are done, you might ask if an algorithm that contains a loop necessarily has to have a branch instruction. The answer is no. It turns out that in many algorithms, especially signal processing algorithms, speed is the most important consideration in its implementation. If any delays can be removed from the code, even at the expense of memory, then sometimes they are. Instructions that are between the start of a loop and the branch back to the beginning can be repeated many times, a process known as unrolling a loop. For example, if you had one instruction that was inside of a for loop, i.e.,

```
        MOV     r1, #10                 ; j=10
Loop
        MLA     r3, r2, r4, r5          ; r3=r2*r4+r5
        SUBS    r1, r1, #1              ; j=j - 1
        BNE     Loop                    ; if j=0, finish
```

you could do away with the for loop entirely by simply repeating the MLA instruction 10 times.

If you recall from the normalization example presented at the beginning of the chapter, a branch forces a pipeline to flush instructions that have already been fetched and decoded; therefore, a routine may spend considerable time just refilling the pipeline. To avoid this, software can simply remove all branches—the routine may be significantly faster but it will occupy more memory because of the repeated instructions. The normalization routine in Section 8.2.1 has been optimized by Symes (Sloss, Symes, and Wright 2004) and is presented below. Notice that the cycle count is fixed for this routine—17 cycles for an ARM7TDMI—due to the conditional execution and lack of branches. The instructions that are not executed, those that fail their condition codes, still have to go through the pipeline and still take a cycle in the execute stage.

```
; Normalization on the ARM7TDMI
; Argument in r0
; Shift count needed for normalization returned in r1
shift   RN      r0
x       RN      r1
        AREA    Prog8d, CODE, READONLY
        ENTRY
        MOV     shift, #0               ; shift=0
        CMP     x, #1<<16               ; if (x<(1<<16))
        LSLCC   x, x, #16               ; {x=x<<16;
        ADDCC   shift, shift, #16       ; shift+=16; }
        TST     x, #0xFF000000          ; if (x<(1<<24))
        LSLEQ   x, x, #8                ; {x=x<<8;
        ADDEQ   shift, shift, #8        ; shift+=8; }
        TST     x, #0xF0000000          ; if (x<(1<<28))
        LSLEQ   x, x, #4                ; {x=x<<4;
        ADDEQ   shift, shift, #4        ; shift+=4; }
        TST     x, #0xC0000000          ; if (x<(1<<30))
        LSLEQ   x, x, #2                ; {x=x<<2;
        ADDEQ   shift, shift, #2        ; shift+=2; }
        TST     x, #0x80000000          ; if (x<(1<<31))
        ADDEQ   shift, shift, #1        ; { shift+=1 ;
```

```
        LSLEQS  x, x, #1              ; x <<= 1;
        MOVEQ   shift, #32            ; if (x==0) shift=32; }
done    B       done
        END
```

As a point of interest, it was mentioned earlier that a new instruction, Count Leading Zeros (CLZ), was added to the v5TE instruction set, and it is included in the v7-M instructions. The entire routine above can be done in two lines of code on the Cortex-M4:

```
; r2 = shift count
; r3 = original value
CLZ    r2, r3
LSL.W  r3, r3, r2   ; r3 << shift count
```

8.6 EXERCISES

1. Code the following IF-THEN statement using Thumb-2 instructions:

```
if (r2 != r7)
    r2 = r2 - r7;
else
    r2 = r2 + r4;
```

2. Write a routine for the ARM7TDMI that reverses the bits in a register, so that a register containing $d_{31}d_{30}d_{29}...d_1d_0$ now contains $d_0d_1...d_{29}d_{30}d_{31}$. Compare this to the instruction RBIT on the Cortex-M4.

3. Code the GCD algorithm given in Section 8.4.1 using Thumb-2 instructions.

4. Find the maximum value in a list of 32-bit values located in memory. Assume the values are in two's complement representations. Your program should have 50 values in the list.

5. Write a parity checker routine that examines a byte in memory for correct parity. For even parity, the number of ones in a byte should be an even number. For odd parity, the number of ones should be an odd number. Create two small blocks of data, one assumed to have even parity and the other assumed to have odd parity. Introduce errors in both sets of data, writing the value 0xDEADDEAD into register r0 when an error occurs.

6. Compare the code sizes (in bytes) for the GCD routines in Section 8.4.1, where one is written using conditional execution and one is written using branches.

7. Digital signal processors make frequent use of Finite Impulse Response filters. The output of the filter, $y(n)$, can be described as a weighted sum of past and present input samples, or

$$y(n) = \sum_{m=0}^{N-1} h(m)x(n-m)$$

where the coefficients $h(m)$ are calculated knowing something about the type of filter you want. A linear phase FIR filter has the property that its coefficients are symmetrical. Suppose that N is 7, and the values for $h(m)$ are given as

$h(0) = h(6) = -0.032$
$h(1) = h(5) = 0.038$
$h(2) = h(4) = 0.048$
$h(3) = -0.048$

Use the sample data $x(n)$ below:

```
SAMPLE    DCW     0x0034,0x0024,0x0012,0x0010
          DCW     0x0120,0x0142,0x0030,0x0294
```

Write an assembly language program to compute just one output value, $y(8)$, placing the result in register r1. You can assume that $x(8)$ starts at the lowest address in memory and that $x(7)$, $x(6)$, etc., follow as memory addresses increase. The coefficients should be converted to Q15 notation, and the input and output values are in Q0 notation.

8. Write a routine to reverse the word order in a block of memory. The block contains 32 words of data.

9. Translate the following conditions into a single ARM instruction:
   a. Add registers r3 and r6 only if N is clear. Store the result in register r7.
   b. Multiply registers r7 and r12, putting the results in register r3 only if C is set and Z is clear.
   c. Compare registers r6 and r8 only if Z is clear.

10. The following is a simple C function that returns 0 if (x + y) < 0 and returns 1 otherwise:

```
int foo(int x, int y){
        if (x+y<0)
        return 0;
        else
        return 1;
        }
```

Suppose that a compiler translated it into the following assembly:

```
foo     ADDS     r0, r0, r1
        BPL      PosOrZ
```

```
done
        MOV     r0, #0
        BX      lr
PosOrZ
        MOV     r0, r1
        B       done
```

This is inefficient. Rewrite the assembly code for the ARM7TDMI using only four instructions (hint: use conditional execution).

11. Write Example 7.10 (finding the absolute value of a number) for the Cortex-M4.

12. What instructions are actually assembled if you type the following lines of code for the Cortex-M4 into the Keil assembler, and why?

```
CMP       r3, #0
ADDEQ     r2, r2, r1
```

9 Introduction to Floating-Point

Basics, Data Types, and Data Transfer

9.1 INTRODUCTION

In Chapter 1 we looked briefly at the formats of the floating-point data types called single-precision and double-precision. These data types are referred to as *float* and *double*, respectively, in C, C++, and Java. These formats have been the standard since the acceptance of the IEEE Standard for Binary Floating-Point Arithmetic (IEEE Standard 1985), known as the IEEE 754-1985 standard, though floating-point was in use long before an effort to produce a standard was considered. Each computer maker had their own data types, rounding modes, exception handling, and odd numeric quirks. In this chapter we take a closer look at the single-precision floating-point data type, the native data type of the Cortex-M4 floating-point unit, and a new format called half-precision. An aim of this chapter is to answer why a programmer would choose to use floating-point over integer in arithmetic computations, and what special considerations are necessary to properly use these data types. This introductory look, here and in Chapters 10 and 11, will let us add floating-point to our programming and make use of a powerful feature of the Cortex-M4.

9.2 A BRIEF HISTORY OF FLOATING-POINT IN COMPUTING

Hardware floating-point is a relatively new part of embedded microprocessors. One of the earliest embedded processors offered with optional floating-point was the ARM10, introduced in 1999. In the last fifteen years embedded processors, such as the ARM11 and Cortex-M4, have been available with hardware floating-point. The adoption of floating-point in the embedded space follows a long tradition of computing features which were first introduced in supercomputer and mainframe computers, and over time migrated to minicomputers, later to desktop processors, and ultimately to the processors which power your smart phone and tablet.

The earliest processor with floating-point capability was the Z3, built by Konrad Zuse in Berlin in the years 1938–1941.* Figure 9.1 shows Dr. Zuse and a reconstruction of the Z3 computer. It featured a 22-bit floating-point unit, with 1 sign bit, 7

* Konrad Zuse's Legacy: The architecture of the Z1 and Z3, *IEEE Annals of the History of Computing*, 19, 2, 1997, pp. 5–16.

FIGURE 9.1 Konrad Zuse with a reconstruction of the Z3 computer.

bits of exponent, and 14 bits of significand. Many of the early machines eschewed floating-point in favor of fixed-point, including the IAS Machine, built by John von Neumann in Princeton, New Jersey. Of the successful commercial computers, the UNIVAC 1100 series and 2200 series included two floating-point formats, a single-precision format using 36 bits and a double-precision format using 72 bits. Numerous machines soon followed with varying data formats. The IBM 7094, shown in Figure 9.2, like the UNIVAC, used 36-bit words, but the IBM 360, which followed in 1964, used 32-bit words, one for single-precision and two for double-precision. The interesting oddity of IBM floating-point was the use of a hexadecimal exponent, that is, the exponent used base 16 rather than base 2, with each increment of the exponent

FIGURE 9.2 IBM 7094 System.

FIGURE 9.3 Seymore Cray and a Cray-1 Computer, circa 1974.

representing 2^4.* In the supercomputer space, machines by Control Data Corporation (CDC) and later by Cray would use a 60-bit floating-point format and be known for their speed of floating-point computation. That race has not stopped. While Cray held the record for years with a speed of 160 million floating-point operations per second (megaflops), modern supercomputers boast speeds in the petaflop (10^{15} flops) range! Figure 9.3 is a photograph of Seymore Cray and the original Cray-1 computer.

However, even with the wide adoption of floating-point there were problems. Companies supported their own formats of floating-point data types, had different models for exceptions, and rounded the results in different ways. While you may not be familiar with floating-point exceptions or rounding just yet, when these concepts are addressed you will see the benefits of a standard that defines the data types, exception handling, and rounding modes. For an example of the problems that arose due to the varied landscape of behaviors, consider the Cray machines. These processors were blazingly fast in their floating-point computations, but they suffered in computational accuracy due to some shortcuts in their rounding logic. They were fast, but not always accurate! In the early 1980s, an IEEE standards committee convened to produce a standard for floating-point which would introduce a consistency to computations done in floating-point, enable work to be performed across a wide variety of computers, and result in a system which could be used by non-numerical experts to produce reliable numerical code. A key leader in this effort was Dr. William Kahan, shown in Figure 9.4, of the University of California at Berkeley, at the time consulting with Intel Corporation on the development of the i8087 floating-point

* See IBM System/360 Principles of Operation, IBM File No. S360-01, pp. 41–42, available from http://bitsavers.informatik.uni-stuttgart.de/pdf/ibm/360/princOps/A22-6821-6_360PrincOpsJan67.pdf.

FIGURE 9.4 Dr. William Kahan.

coprocessor. The specification defined the format of the data types, including special values such as infinities and not-a-numbers (NaNs, to be considered in a later section); how rounding was to be done; what conditions would result in exceptions; and how exceptions would be handled and reported. In 2008, a revision of the standard, referred to as IEEE 754-2008, was released, adding decimal data types and addressing a number of issues unforeseen 25 years ago. Most processors with floating-point hardware, from supercomputers to microcontrollers, implement some subset of the IEEE 754 standard.

9.3 THE CONTRIBUTION OF FLOATING-POINT TO THE EMBEDDED PROCESSOR

The cost of an integrated circuit is directly related to the size of the die. The larger the size of the die, the fewer of them that can be put on a wafer. With constant wafer costs, the more die on the wafer, the lower the cost of each die. So it is a reasonable question to ask why manufacturers spend the die area on an FPU, or, more specifically, what value does the floating-point unit of the Cortex-M4 bring? To answer these questions it is necessary to first consider how floating-point computations differ from integer computations. As we saw in Chapter 2, the integer data types are commonly in three formats:

- Byte, or 8 bits
- Halfword, or 16 bits
- Word, or 32 bits

Each of these formats may be treated as signed or unsigned. For the moment we will consider only 32-bit words, but each data type shares these characteristics. The range of an unsigned word value is 0 to 4,294,967,295 (2^{32}–1). Signed word values are in the range –2,147,483,648 to 2,147,483,647, or -2^{31} to 2^{31}–1. While these are large numbers, many fields of study cannot live within these bounds. For example, the national debt is $17,320,676,548,008.59 (as of January 4, 2014), a value over 2000 times larger than can be represented in a 32-bit unsigned word. In the field of astronomy, common distances are measured in parsecs, with one parsec equal to 3.26 light years, or about 30,856,780,000,000 km. While such a number cannot be represented in a 32-bit unsigned word, it is often unnecessary to be as precise as financial computations. Less precise values will often suffice. So, while the charge on an electron, denoted e, is $1.602176565 \times 10^{-19}$ coulombs, in many instances computations on e can tolerate reduced precision, perhaps only a few digits. So 1.60×10^{-19} may be precise enough for some calculations. Floating-point enables us to trade off precision for range, so we can represent values larger than 32-bit integers and also much smaller than 1, but frequently with less than full precision.

Could this mean floating-point is always the best format to use? Simply put, no. Consider the 32-bit integer and 32-bit single-precision floating-point formats. Both have the same storage requirements (32 bits, or one word), both have the same number of unique bit patterns (2^{32}), but integers have a fixed numeric separation, that is, each integer is exactly the same distance from the integer just smaller and the integer just larger. That numeric separation is exactly and always 1. Consider that we represent the decimal value 1037 in 32-bit binary as

```
0000 0000 0000 0000 0000 0100 0000 1101;
```

the integer value just smaller is 1036, represented in 32-bit binary as

```
0000 0000 0000 0000 0000 0100 0000 1100;
```

and the integer value just larger is 1038, and it is represented in 32-bit binary as

```
0000 0000 0000 0000 0000 0100 0000 1110.
```

In each case the difference between sequential values is exactly 1 (verify you believe this from the last 4 bits). This is why integers make a great choice for counters and address values, but not always for arithmetic calculations. Why would a 32-bit floating-point value be better for arithmetic? As we showed above, the range of a 32-bit integer is insufficient for many problems—it is simply not big enough on the end of the number curve, and not small enough to represent values between 0 and 1. If we use 64-bit integers we extend the range significantly, but again not enough for all problems. We will see that floating-point values have a much greater range than even 64-bit integers. They accomplish this by not having a fixed numeric separation, but a variable one that depends on the value of the exponent. We'll explain this in Section 9.5.

So, back to our question—Why include floating-point capability in the processor? To begin our evaluation, let's ask some questions. First, does the application have inputs, outputs, or intermediate values larger than representable by the available integers? Second, does the application have inputs, outputs, or intermediate values between 0 and 1? If either of these questions is yes, can we use fixed-point representations to satisfy the range required? Third, do any of the algorithms in the application require correct rounding, rather than truncation? Fourth, how easy is it to ensure that no intermediate value is outside the range of the available integers? In many cases the analysis required to ensure that all inputs, outputs, and intermediate values remain in the range available is not trivial, but can be quite difficult. The answers to these questions will point the system designer to one of two conclusions–that the integer formats are sufficient, or the problems are better processed with floating-point. The following chapters introduce the key elements of floating-point, where floating-point differs from integer and fixed-point computation, and what benefits come naturally to computations in floating-point. Knowing this will make the decision easier for the system designer.

9.4 FLOATING-POINT DATA TYPES

The IEEE 754-2008 specification defines four binary floating-point formats: 16-bit, 32-bit, 64-bit, and 128-bit, commonly referred to as *half-precision, single-precision, double-precision*, and *quad-precision*, respectively. C, C++, and Java refer to the 32-bit format as *float* and the 64-bit format as *double*. The Cortex-M4 does not support the two larger formats, but does support a half-precision floating-point format for data storage and the single-precision data type for computation. Figure 9.5 shows the half-precision, single-precision, and double-precision data formats.

From Figure 9.5 you can see the floating-point formats are composed of three component parts: the *sign bit*, represented by *s*; the *exponent*, typically in a biased form (see the explanation of bias below); and the *fraction*. The value of a floating-point data value is computed according to the formula for *normal values*, covered in Section 9.6.1. We will consider special values in a later sections. This format is called *sign magnitude* representation, since the sign bit is separate from the bits that comprise the magnitude of the value. The equation for normal values in a floating-point format is given by*

$$F = (-1)^s \times 2^{(\text{exp}-\text{bias})} \times 1.\text{f} \tag{9.1}$$

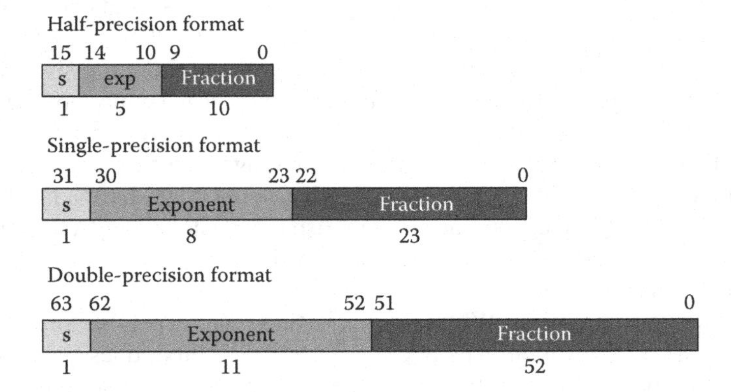

FIGURE 9.5 IEEE 754-2008 data formats.

* We will consider values, or encodings, for values that are not in the space of normal values in Section 9.6.

where:

- *s* is the *sign*,
  - 0 for positive
  - 1 for negative
- *exp* is the *exponent*,
  - The bias is a constant specified in the format
  - The purpose is to create a positive exponent
- *f* is the *fraction*, or sometimes referred to as the *mantissa*

We refer to the value 1.f as the *significand*, and this part of the equation is always in the range [1.0, 2.0) (where the value may include 1.0 but not 2.0). The set of possible values is referred to as the *representable values*, and each computation must result in either one of these representable values or a special value. The bias is a constant added to the true exponent to form an exponent that is always positive. For the single-precision format, the bias is 127, resulting in an exponent range of 1 to 254 for normal numbers. The exponent values 0 and 255 are used for special formats, as will be considered later. Table 9.1 shows the characteristics of the three standard data types.

EXAMPLE 9.1

Form the single-precision representation of 6.5.

SOLUTION

The sign is positive, so the sign bit will be 0. The power of 2 that will result in a significand between 1 and almost 2 is 4.0 (2^2), resulting in a significand of 1.625. Expressed in floating-point representation, the value 6.5 is

$$6.5 = -1^0 \times 2^2 \times 1.625$$

TABLE 9.1
Floating-Point Formats and Their Characteristics

	Format		
	Half-Precision[a]	Single-Precision	Double-Precision
Format width in bits	16	32	64
Exponent width in bits	5	8	11
Fraction bits	10	23	52
Exp maximum	+15	+127	+1023
Exp minimum	−14	−126	−1022
Exponent bias	15	127	1023

[a] The Cortex-M4 has an alternative format for half-precision values. This format may be selected by setting the AHF bit in the FPSCR, and the format will be interpreted as having an exponent range that includes the max exponent, 2^{16}. This precision does not support NaNs or infinities. The maximum value is $(2-2^{-10}) \times 2^{16}$ or 131008.

S	Exponent								Fraction																						
31	30	29	28	27	26	25	24	23	22	21	20	19	18	17	16	15	14	13	12	11	10	9	8	7	6	5	4	3	2	1	0
0	1	0	0	0	0	0	0	1	1	0	1	0	0	0	0	0	0	0	0	0	0	0	0	0	0	0	0	0	0	0	0
4				0				D				0				0				0				0				0			

FIGURE 9.6 Result of Example 9.1.

To finish the example, convert the resulting factor to a significand in binary.

$$1.625 = 1 + ½ + ⅛\text{, or in binary, } 1.101.$$

The exponent is 2, and when the bias is added to form the exponent part of the single-precision representation, the biased exponent becomes 129, or 0x81.

The resulting single-precision value is 0x40D00000, shown in binary and hexadecimal in Figure 9.6.

EXAMPLE 9.2

Form the single-precision representation of −0.4375.

Solution

The sign is negative, so the sign bit will be 1. The power of 2 that will result in a significand between 1 and almost 2 is 2^{-2} (0.25), giving a significand of 1.75.

$$-0.4375 = -1^{1} \times 2^{-2} \times 1.75$$

$$1.75 = 1 + ½ + ¼\text{, or in binary, } 1.11.$$

The exponent is −2, and when the bias is added to form the exponent of the single-precision representation, the biased exponent becomes 125, or 0x7D. The resulting single-precision value is 0xBEE00000. See Figure 9.7.

It's unlikely you will ever have to do these conversions by hand. The assembler will perform the conversion for you. Also, a number of useful websites will do the conversions for you. See, e.g., (http://babbage.cs.qc.cuny.edu/IEEE-754.old/Decimal.html) for conversions from decimal to floating-point. See also (http://babbage.cs.qc.cuny.edu/IEEE-754.old/32bit.html) for an excellent website that has a very useful calculator to perform the conversion from single-precision floating-point to decimal. Also, the website at (http://www.h-schmidt.net/FloatConverter) allows you to

S	Exponent								Fraction																						
31	30	29	28	27	26	25	24	23	22	21	20	19	18	17	16	15	14	13	12	11	10	9	8	7	6	5	4	3	2	1	0
1	0	1	1	1	1	1	0	1	1	1	0	0	0	0	0	0	0	0	0	0	0	0	0	0	0	0	0	0	0	0	0
B				E				E				0				0				0				0				0			

FIGURE 9.7 Result of Example 9.2.

set each bit separately in a single-precision representation and see immediately the contribution to the final value.

9.5 THE SPACE OF FLOATING-POINT REPRESENTABLE VALUES

In school we learned about the number line and the whole numbers. On this number line, each whole number was separated from its neighbor whole number by the value 1. Regardless of where you were on the number line, any whole number was 1 greater than the whole number to the left and 1 less than the whole number to the right. Such is not the case for the floating-point number line. Recall from Equation 9.1 above that the significand is multiplied by a power of 2. The larger the exponent, the greater the multiplication factor applied to the significand. Two significands that are contiguous, i.e., the larger significand is the next higher value, would differ by a factor of the exponent rather than a fixed value. Let's represent this idea using a simple format with 2 bits of fraction and an exponent range of $-3 \leq E \leq 0$. The floating-point number line looks like Figure 9.8.

There are several things to notice in the number line in Figure 9.8. First, the number of representable values associated with each exponent is fixed at 2^n, where n is the number of bits in the fraction. In this example, two bits give four representable values for each exponent. Notice that four values exist with an exponent of –1 using our format: ½, ⅝, ¾, and ⅞. Second, notice the numeric separation between each representable value is a function of the exponent value, and as the exponent increases by one, the numeric separation doubles. The only exception is in the *subnormal* range, and we will discuss subnormals in Section 9.6.2. If we consider a single-precision data value with the exponent equal to 0 (a biased exponent of 127), the range of values with this exponent are:

$$1.0 \ldots 1.99999998808 \ (2^1 - 2^{-23})$$

That is, the minimum value representable is 1.0, while the maximum value is just less than 2.0. With a fraction of 23 bits, the numeric separation between representable values is 2^{-23}, or $\sim 1.192 \times 10^{-7}$, a fairly small amount.

Contrast this to an exponent of 23 (a biased exponent of 150). Now each value will be in the range

$$8388608 \ldots 16777215$$

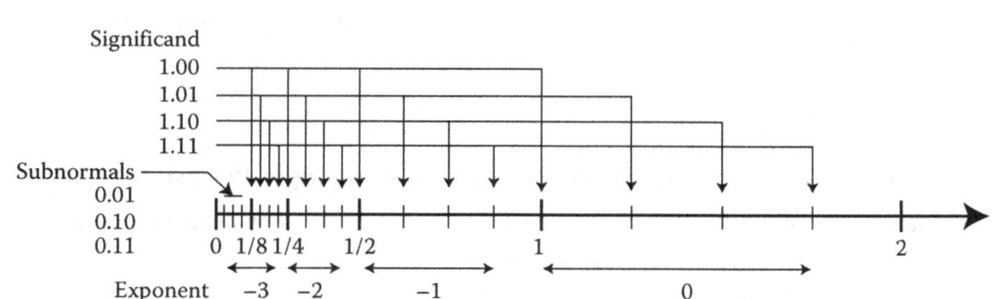

FIGURE 9.8 Floating-point number line for positive values, 2 exponent bits and 2 fractional bits (see Ercegovac and Lang 2004).

In this instance, the numeric separation between representable values is 1.0, much larger than the 1.192×10^{-7} of the previous example.

If we continue this thought with an exponent closer to the maximum, say 73 (a biased exponent of 200), we have this range of values:

$$9.445 \times 10^{21} \ldots 1.889 \times 10^{22}$$

Here the numeric separation between representable values is roughly 1.126×10^{15}! If we go in the other direction, say with an exponent value of –75 (a biased exponent of 52), the range becomes

$$2.647 \times 10^{-23} \ldots 5.294 \times 10^{-23}$$

with a numeric separation of 3.155×10^{-30}! Table 9.2 is a summary of the findings.

From Table 9.2 it is evident that the range of single-precision values and the numeric separation vary a great deal. Notice that the numeric separation between values for an exponent of 73 is greater than the total range for values with an exponent of 23. The key to understanding floating-point as a programmer is that *floating-point precision is not fixed but a function of the exponent.* That is, while the numeric separation in an integer data type is always 1, the numeric separation of a floating-point data type varies with the exponent. This is rarely a problem for scientific computations—we typically are interested in only a few digits regardless of the magnitude of the results. So if we specify the precision of our results is to be 4 digits, 1 to the left of the decimal point and three to the right, we may compute

$$5.429 \times 10^{15}$$

but another calculation may result in

$$-2.907 \times 10^{-8}$$

and we would not consider this in error even though the value of the second calculation is *much smaller* than the *smallest variation we are interested in* of the first result (a factor of 10^{12}). Rather, the precision of each of the calculations is the same—4 digits. Thinking of floating-point as a base-2 version of scientific notation will help in grasping the useful properties of floating-point, and in using them properly.

TABLE 9.2
Examples of the Range of Numeric Separation in Single-Precision Values

Exponent	exp-bias	Range	Numeric Separation
52	–75	$2.647 \times 10^{-23} \ldots 5.294 \times 10^{-23}$	3.155×10^{-30}
127	0	1.0 … 1.9999998	1.192×10^{-7}
150	23	8388608 … 16777215	1.0
200	73	$9.445 \times 10^{21} \ldots 1.889 \times 10^{22}$	1.126×10^{15}

9.6 FLOATING-POINT REPRESENTABLE VALUES

All representable values have a single encoding in each floating-point format, but not all floating-point encodings represent a number. This is another difference between floating-point and integer representation. The IEEE 754-2008 specification defines five classes of floating-point encodings: normal numbers, subnormal numbers, zeros, NaNs, and infinities. Each class has some shared properties and some unique properties. Let's consider each one separately.

9.6.1 Normal Values

We use the term *normal value* to define a floating-point value that satisfies the equation

$$F = (-1)^s \times 2^{(\text{exp}-\text{bias})} \times 1.\text{f} \tag{9.1}$$

which we saw earlier in Section 9.4. In the space of normal values, each floating-point number has a single encoding, that is, an encoding represents only one floating-point value and each representable value has only one encoding. Put another way, no aliasing exists within the single-precision floating-point data type. It is possible to have multiple encodings represent a single value when represented in *decimal floating-point* formats, but this is beyond the scope of this text. See the IEEE 754-2008 specification for more on this format.

Recall that a 32-bit signed integer has a range of –2,147,483,648 to 2,147,483,647 ($+/-2.147 \times 10^9$). Figure 9.9 shows the range of signed 32-bit integers, half-precision (16-bit) and single-precision (32-bit) floating-point data types for the normal range. Notice the range of the signed 32-bit integer and the half-precision data types is roughly the same; however, notice the much greater range available in the single-precision floating-point data type.

Remember, the tradeoff between the integer data types and the floating-point data types is in the precision of the result. In short, as we showed in Figure 9.8, *the precision of a floating-point data value is a function of the exponent.* As the exponent

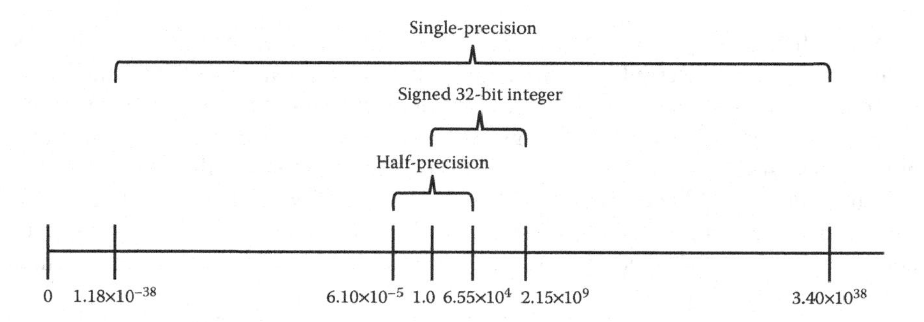

FIGURE 9.9 Relative normal range for signed 32-bit integer, half-precision floating-point, and single-precision floating-point data types.

TABLE 9.3
Several Normal Half-Precision and Single-Precision Floating-Point Values

	Format	
	Half-Precision	**Single-Precision**
1.0	0x3C00	0x3F800000
2.0	0x4000	0x40000000
0.5	0x3800	0x3F000000
1024	0x6400	0x44800000
0.005	0x1D1F	0x3BA3D70A
6.10×10^{-5}	0x0400	0x38800000
6.55×10^{4}	0x7BFF	0x477FE000
1.175×10^{-38}	*Out of range*	0x00800000
3.40×10^{38}	*Out of range*	0x7F7FFFFF

increases, the precision decreases, resulting in an increased numeric separation between representable values.

Table 9.3 shows some examples of normal data values for half-precision and single-precision formats. Note that each of these values may be made negative by setting the most-significant bit. For example, –1.0 is `0xBF800000`. Using the technique shown in Section 9.4, try out some of these. You can check your work using the conversion tools listed in the References.

9.6.2 Subnormal Values

The inclusion of *subnormal values** was an issue of great controversy in the original IEEE 754-1985 deliberations. When a value is non-zero and too small to be represented in the normal range, it value may be represented by a subnormal encoding. These values satisfy Equation 9.2:

$$F = (-1)^s \times 2^{-126} \times 0.\text{f} \tag{9.2}$$

Notice first the exponent value is fixed at *–126*, one greater than the negative bias value. This value is referred to as *emin*, and is the exponent value of the smallest normal representation. Also notice that the 1.0 factor is missing, changing the significand range to [0.0, 1.0). The subnormal range extends the lower bounds of the representable numbers by further dividing the range between zero and the smallest normal representable value into 2^{23} additional representable values. If we look again at Figure 9.8, we see in the region marked *Subnormals* that the range between 0 and the minimum normal value is represented by *n* values, as in each exponent range of the normal

* The ARM documentation in the ARM v7-M Architecture Reference Manual uses the terms "denormal" and "denormalized" to refer to subnormal values. The ARM Cortex-M4 Technical Reference Manual uses the terms "denormal" and "subnormal" to refer to subnormal values.

TABLE 9.4
Subnormal Range for Half-Precision and Single-Precision

	Format	
	Half-Precision	**Single-Precision**
Minimum	$+/-5.96 \times 10^{-8}$	$+/-1.45 \times 10^{-45}$
Maximum	$+/-6.10 \times 10^{-5}$	$+/-1.175 \times 10^{-38}$

values. The numeric separation in the subnormal range is equal to that of the normal values with *minimum normal exponent.* The minimum value in the normal range for the single-precision floating-point format is 1.18×10^{-38}. The subnormal values increase the minimum range to 1.4×10^{-45}. Be aware, however, when an operand in the subnormal range decreases toward the minimum value, the number of significant digits decreases. In other words, the precision of subnormal values may be significantly less than the precision of normal values, or even larger subnormal values. The range of subnormal values for the half-precision and single-precision data types is shown in Table 9.4. Table 9.5 shows some examples of subnormal data values. As with the normal values, each of these values may be made negative by setting the most significant bit.

EXAMPLE 9.3

Convert the value -4.59×10^{-41} to single-precision.

SOLUTION

The value is below the minimum threshold representable as a normal value in the single-precision format, but is greater than the minimum representable subnormal value and is in the subnormal range for the single-precision format.

Recalling our conversion steps above, we can use the same methodology for subnormal values so long as we recall that the exponent is fixed at the value 2^{-126} and no implicit 1 is present.

TABLE 9.5
Examples of Subnormal Values for Half-Precision and Single-Precision

	Format	
	Half-Precision	**Single-Precision**
6.10×10^{-5}	0x03FF	
1.43×10^{-6}	0x0018	
5.96×10^{-8}	0x0001	
1.175×10^{-38}		0x007FFFFF
4.59×10^{-41}		0x00008000
1.45×10^{-45}		0x00000001

S	Exponent								Fraction																						
31	30	29	28	27	26	25	24	23	22	21	20	19	18	17	16	15	14	13	12	11	10	9	8	7	6	5	4	3	2	1	0
1	0	0	0	0	0	0	0	0	0	0	0	0	0	0	0	1	0	0	0	0	0	0	0	0	0	0	0	0	0	0	0
8				0				0				0				8				0				0				0			

FIGURE 9.10 Single-precision representation of -4.592×10^{-41}.

First, divide -4.592×10^{-41} by 2^{-126} and we have −0.00390625, which is equal to 2^{-8}. This leaves us with

$$-4.592 \times 10^{-41} = -1^1 \times 2^{-126} \times 0.00390625$$

The result single-precision value is 0x80008000, shown in binary and hexadecimal in Figure 9.10.

The conversion to and from half-precision is done in an identical manner, but remember the subnormal exponent for the half-precision format is −14 and the format is only 16 bits.

A computation that results in a subnormal value may set the *Underflow* flag and may signal an exception. We will address exceptions in a later chapter.

9.6.3 Zeros

It's odd to think of zero as anything other than, well, zero. In floating-point zeros are signed. You may compute a function and see a negative zero as a result! Zeros are formed by a zero exponent and zero fraction. A critical bit of information here—if the fraction is *not* zero, the value is a subnormal, as we saw above. While numerous subnormal encodings are possible, only two zero encodings, a positive zero with a sign bit of zero, and a negative zero with a sign bit of one, are possible. How is it possible to have a negative zero? There are several ways outlined in the IEEE 754-2008 specification. One way is to be in Round to Minus Infinity mode (we will consider rounding in Chapter 10) and sum two equal values that have opposite signs.

EXAMPLE 9.4

Add the two single-precision values 0x3F80000C and 0xBF80000C with different rounding modes.

Solution

Let register s0 contain 0x3F80000C and register s1 contain 0xBF80000C. The two operands have the same magnitude but opposite sign, so the result of adding the two operands using the Cortex-M4 VADD instruction (we will consider this instruction in Chapter 11)

```
VADD s2, s0, s1
```

in each case is zero. But notice that the sign of the zero is determined by the rounding mode. We will consider rounding modes in detail in Chapter 10, but

TABLE 9.6
Operations with Zero Result in Each Rounding Mode

Rounding Mode	Result	
roundTiesToEven	0x00000000	Positive Zero
roundTowardPositive	0x00000000	Positive Zero
roundTowardNegative	0x80000000	Negative Zero
roundTowardZero	0x00000000	Positive Zero

TABLE 9.7
Format of Signed Zero in Half-Precision and Single-Precision

	Format	
	Half-Precision	Single-Precision
+0.0	0x0000	0x00000000
–0.0	0x8000	0x80000000

for now consider the four in Table 9.6. (The names give a clue to the rounding that is done. For example, roundTowardPositive always rounds up if the result is not exact. The rounding mode roundTiesToEven uses the method we learned in school—round to the nearest valid number, and if the result is exactly halfway between two valid numbers, pick the one that is even.)

Likewise, a multiplication of two values, one positive and the other negative, with a product too small to represent as a subnormal, will return a negative zero. And finally, the square root of –0 returns –0. Why bother with signed zeros? First, the negative zero is an artifact of the sign-magnitude format, but more importantly, the sign of zero is an indicator of the direction of the operation or the sign of the value before it was rounded to zero. This affords the numeric analyst with information on the computation, which is not obvious from an unsigned zero result, and this may be useful even if the result of the computation is zero.

The format of the two zeros for half-precision and single-precision are shown in Table 9.7.

9.6.4 Infinities

Another distinction between floating-point and integer values is the presence of an *infinity* encoding in the floating-point formats. A floating-point infinity is encoded with an exponent of all ones and a fraction of all zeros. The sign indicates whether it is a positive or negative infinity. While it is tempting to consider the positive infinity as the value just greater than the maximum normal value, it is best considered as a mathematical symbol and not as a number. In this way computations involving infinity will behave as would be expected. In other words, any operation computed with an

TABLE 9.8
Format of Signed Infinity in Half-Precision and Single-Precision

	Format	
	Half-Precision	**Single-Precision**
–Infinity	0xFC00	0xFF800000
+Infinity	0x7C00	0x7F800000

infinity value by a normal or subnormal value will return the infinity value. However, some operations are *invalid*, that is, there is no generally accepted result value for the operation. An example is multiplication of infinity by zero. We note that the IEEE 754-2008 specification defines the nature of the infinity in an *affine sense*, that is,

$$-\infty < \textit{all finite numbers} < +\infty$$

Recall from Section 7.2.2 that overflow in an integer computation produces an incorrect value and sets a hardware flag. To determine whether overflow occurred, a check on the flags in the status register must be made before you can take appropriate action. Multiplying two very large values that result in a value greater than the maximum for the floating-point format will return an infinity,* and further calculations on the infinity will indicate the overflow. While there is an overflow flag (more on this in Chapter 10), in most cases the result of a computation that overflows will indicate as much without requiring the programmer to check any flags. The result will make sense as if you had done it on paper. The format of the half-precision and single-precision infinities is shown in Table 9.8.

9.6.5 Not-a-Numbers (NaNs)

Perhaps the oddest of the various floating-point classes is the not-a-number, or NaN. Why would a numerical computation method include a data representation that is "not a number?" A reasonable question, certainly. They have several uses, and we will consider two of them. In the first use, a programmer may choose to return a NaN with a unique *payload* (the bits in the fraction portion of the format) as an indicator that a specific, typically unexpected, condition existed in a routine within the program. For instance, the programmer believes the range of data for a variable at a point in the program should not be greater than 100. But if it is, he can use a NaN to replace the value and encode the payload to locate the line or algorithm in the routine that caused the behavior. Secondly, NaNs have historically found use as the default value put in registers or in data structures. Should the register or data structure be *read* before it is written with valid data, a NaN would be returned. If

* In some rounding modes, a value of *Maximum Normal* will be returned. We will consider this case in the section in our discussion of exceptions.

the NaN is of a type called *signaling NaNs*, the Invalid Operation exception would be signaled, giving the programmer another tool for debugging. This use would alert the programmer to the fact that uninitialized data was used in a computation, likely an error. The Motorola MC68881 and later 68K floating-point processors initialized the floating-point register file with signaling NaNs upon reset for this purpose. Both signaling NaNs, and a second type known as *quiet* NaNs, have been used to represent non-numeric data, such as symbols in a symbolic math system. These programs operate on both numbers and symbols, but the routines operating on numbers can't handle the symbols. NaNs have been used to represent the symbols in the program, and when a symbol is encountered it would cause the program to jump to a routine written specifically to perform the needed computation on symbols rather than numbers. This way it would be easy to intermix symbols and numbers, with the arithmetic of the processor operating on the numbers and the symbol routines operating whenever an operand is a symbol.

How does one use NaNs? One humorous programmer described NaNs this way: when you think of computing with NaNs, replace the NaN with a "Buick" in a calculation.* So, what is a NaN divided by 5? Well, you could ask instead, "What is a Buick divided by 5?" You quickly see that it's not possible to reasonably answer this question, since a Buick divided by 5 is *not-a-number*, so we will simply return the Buick (unscratched, if we know what's good for us). Simply put, in an operation involving a NaN, the NaN, or one of the NaNs if both operands are NaN, is returned. This is the behavior of an IEEE 754-2008-compliant system in most cases when a NaN is involved in a computation. The specification does not direct which of the NaNs is returned when two or more operands are NaN, leaving it to the floating-point designer to select which is returned.

A NaN is encoded with an exponent of all ones and a non-zero fraction. Note that an exponent of all ones with a zero fraction is an infinity encoding, so to avoid confusing the two representations, a NaN must not have a zero fraction. As we mentioned above, NaNs come in two flavors: signaling NaNs (sNaN) and non-signaling, or quiet, NaNs (qNaN). The difference is the value of the first, or most significant, of the fraction bits. If the bit is a one, the NaN is quiet. Likewise, if the bit is a zero, the NaN is signaling, but only if at least one other fraction bit is a one. In the half-precision format, bit 9 is the bit that identifies the NaN type; in the single-precision format it's bit 22. The format of the NaN encodings for the half-precision format and the single-precision format is shown in Table 9.9.

Why two encodings? The signaling NaN will cause an Invalid Operation exception (covered in Section 10.3.4) to be set, while a quiet NaN will not. What about the fraction bits when a NaN is an operand to an operation? The specification requires that the fraction bits of a NaN be preserved, that is, returned in the NaN result, if it is the only NaN in the operation and if preservation is possible. (An example when it would not be possible to preserve the fraction is the case of a format conversion in which the fraction cannot be preserved because the final format lacks the necessary number of bits.) If two or more NaNs are involved in an operation, the fraction of one of them is to be preserved, but which is again the decision of the processor designer.

* Buick is a brand of General Motors vehicle popular in the 1980s.

TABLE 9.9
Format of NaN Encodings in Half-Precision and Single-Precision

	Format	
	Half-Precision	**Single-Precision**
Sign bit	0/1	0/1
Exponent bits	Must be all ones, 0x1F	Must be all ones, 0xFF
NaN type bit	Bit 9	Bit 22
Payload bits	Bits 8-0	Bits 21-0

TABLE 9.10
Examples of Quiet and Signaling NaNs in Half-Precision and Single-Precision Formats

	Format	
	Half-Precision	**Single-Precision**
Quiet NaN, 0x01	`0x7D01`	`0x7FC00001`
Quiet NaN, 0x55	`0x7D55`	`0x7FC00055`
Signalling NaN, 0x01	`0x7C01`	`0x7F800001`
Signalling NaN, 0x55	`0x7C55`	`0x7F800055`

The sign bit of a NaN is not significant, and may be considered as another payload bit. Several of the many NaN values are shown in Table 9.10, with payloads of 0x01 and 0x55. Notice how the differentiator is the most-significant fraction bit.

9.7 THE FLOATING-POINT REGISTER FILE OF THE CORTEX-M4

Within the floating-point unit of the Cortex-M4 is another register file made up of 32 single-precision registers labeled s0 to s31. One difference to note between the ARM registers and the FPU registers is that none of the FPU registers are banked, as are some of the ARM registers. The Cortex-M4 can also address registers as double-precision registers for loads and stores even without specific instructions which operate on double-precision data types. Likewise, half-precision and integer data can be stored in the FPU registers in either the upper or lower half of the register. The register file is shown in Figure 9.11.

Each single-precision register may be used as a source or destination, or both, in any instruction. There are no limitations on the use of the registers, unlike register r13, register r14, and register r15 in the integer register file. This is referred to as a *flat register file*, although some restrictions do exist when a standard protocol, such as the ARM Architecture Procedure Call Standard (AAPCS), is in place for passing operands and results to subroutines and functions. The FPU registers are aliased,

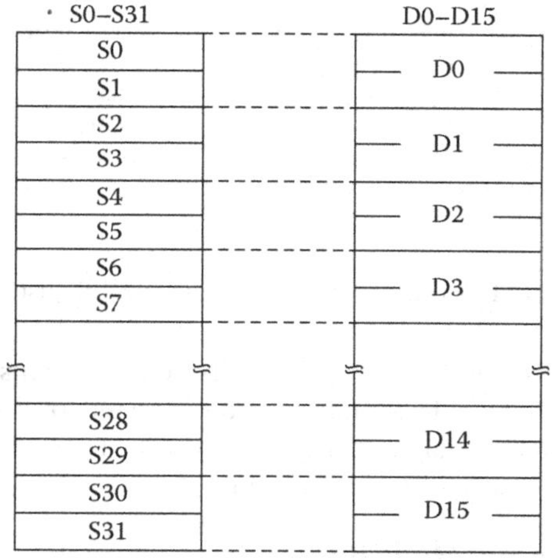

FIGURE 9.11 Cortex-M4 floating-point register file.

such that two single-precision registers may be referenced as a double-precision register. The aliasing follows the relation shown below.

$$d[x] \Leftrightarrow \{s[(2x)+1], s[2x]\}$$

For example, register d[6] is aliased to the register pair {s13, s12}. In several of the load and store instructions, the FPU operand may be either a single-precision or double-precision register. This enables 64-bit data transfers with memory and with the ARM register file. It's important to ensure that you know which single-precision registers are aliased to a double-precision register, so you don't accidently overwrite a single-precision register with a load to a double-precision register.

9.8 FPU CONTROL REGISTERS

Two control registers are of immediate importance, and they are the FPSCR and the CPACR. The first controls the internal workings of the FPU, while the second enables the FPU. If the FPU is not enabled, any access to the FPU will result in a fault. This will be covered in more detail in Chapter 15, but for now we need to know that the FPU must be enabled or our programs will not work.

9.8.1 The Floating-Point Status and Control Register, FPSCR

In Chapter 7, we became familiar with the various status registers, e.g., the CPSR and APSR. We also examined the use of the register to hold condition code flags and to specify various options and modes of operation. The equivalent register in the FPU is the Floating-Point Status and Control Register (FPSCR), shown in Figure 9.12. Reading and writing the FPSCR is covered in Chapter 11. Notice that the APSR

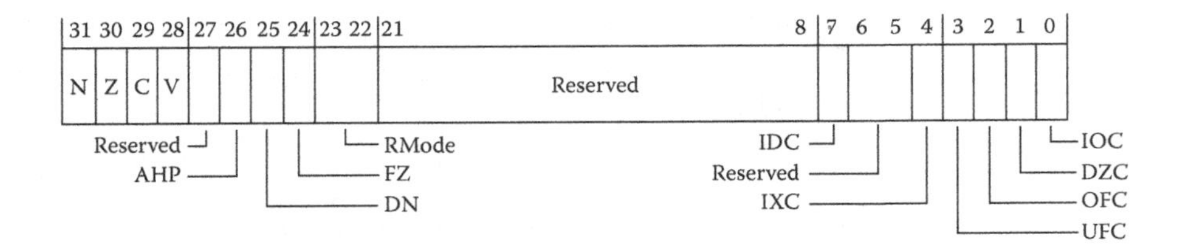

FIGURE 9.12 Cortex-M4 Floating-Point Status and Control Register.

and the FPSCR are alike in that the upper 4 bits hold the status of the last comparison, the N, Z, C, and V bits. These bits record the results of floating-point compare instructions (considered in Chapter 11) and can be transferred to the APSR for use in conditional execution and conditional branching.

9.8.1.1 The Control and Mode Bits

The bits following the status bits are used to specify modes of operation. The AHP bit specifies the "alternative half-precision format" to select the format of the half-precision data type. If set to zero, the IEEE 754-2008 format is selected, and if set to 1, the ARM alternative format is selected. The DN bit selects whether the FPU is in "default NaN" mode. When not in default NaN mode (the common case), operations with NaN input values preserve the NaN (or one of the NaN values, if more than one input operand is a NaN) as the result. When in default NaN mode any operation involving a NaN returns the *default NaN* as the result, regardless of the NaN payload or payloads. The default NaN is a qNaN with an all-zero payload, as in Table 9.11.

The FZ bit selects whether the processor is in *flush-to-zero* mode. When set, the processor ignores subnormal inputs, replacing them in computations with signed zeroes, and *flushes* a result in the subnormal range to a signed zero. Both the DN and FZ bits are discussed in greater detail in Chapter 11. Bits 23 and 22 contain the RMode bits. These bits specify the *rounding mode* to be used in the execution of most operations. The default rounding mode is *roundTiesToEven*, also known as *Round to Nearest Even*. It's important to know where these bits may be found, but we will not take up rounding until Chapter 10. The rounding mode is selected by setting the RMode bits to one of the bit patterns shown in Table 9.12.

TABLE 9.11
Format of the Default Nan for Half-Precision and Single-Precision Data Types

	Format	
	Half-Precision	**Single-Precision**
Sign bit	0	0
Exponent	0x1F	0xFF
Fraction	bit [9] = 1, bits [8:0] = 0	bit [22] = 1, bits [21:0] = 0

TABLE 9.12
Rounding Mode Bits

Rounding Mode	Setting in FPSCR[22:23]
roundTiesToEven	0b00 (default)
roundTowardPositive	0b01
roundTowardNegative	0b10
roundTowardZero	0b11

9.8.1.2 The Exception Bits

The status bits in the lower 8 bits of the FPSCR indicate when an exceptional condition has occurred. We will examine exceptions in Chapter 10, but here we only need to know that these bits are set by hardware and cleared only by a reset or a write to the FPSCR. The Cortex-M4 does not trap on any exceptional conditions, so these bits are only useful to the programmer to identify an exceptional condition has occurred since the bit was last cleared.

The exception bits are shown in the Table 9.13. Each of these bits is "sticky", that is, they are set on the first instance of the condition, and remain set until cleared by a write to the FPSCR. If the bits are cleared before a block of code, they will indicate whether their respective condition occurred in that block. They won't tell you what instruction or operand(s) caused the condition, only that it occurred somewhere in the block of code. To learn this information more precisely you can step through the code and look for the instruction that set the exception bit of interest.

TABLE 9.13
FPSCR Exception Bits

FPSCR Bit Number	Bit Name	This Bit Is Set When
7	IDC Input Denormal	An input to an operation was subnormal and was flushed to zero before used in the operation. *Valid only in flush-to-zero mode.*
4	IXC Inexact	An operation returned a result that was not representable in the single-precision format, and a rounded result was written to the register file.
3	UFC Underflow	An operation returned a result that, in absolute value, was smaller in magnitude than the positive minimum normalized number *before rounding*, and was not exact.
2	OFC Overflow	An operation returned a result that, in absolute value, was greater in magnitude than the positive maximum number *after rounding*.
1	DZC Division by Zero	A divide had a zero divisor and the dividend was not zero, an infinity or a NaN.
0	IOC Invalid Operation	An operation has no mathematical value or cannot be represented.

9.8.2 The Coprocessor Access Control Register, CPACR

The Coprocessor Access and Control Register, known as the CPACR, controls the access rights to all implemented coprocessors, including the FPU. Coprocessors are addressed by coprocessor number, a four-bit field in coprocessor instructions that identifies to the coprocessor whether it is to handle this instruction or to ignore it. Coprocessors are identified by CPn, where n is a number from 0 to 15. Coprocessors CP8 to CP15 are reserved by ARM, allowing system-on-chip designers to utilize CP0-CP7 for special function devices that can be addressed by coprocessor instructions. ARM processors have supported user coprocessors from the ARM1, but designing and incorporating custom coprocessors is not a trivial exercise, and is beyond the scope of this book. The FPU in ARM processors uses coprocessor numbers CP10 and CP11. The two coprocessor numbers are part of each FPU instruction, and specify the precision of the instruction, with CP10 specifying single-precision execution and CP11 specifying double-precision execution. Since the Cortex-M4 executes instructions operating on single-precision operands only, CP10 must be enabled. However, some of the instructions which load and store 64-bit double-precision data are in CP11 space, so it makes sense to enable both CP10 and CP11.

To enable the FPU the two bits corresponding to CP10 and CP11, bits 23:22 and 21:20, must be set to either 01 or 11. If CP10 and CP11 are each set to 01, the FPU may be accessed only in a privileged mode. If code operating in unprivileged Thread mode attempts to execute a FPU instruction, a UsageFault will be triggered and execution will transfer to a handler routine. For more information on exceptions and exception handling, see Chapter 15. If the bits are set to 11, the FPU is enabled for operations in privileged and unprivileged modes. This is the mode in which we will operate for our examples, but if you were designing a system you would have the flexibility to utilize the privileged and unprivileged options in your system code. The format of the CPACR is shown in Figure 9.13.

The following code may be used to enable CP10 and CP11 functionality in both privileged and unprivileged modes. The CPACR is a memory-mapped register, that is, it is addressed by a memory address rather than by a register number. In the Cortex-M4 the CPACR is located at address 0xE000ED88.

```
; Enable the FPU, both CP10 and CP11, for
; privileged and unprivileged mode accesses
; CPACR is located at address 0xE000ED88
LDR.W    r0, =0xE000ED88
; Read CPACR
LDR      r1, [r0]
; Set bits 20-23 to enable CP10 and CP11 coprocessors
ORR      r1, r1, #(0xF << 20)
```

31 30 29 28 27 26 25 24	23 22	21 20	19 18 17 16	15 14	13 12	11 10	9 8	7 6	5 4	3 2	1 0
Reserved	CP11	CP10	Reserved	CP7	CP6	CP5	CP4	CP3	CP2	CP1	CP0

FIGURE 9.13 Cortex-M4 Coprocessor Access Control Register.

```
; Write back the modified value to the CPACR
STR      r1, [r0]
; Wait for store to complete
DSB
```

It is necessary to execute this code or some code that performs the same functions before executing any code that loads data into the FPU or executes any FPU operations.

9.9 LOADING DATA INTO FLOATING-POINT REGISTERS

We have seen the various data types and formats available in the Cortex-M4 FPU, but how is data loaded into the register file and stored to memory? Fortunately, the instructions for loading and storing data to the FPU registers share features with the integer instructions seen in Chapter 5. We will first consider transfers to and from memory, then with the integer register file, and finally between FPU registers.

9.9.1 Floating-Point Loads and Stores: The Instructions

Memory is accessed in the same way for floating-point data and integer data. The instructions and the format for floating-point loads and stores is given below.

VLDR|VSTR{<cond>}.32 <Sd>, [<Rn>{, #+/ – <imm>}]
VLDR|VSTR{<cond>}.64 <Dd>, [<Rn>{, #+/ – <imm>}]

The <cond> is an optional condition field, as discussed in Chapter 8. Notice that these instructions do not follow the convention of naming the destination first. For both loads and stores the FPU register is named first and the addressing follows. All FPU instructions may be predicated by a condition field; however, as described in Chapter 8, selecting a predicate, such as NE, introduces an IT instruction to affect the predicated execution. The <Sd> value is a single-precision register, the <Dd> register is a pair of single-precision registers, the <Rn> register is an integer register, and the <imm> field is an 8-bit signed offset field. This addressing mode is referred to as *pre-indexed addressing*, since the offset is added to the address in the index register to form the effective address. For example, the instruction

```
VLDR s5, [r6, #08]
```

loads the 32-bit value located in memory into FPU register s5. The address is created from the value in register r6 plus the offset value of 8. Only fixed offsets and a single-index register are available in the FPU load and store instructions. An offset from an index register is useful in accessing constant tables and stacked data. Stacks will be covered in Chapter 13, and we will see an example of floating-point tables in Chapter 12.

VLDR may also be used to create literal pools of constants. This use is referred to as a *pseudo-instruction*, meaning the instruction as written in the source file is not a valid Cortex-M4 instruction, but is used by the assembler as a shortcut. The VLDR pseudo-instruction used with immediate data creates a constant table and generates VLDR PC-relative addressed instructions. The format of the instruction is:

VLDR{<cond>}.F32 Sd, =constant
VLDR{<cond>}.F64 Dd, =constant

Any value representable by the precision of the register to be loaded may be used as the constant. The format of the constants in the Keil tools may be any of the following:

[+/–]number.number (e.g., –5.873, 1034.77)
[+/–]number[e[+/–]number] (e.g., 6e-5, –123e12)
[+/–]number.number[e[+/–]number] (e.g., 1.25e-18, –5.77e8)

For example, to load Avogadro's constant, the molar gas constant, and Boltzmann's constant in single-precision, the following pseudo-instructions are used to create a literal pool and generate the VLDR instructions to load the constant into the destination registers.

```
VLDR.F32 s14, =6.0221415e23        ; Avogadro's number
VLDR.F32 s15, =8.314462            ; molar gas constant
VLDR.F32 s16, =1.3806505e-23       ; Boltzmann's constant
```

The following code is generated:

```
   41:            VLDR.F32 s14, = 6.0221415e23    ; Avogadro's number
0x0000001C ED9F7A03   VLDR           s14, [pc,#0x0C]
   42:            VLDR.F32 s15, = 8.314462        ; molar gas constant
0x00000020 EDDF7A03   VLDR           s15, [pc,#0x0C]
   43:            VLDR.F32 s16, = 1.3806505e-23   ; Boltzmann's constant
0x00000024 ED9F8A03   VLDR           s16, [pc,#0x0C]
```

The memory would be populated as shown below.

```
0x0000002C 0C30      DCW      0x0C30
0x0000002E 66FF      DCW      0x66FF
0x00000030 0814      DCW      0x0809
0x00000032 4105      DCW      0x4105
0x00000034 8740      DCW      0x8740
0x00000036 1985      DCW      0x1985
```

You should convince yourself these constants and offsets are correct.

For hexadecimal constants, the following may be used:

VLDR{<cond>}.F32 Sd, =0f_xxxxxxxx

where xxxxxxxx is an 8 character hex constant. For example,

```
VLDR.F32 s17, =0f_7FC00000
```

will load the default NaN value into register s17.

Note that Code Composer Studio does not support VLDR pseudo-instructions. See Section 6.3.

9.9.2 The VMOV Instruction

Often we want to copy data between ARM registers and the FPU. The VMOV instruction handles this, along with moving data between FPU registers and loading constants into FPU registers. The first of these instructions transfers a 32-bit operand between an ARM register and an FPU register; the second between an FPU register and an ARM register:

```
VMOV{<cond>}.F32 <Sd>, <Rt>
VMOV{<cond>}.F32 <Rt>, <Sn>
```

The format of the data type is given in the .F32 extension. When it could be unclear which data format the instruction is transferring, the data type is required to be included. The data type may be one of the following shown in Table 9.14.

We referred to the operand simply as a 32-bit operand because what is contained in the source register could be any 32-bit value, not necessarily a single-precision operand. For example, it could contain two half-precision operands. However, it does not have to be a floating-point operand at all. The FPU registers could be used as temporary storage for any 32-bit quantity.

The VMOV instruction may also be used to transfer data between FPU registers. The syntax is

```
VMOV{<cond>}.F32 <Sd>, <Sn>
```

One important thing to remember in any data transfer operation is that the content of the source register is ignored in the transfer. That is, the data is simply transferred bit by bit. This means that if the data in the source register is an sNaN, the IOC flag will not be set. This is true for any data transfer operation, whether between FPU registers, or between an FPU register and memory, or between an FPU register and an ARM register.

As a legacy of the earlier FPUs that processed double-precision operands, the following VMOV instructions transfer to or from an ARM register and the upper or lower half of a double-precision register. The x is replaced with either a 1, for the top half, or a 0, for the lower half. This is necessary to identify which half of the double-precision register is being transferred.

TABLE 9.14
Data Type Identifiers

Data Type	Identifier
Half-precision	.F16
Single-precision	.F32 or .F
Double-precision	.F64 or .D

```
VMOV{<cond>}.F32 <Dd[x]>, <Rt>
VMOV{<cond>}.F32 <Rt>, <Dn[x]>
```

It is not necessary to include the .F32 in the instruction format above, but it is good practice to make the data type explicit whenever possible. The use of this form of the VMOV instruction is common in routines which process double-precision values using integer instructions, such as routines that emulate double-precision operations. You may have access to integer routines that emulate the double-precision instructions that are defined in the IEEE 754-2008 specification but are not implemented in the Cortex-M4.

Two sets of instructions allow moving data between two ARM registers and two FPU registers. One key thing to note is that the ARM registers may be independently specified but the FPU registers must be contiguous. As with the instructions above, these are useful in handling double-precision operands or simply moving two 32-bit quantities in a single instruction. The first set is written as

```
VMOV{<cond>} <Sm>, <Sm1>, <Rt>, <Rt2>
VMOV{<cond>} <Rt>, <Rt2>, <Sm>, <Sm1>
```

The transfer is always between Sm and Rt, and Sm1 and Rt2. Sm1 must be the next contiguous register from Sm, so if Sm is register s6 then Sm1 is register s7. For example, the following instruction

```
VMOV s12, s13, r6, r11
```

would copy the contents of register r6 into register s12 and register r11 into register s13. The reverse operation is also available. The second set of instructions substitutes the two single-precision registers with a reference to a double-precision register. This form is a bit more limiting than the instructions above, but is often more useful in double-precision emulation code. The syntax for these instructions is shown below.

```
VMOV{<cond>} <Dm>, <Rt>, <Rt2>
VMOV{<cond>} <Rt>, <Rt2>, <Dm>
```

One final VMOV instruction is often very useful when a simple constant is needed. This is the immediate form of the instruction,

```
VMOV{<cond>}.F32 <Sd>, #<imm>
```

For many constants, the VMOV immediate form loads the constant without a memory access. Forming the constant can be a bit tricky, but fortunately for us, the assembler will do the heavy lifting. The format of the instruction contains two immediate fields, imm4H and imm4L, as we see in Figure 9.14.

The destination must be a single-precision register, meaning this instruction cannot be used to create half-precision constants. It's unusual for the programmer to need to determine whether the constant can be represented, but if code space or speed is an issue, using immediate constants saves on area and executes faster than the PC-relative loads generated by the VLDR pseudo-instruction.

15	14	13	12	11	10	9	8	7	6	5	4	3	2	1	0	15	14	13	12	11	10	9	8	7	6	5	4	3	2	1	0
1	1	1	0	1	1	1	0	1	D	1	1	imm4H				Vd				1	0	1	0	(0)	0	(0)	0	imm4L			

FIGURE 9.14 VMOV immediate instruction.

	31	30	29	28	27	26	25	24	23	22	21	20	19	18 ... 0
	Sign	E[7]	E[6]	E[5]	E[4]	E[3]	E[2]	E[1]	E[0]	F[22]	F[21]	F[20]	F[19]	F[18–0]
Imm[x]	[7]	~[6]	[6]	[6]	[6]	[6]	[6]	[5]	[4]	[3]	[2]	[1]	[0]	19'b0

FIGURE 9.15 Formation of constants using the VMOV immediate instruction.

	31	30	29	28	27	26	25	24	23	22	21	20	19	18 ... 0
	Sign	E[7]	E[6]	E[5]	E[4]	E[3]	E[2]	E[1]	E[0]	F[22]	F[21]	F[20]	F[19]	F[18–0]
Imm[x]	[7]	~[6]	[6]	[6]	[6]	[6]	[6]	[5]	[4]	[3]	[2]	[1]	[0]	19'b0
Binary	0	0	1	1	1	1	1	1	1	0	0	0	0	0
Hexa-decimal	3				F				8				00000	

FIGURE 9.16 Formation of 1.0 using VMOV immediate instruction.

The single-precision operand is formed from the eight bits contained in the two 4-bit fields, imm4H and imm4L. The imm4H contains bits 7-4, and imm4L bits 3-0. The bits contribute to the constant as shown in Figure 9.15.

While at first glance this does look quite confusing, many of the more common constants can be formed this way. The range of available constants is

$$+/- (1.0 \ldots 1.9375) \times 2^{(-3 \ldots +4)}$$

For example, the constant 1.0, or 0x3F800000, is formed when the immediate field is imm4H = 0111 and imm4L = 0000. When these bits are inserted as shown in Figure 9.15, we have the bit pattern shown in Figure 9.16.

Some other useful constants suitable for the immediate VMOV include those listed in Table 9.15. Notice that 0 and infinity cannot be represented, and if the constant cannot be constructed by this instruction, the assembler will create a literal pool.

9.10 CONVERSIONS BETWEEN HALF-PRECISION AND SINGLE-PRECISION

A good way to reduce the memory usage in a design is to use the smallest format that will provide sufficient range and precision for the data. As we saw in Section 9.6.1, the half-precision data type has a range of $+/- 6.10 \times 10^{-5}$ to $+/- 6.55 \times 10^{4}$, with 10 fraction bits, giving roughly 3.3 digits of precision. When the data can be represented in this format, only half the memory is required as compared to using single-precision data for storage.

The instructions VCVTB and VCVTT convert a half-precision value in either the lower half or upper half of a floating-point register, respectively, to a single-precision

TABLE 9.15
Useful Floating-Point Constants

Constant Value	imm4H	Imm4L
0.5	0110	0000
0.125	0100	0000
2.0	0000	0000
31	0011	1111
15	0010	1110
4.0	0001	0000
–4.0	1001	0000
1.5	0111	1000
2.5	0001	0100
0.75	0110	1000

value, or convert a single-precision value to a half-precision value and store it in either the lower half or upper half of the destination floating-point register. The syntax of these instructions is

```
VCVTB{<cond>}.F32.F16  <Sd>, <Sm>
VCVTT{<cond>}.F32.F16  <Sd>, <Sm>
VCVTB{<cond>}.F16.F32  <Sd>, <Sm>
VCVTT{<cond>}.F16.F32  <Sd>, <Sm>
```

The B variants operate on the lower 16 bits of the Sm or Sd register, while the T variants operate on the upper 16 bits. These instructions provide a means of storing table data that does not require the precision or range of single-precision floating-point but can be represented sufficiently in the half-precision format.

9.11 CONVERSIONS TO NON-FLOATING-POINT FORMATS

Often data is input to a system in integer or fixed-point formats and must be converted to floating-point to be operated on. For example, the analog-to-digital converter in the TM4C1233H6PM microcontroller from Texas Instruments outputs a 12-bit digital conversion in the range 0 to the analog supply voltage, to a maximum of 4 volts. Using the fixed-point to floating-point conversion instructions, the conversion from a converter output to floating-point is possible in two instructions—one to move the data from memory to a floating-point register, and the second to perform the conversion. The range of options in the fixed-point conversion instructions makes it easy to configure most conversions without any scaling required. In Chapter 18, we will look at how to construct conversion routines using these instructions, which may be easily called from C or C++.

In the following sections, we will look at the instructions for conversion between 32-bit integers and floating-point single-precision, and between 32-bit and 16-bit fixed-point and floating-point single-precision.

9.11.1 Conversions between Integer and Floating-Point

The Cortex-M4 has two instructions for conversion between integer and floating-point formats. The instructions have the format

```
VCVT{R}<c>.<T32>.F32  <Sd>, <Sm>
VCVT<c>.F32.<T32>     <Sd>, <Sm>
```

The <T32> may be replaced by either S32, for 32-bit signed integer, or U32, for 32-bit unsigned integer. Conversions to integer format commonly use the roundTowardZero (RZ) format. This is the behavior seen in the C and C++ languages; conversion of a floating-point value to an integer always truncates any fractional part. For example, each of the following floating-point values, 12.0, 12.1, 12.5, and 12.9, will return 12 when converted to integer. Likewise, –12.0, –12.1, –12.5, and –12.9 will return –12. To change this behavior, the R variant may be used to perform the conversion using the rounding mode in the FPSCR. When the floating-point value is too large to fit in the destination precision, or is an infinity or a NaN, an Invalid Operation exception is signaled, and the largest value for the destination type is returned. Exceptions are covered in greater detail in Chapter 10.

A conversion from integer to floating-point always uses the rounding mode in the FPSCR. If the conversion is not exact, as in the case of a very large integer that has more bits of precision than are available in the single-precision format, the Inexact exception is signaled, and the input integer is rounded. For example, the value 10,000,001 cannot be precisely represented in floating-point format, and when converted to single-precision floating-point will signal the Inexact exception.

9.11.2 Conversions between Fixed-Point and Floating-Point

The formats of the fixed-point data type in the Cortex-M4 can be either 16 bits or 32 bits, and each may be signed or unsigned. The position of the binary point is identified by the <fbits> field, which specifies the number of *fractional bits* in the format. For example, let us specify an unsigned, 16-bit, fixed-point format in which there are 8 bits of integer data and 8 bits of fractional data. So the range of this data type is [0, 128), with a numeric separation of 1/256, or 0.00390625. That is, the value increments by 1/256 as one is added to the least-significant bit.

The instructions have the format

```
VCVT{<cond>}.<Td>.F32  <Sd>, <Sd>, #<fbits>
VCVT{<cond>}.F32.<Td>  <Sd>, <Sd>, #<fbits>
```

The <Td> value is the format of the fixed-point value, one of U16, S16, U32, or S32. Rounding of the conversions depends on the direction. Conversions from fixed-point to floating-point are always done with the *roundTiesToEven* rounding mode, and conversions from floating-point to fixed-point use the *roundTowardZero* rounding mode. We will consider these rounding modes in Chapter 10. One thing to notice in these instructions is the reuse of the source register for the destination register. This is due to the immediate <fbits> field. Simply put, there is not room in the instruction word for two registers, so the source register is overwritten. This should not be an issue; typically this instruction takes a fixed-point value and converts it, and the fixed-point value is needed only for the conversion. Likewise, when a floating-point value is converted to a fixed-point value, the need for the floating-point value is often gone.

EXAMPLE 9.5

Convert the 16-bit value 0x0180 in U16 format with 8 bits of fraction to a single-precision floating-point value.

SOLUTION

```
            ADR     r1, DataStore
            LDRH    r2, [r1]

            ; Convert each of the 16-bit data to single-precision with
            ; different <fbits> values

            VMOV.U16       s7, r2        ; load the 16-bit fixed-pt to s reg
            VCVT.F32.U16   s7, s7, #8    ; convert the fixed-pt to SP with
                                         ; 8 bits of fraction
loop        B  loop

            ALIGN
DataStore
            DCW            0x0180
```

The value in register s7 after this code is run is 0x3FC00000, which is 1.5. How did the Cortex-M4 get this value? Look at Table 9.16.

Notice that we specified 8 bits of fraction (here 8′b10000000, representing 0.5 in decimal) and 8 bits of integer (here 8′b00000001, representing 1.0), hence the final value of 1.5. In this format, the smallest representable value would be 0x0001 and would have the value 0.00390625, and the largest value would be 0xFFFF, which is 255.99609375 (256 – 0.00390625). Any multiple of 0.00390625 between these two values may be represented in 16 bits. If we wanted to do this in single-precision, each value would require 32 bits. With the U16 format we can represent each in only 16 bits.

There are valid uses for this type of conversion. The cost of memory is often a factor in the cost of the system, and minimizing memory usage, particularly ROM storage, will help. Another use is generating values that may be used by peripherals that expect outputs in a non-integer range. If we want to control a motor and the motor control

TABLE 9.16
Output of Example 9.5

Format U/S, <fbits >	Hex Value	Binary Value	Decimal Value	Single-Precision Floating-Point Value
U16, 8	0x0180	00000001.10000000	1.5	0x3FC00000

inputs are between 0 to almost 10, with 4 bits of fraction (so we can increment by 1/16, i.e., 0, 1/16, 1/8, 3/16, ... 9.8125, 9.875) the same instruction can be used to convert from floating-point values *to* U16 values. Conversion instructions are another tool in your toolbox for optimizing your code for speed or size, and in some cases, both.

The 16-bit formats may also be interpreted as signed when the S16 format is used, and both signed and unsigned fixed-point 32-bit values are available. Table 9.17 shows how adjusting the #fbits value can change how a 16-bit hex value is interpreted. If the #fbits value is 0, the 16 bits are interpreted as an integer, either signed

TABLE 9.17
Ranges of Available 16-Bit Fixed-Point Format Data

fbits	Integer Bits: Fraction Bits	Numeric Separation	Range Unsigned Range Signed
0	16:0	2^{0}, 1	0 ... 65,535 –32,768 ... 32,767
1	15:1	2^{-1}, 0.5	0 ... 32,767.5 –16,384 ... 16,383.5
2	14:2	2^{-2}, 0.25	0 ... 16,383.75 –8,192 ... 8,191.75
3	13:3	2^{-3}, 0.125	0 ... 8,191.875 –4,096 ... 4,047.875
4	12:4	2^{-4}, 0.0625	0 ... 4,095.9375 –2,048 ... 2,023.9375
5	11:5	2^{-5}, 0.03125	0 ... 2,047.96875 –1,024 ... 1,023.96875
6	10:6	2^{-6}, 0.015625	0 ... 1,023.984375 –512 ... 511.984375
7	9:7	2^{-7}, 0.0078125	0 ... 511.9921875 –256 ... 255.9921875
8	8:8	2^{-8}, 0.00390625	0 ... 255.99609375 –128 ... 127.99609375
9	7:9	2^{-9}, 0.001953125	0 ... 127.998046875 –64 ... 63.998046875
10	6:10	2^{-10}, 0.000976563	0 ... 63.999023438 –32 ... 31.999023438
11	5:11	2^{-11}, 0.000488281	0 ... 31.99951171875 –16 ... 15.99951171875
12	4:12	2^{-12}, 0.000244141	0 ... 15.999755859375 –8 ... 7.999755859375
13	3:13	2^{-13}, 0.00012207	0 ... 7.9998779296875 4 ... 3.9998779296875
14	2:14	2^{-14}, 6.10352E-05	0 ... 3.99993896484375 2 ... 1.99993896484375
15	1:15	2^{-15}, 3.05176E-05	0 ... 1.999969482421875 –1 ... 0.999969482421875
16	0:16	2^{-16}, 1.52588E-05	0 ... 0.999984741210937 –0.5 ... 0.499984741210937

or unsigned, and the numeric separation is 1, as we expect in the integer world. However, if we choose #fbits to be 8, the 16 bits are interpreted as having 8 integer bits and 8 fraction bits, and the range is that of an 8-bit integer, but with a numeric separation of 2^{-8}, or 0.00390625, allowing for a much higher precision than is available with integers by trading off range.

When the range and desired precision are known, for example, for a sensor attached to an analog-to-digital converter (ADC) or for a variable speed motor, the fixed-point format can be used to input the data directly from the converter without having to write a conversion routine. For example, if we have an ADC with 16-bit resolution over the range 0 to $+V_{REF}$, we could choose a V_{REF} value of 4.0 V. The U16 format with 14 fraction bits has a range of 0 up to 4 with a resolution of 2^{-14}. All control computations for the motor control could be made using a single-precision floating-point format and directly converted to a control voltage using

```
VCVT.U16.F32 s9, s9, #14
```

The word value in the s9 register could then be written directly to the ADC buffer location in the memory map. If the conversion is not 16 bits, but say 12 bits, conversion with the input value specified to be the format U16 with 10 fraction bits would return a value in the range 0 to 4 for all 12-bit inputs. Similarly, if V_{REF} is set to 2 V, the U16 format with 15 fraction bits would suffice for 16-bit inputs and the U16 with 11 fraction bits for 12-bit inputs. The aim of these instructions is to eliminate the need for a multiplier step for each input sampled or control output. Careful selection of the V_{REF} and the format is all that is required. Given the choice of signed and unsigned formats and the range of options available, these conversion instructions can be a powerful tool when working with physical input and output devices.

9.12 EXERCISES

1. Represent the following values in half-precision, single-precision, and double-precision.
   a. 1.5
   b. 3.0
   c. −4.5
   d. −0.46875
   e. 129
   f. −32768

2. Write a program in a high-level language to take as input a value in the form (−)x.y and convert the value to single-precision and double-precision values.

3. Using the program from Exercise 2 (or a converter on the internet), convert the following values to single-precision and double-precision.
   a. 65489
   b. 2147483648

c. 2^{29}
d. −0.38845
e. 0.0004529
f. 11406
g. −57330.67

4. Expand the program in Exercise 2 to output half-precision values. Test your output on the values from Exercise 3. Which would fit in half-precision?

5. Write a program in a high-level language to take as input a single-precision value in the form `0xXXXXXXXX` (where X is a hexadecimal value) and convert the input to decimal.

6. Using the program from Exercise 5 (or a converter on the internet), convert the following single-precision values to decimal. Identify the class of value for each input. If the input is a NaN, give the payload as the value, and NaN type in the class field.

Single-Precision Value	Value	Class
a. `0x3fc00000`		
b. `0x807345ff`		
c. `0x7f350000`		
d. `0xffffffff`		
e. `0x20000000`		
f. `0x7f800000`		
g. `0xff800ffe`		
h. `0x42c80000`		
i. `0x4d800000`		
j. `0x80000000`		

7. What value would you write to the FPSCR to set the following conditions?
   a. FZ unset, DN unset, roundTowardZero rounding mode
   b. FZ set, DN unset, roundTowardPositive rounding mode
   c. FZ set, DN set, roundTiesToEven rounding mode

8. Complete the following table for each of the FPSCR values shown below.

	N	Z	C	V	DN	FZ	RMode	IDC	IXC	UFC	OFC	DZC	IOC
0x41c00010													
0x10000001													
0xc2800014													

9. Give the instructions to load the following values to FPU register s3.
   a. 5.75×10^3
   b. 147.225
   c. −9475.376
   d. -100.6×10^{-8}

10. Give the instructions to perform the following load and store operations.
    a. Load the 32-bit single-precision value at the address in register r4 into register s12.
    b. Load the 32-bit single-precision value in register r6 to register s12. Repeat for a store of the value in register s15 to register r6.
    c. Store the 32-bit value in register s4 to memory at the address in register r8 with an offset of 16 bytes.
    d. Store the 32-bit constant 0xffffffff to register s28.

11. Give the instructions to perform a conversion of four fixed-point data in unsigned 8.8 format stored in register s8 to register s11 to single-precision format.

12. What instruction would you use to convert a half-precision value in the lower half of register s5 to a single-precision value, and store the result in register s2?

13. Give the instructions to load 8 single-precision values at address 0x40000100 to FPU registers s8 to s15.

14. How many subnormal values are there in a single-precision representation? Is this the same number as values for any non-zero exponent?

10 Introduction to Floating-Point
Rounding and Exceptions

10.1 INTRODUCTION

Rounding is one of the most important but confusing aspects of floating-point. We learned this early in school with problems asking for the nearest whole number when we divided 9 by 4. The pencil and paper result is 2.25, but what whole number do we give in the answer? The solution, we were taught, is to add 0.5 to the result and drop the fraction. So,

$$9/4 + 0.5 = 2.75,$$

and dropping the fraction gives 2. What if the problem was 9 divided by 2? We get 4.5, and adding 0.5 gives us 5. Is this the best we can do, since the computed value is exactly halfway between two whole numbers? We have the same issue in floating-point. The result of each operation must be a representable value, but what if the intermediate result of the operation was not? We have to *round the intermediate result* to a representable value. In this chapter we will look carefully at rounding and the various *rounding modes* specified by the IEEE 754-2008 specification.

A second important issue concerns what we do when an operation has no mathematically agreed upon answer, such as 0/0, or if some unusual event occurred in our computation. We call these situations *exceptions*, and while they are often not problematic, sometimes it can signal a situation that may require attention. We will consider each of these exceptions first generally, and then the specific response of the Cortex-M4 with floating-point hardware to the situations that signal each exception. Next we will consider whether we can count on some of the mathematical laws we learned in school, and finish the chapter looking at normalization and cancelation, two steps in floating-point computation and how they impact rounding and exceptions.

10.2 ROUNDING

Since only a finite set of representable values exists for each floating-point data type, we must have a method to deal with a computation that does not result in a representable value. For example, when we add

$$1.0 \text{ (0x3F800000)} + 2^{24} \text{ (0x4B800000)}$$

on a Cortex-M4, with the default rounding mode (roundTiesToEven), the answer is

$$2^{24} \text{ (0x4B800000).}$$

Why is the result the same as the second operand? Where is the contribution of the first operand? When we add a 1 to an integer, we expect the result to be the input integer incremented by one. On first glance, this seems to be an error. However, in this floating-point example, it makes no difference to the final result that we added 1.0 to 2^{24}. In fact, it makes no difference whether we add 1.0 to 2^{24} once, or whether we add it a million times. Each time this instruction is executed we will get the same result. How can this be? The answer is one of the most important features, and programming landmines, in using floating-point arithmetic, namely the frequent need to round the computed result to a representable value. The IEEE 754-2008 standard requires that each computation be computed *as if to infinite precision** and then rounded to a representable value or a special value. Internal to the Cortex-M4, the computation is performed to an intermediate precision larger than single-precision, which represents the *infinitely precise* internal sum, of

$$1.0 + 2^{24} = 16{,}777{,}217.0$$

as we would expect. However, this value is not a representable value for the single-precision data type. Recalling the formula for single-precision values

$$F = (-1)^s \times 2^{(\text{exp}-\text{bias})} \times 1.\text{f} \tag{9.1}$$

which we saw earlier in Sections 1.5.2 and 9.4, the value $2^{24} + 1.0$, results in the following floating-point component parts according to our formula (the significand part is represented in binary and the exponent in decimal):

$$16{,}777{,}217.0 = (-1)^0 \times 2^{(151-127)} \times 1.000000000000000000000001$$

Recall a single-precision value has only 23 fraction bits (bits [22:0] in a representation with the least-significant bit numbered 0). There are 24 bits in the significand of our example after the binary point, and only 23 of them can fit in the final significand. So we must select a representable value to return in place of the infinitely precise internal value. Every computation results in either exactly a representable value or one between two representable values, as shown in Figure 10.1. In this figure, values n_1 and n_2 are representable values in single-precision. Result A is exactly the representable value n_1; result B is exactly halfway between representable values n_1 and n_2; result C is closer to representable value n_1, while result D is closer to representable

* It doesn't really keep an infinite set of bits from a computation. That would make for a really large processor. Rather, the computation is done as if to infinite precision, meaning that internal to the Cortex-M4 the computation retains enough information to round the result correctly. See the IEEE 754-2008 standard section 5.1, p. 17.

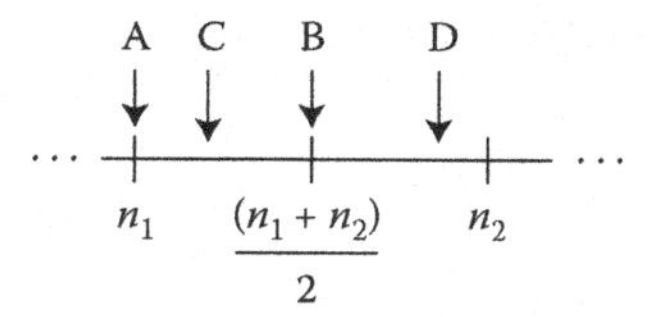

FIGURE 10.1 Possible results between two representable values.

value n_2. In each case, the representable value to be returned is determined by the current rounding mode and the bits in the infinitely precise internal value.

Each of the available rounding modes is the subject of the following sections, which describe the rounding modes defined in the IEEE 754-2008 standard, and those that are available in the Cortex-M4.

10.2.1 Introduction to Rounding Modes in the IEEE 754-2008 Specification

The IEEE 754-2008 standard specifies that the computed infinitely precise internal value be *rounded* to a representable value according to a selected *rounding mode.* Five rounding modes are specified by the standard:

roundTiesToEven
roundTiesToAway
roundTowardPositive
roundTowardNegative
roundTowardZero

We will focus our attention on four of these: *roundTiesToEven*, which we sometimes refer to as *Round to Nearest Even*, or RNE*; *roundTowardPositive*, also known as *Round to Plus Infinity*, or RP; *roundTowardNegative*, also known as *Round to Minus Infinity*, or RM; and *roundTowardZero*, also known as *Round to Zero*, or RZ.† Recall from Chapter 9, the rounding mode is set in the FPSCR in the Cortex M4, and the VMSR and VMRS instructions, covered in Chapter 11, enable the reading and writing of the FPSCR.

Inside the processor, a computed result will have additional bits beyond the 23 bits of the fraction. These bits are computed faithfully, that is, they are correct for the operation; however, they are not simply more bits of precision. Rather, two additional bits are computed—the *guard bit* and the *sticky bit*, shown in Figure 10.2. The guard bit is the bit immediately lower in rank than the least-significant bit position in the final result. If the infinitely precise internal significand were normalized to the range [1.0, 2.0), this would be the 25th bit of the significand, counting from left to right. The sticky bit is formed by ORing all bits with lower significance than the guard bit.

* In the IEEE 754-1985 specification this referred to as *Round to Nearest Even*, hence RNE. The abbreviations of the other rounding modes should be self-explanatory.

† The IEEE 754-2008 Standard does not require *roundTiesToAway* (RNA). See Clause 4.3.3, p. 16.

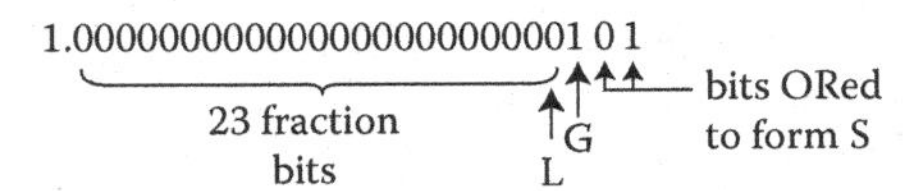

FIGURE 10.2 Internal representation of $2^{24} + 1.25$.

In other words, if the final result was computed to 40 bits, the upper 24 bits would be the significand of the pre-rounded, final result; the next bit would be the guard bit; and the OR of the final 15 bits would be the sticky bit.

EXAMPLE 10.1

Add 2^{24} (0x4B800000) + 1.25 (0x3FA00000)

Solution

The value 2^{24} is much larger than 1.25, and when represented in single-precision format, the 1.25 only contributes in the guard and sticky bits. When represented in infinite precision, we can see all the bits in the intermediate sum.

$$1.25 + 2^{24} = (-1)^0 \times 2^{(151-127)} \times 1.00000000000000000000000101$$

The guard bit is the 1 in bit position 2 (counting from the right side), while the sticky bit would be the OR of bits 0 (bit position 1) and 1 (bit position 0) and would be 1. We also refer to the least-significant bit of the pre-rounded significand as the *L bit* (for least-significant bit, or LSB). See Table 10.1.

10.2.2 The roundTiesToEven (RNE) Rounding Mode

The *roundTiesToEven* (RNE) rounding mode is the default rounding mode in the Cortex M4. The following equation governs the decision to increment the significand:

$$\text{Increment} = (\text{L} \ \& \ \text{G})|(\text{G} \ \& \ \text{S})$$

In truth table form this equation looks like Table 10.2.

The *roundTiesToEven* rounding mode causes the significand to be incremented whenever the bits not part of the pre-rounded significand would contribute greater than 1/2 the LSB value to the final result, and never when the bits would contribute less than 1/2 the LSB value. In the case of the bits contributing exactly 1/2 of the LSB, the pre-rounded significand is incremented when it is odd, that is, the L bit is

TABLE 10.1
Significand with Guard and Sticky Bits

Value	2^0	$2^{-1}\ldots2^{-22}$	2^{-23} ("L")	Guard ("G")	Bits Contributing to the Sticky Bit ("S")	
Bit position	25	24…4	3	2	1	0
Bit value	1.	0…0	0	1	0	1

TABLE 10.2
roundTiesToEven Rounding Summary

L LSB	G Guard	S Sticky	Increment?	Note
0	0	0	No	Pre-rounded result is exact, no rounding necessary
0	0	1	No	Only sticky set
0	1	0	No	Tie case, L bit not set
0	1	1	Yes	Guard and Sticky set—rounding bits >1/2 LSB
1	0	0	No	Pre-rounded result is exact, no rounding necessary
1	0	1	No	Only L bit and Sticky set
1	1	0	Yes	Tie case, L-bit set
1	1	1	Yes	Guard and sticky set—rounding bits are >1/2 LSB

set, and not when it is even. This is what the *Even* signifies in the name *roundTiesToEven*. This paradigm for rounding results is the statistically most accurate results for a random sample of operations and operands.

So let's return to our example case of adding 1.25 to 2^{24}. We know that the significand of the final value is 1.00000000000000000000000101. In Figure 10.2 we identified the L, G, and S bits.

The guard bit adds exactly 1/2 the value of the L bit to the significand, and the sticky bit increases the rounding contribution to greater than 1/2 of the L bit value. The significand does get incremented, and we have $2^{24} + 2$ as a final result.

EXAMPLE 10.2

Show a pair of operands that, when added, demonstrate the tie case without an increment.

Solution

The two operands could be 16,777,216 (2^{24}) and 1.0.

From Figure 10.2, the contribution of the 1.0 term would be only the G bit. L and S are each zero. From Table 10.2, we would not increment in this case (see the third line from the top of the table.)

EXAMPLE 10.3

Show a pair of operands that, when added, demonstrate the tie case with an increment.

Solution

The two operands could be 16,777,218 ($2^{24} + 2.0$) and 1.0.

From Figure 10.2, the contribution of the 1.0 term would be only the G bit. L is set to a one and S is zero. From Table 10.2, we would increment in this case (see the seventh line from the top of the table.)

EXAMPLE 10.4

Show a pair of operands that, when added, demonstrate an increment due to G and S.

SOLUTION

The two operands could be 16,777,216 (2^{24}) and 1.25.

From Figure 10.2, the contribution of the 1.25 term would be the G and S bits each set. L is zero. From Table 10.2, we would increment in this case. (See the fourth line from the top of the table.)

10.2.3 THE DIRECTED ROUNDING MODES

The other three rounding modes are called *directed rounding modes*. These find use in specialized mathematical operations in which creating a bound of a problem is more useful than computing a single result. In these instances, knowing the bounds of a function with a specific data set conveys more useful information about the error range than would a single result of unknown accuracy. One such area of mathematics is called *interval arithmetic*. In interval arithmetic, each value is represented as a *pair of bounds*, representing the *range* of possible numerical values for a result rather than a single result value. These computations are useful when the true value cannot be known due to measurement inaccuracies, rounding, or limited precision.

For example, we may say a friend's height is between 6 feet and 6 feet two inches. While we don't know exactly the height of our friend, we are sure he is at least 6 feet but no more than 6 feet 2 inches tall. If we were to measure the average height of a class of boys, we could measure each boy using a tape measure, but each of these measurements may not be accurate. For example, we measure Tom to be 6 foot 3/4 inches. But is this his true height? Perhaps he has let his hair grow and this added a 1/4 inch, or his shoes or socks are contributing to our measurement. Next week he will get his hair cut, and he would measure only 6 foot 1/2 inches. If we record the boys' heights to a precision of 1 inch, rounding any fraction down for a lower value and rounding up for a higher value, we could create a pair of bounding values, each with a precision of 1 inch, one lower than the measured value, and one higher than the measured value. With the measurement we have for Tom, we could record Tom's height as (6′ 0″, 6′ 1″). This way we give accurate bounds for his height, but not a specific number. In floating-point interval arithmetic, we would round any imprecise value, both down, for a lower bound, and up, for a higher bound, creating a bounding pair for all computations. When we have all the measurement pairs for the class, we would compute an average of the lower entries in each pair, and an average of the upper entries in the pairs, producing again another pair. Now we can say with some certainty that the average height of the boys in the class is between the lower bound and the upper bound. As you can see, while this is *imprecise*, in that we don't have a single value, it is more *accurate* than any single number could be.

The three directed rounding modes are often used in interval arithmetic, and are discussed below.

10.2.3.1 The roundTowardPositive (RP) Rounding Mode

This rounding mode will increment any *imprecise positive result*, and not increment any precise or negative result. The inputs to the rounding equation are G and S bits, the sign bit, and incrementing is done if both the pre-rounded result is positive and if either G or S is set. If the pre-rounded result is negative, no incrementing is done. In this mode, the increment equation is

$$\text{Increment} = \sim\text{sign} \ \& \ (G|S)$$

Recall the sign bit in a positive floating-point number is 0. If we consider Example 10.4, we see the S bit is set and the sign is positive (sign = 0). In RP mode the final significand would be incremented.

10.2.3.2 The roundTowardNegative (RM) Rounding Mode

Some explanation is useful here. When we say a value in incremented, we are referring to the significand regardless of the sign off the value. For example, if we have in –1.75 in decimal and we round this value up, we would have –2.0. Simply put, an increment always causes the result to be further from zero, regardless of the sign of the result.

The RM rounding mode will increment any *imprecise negative result*, and not increment any precise or positive result. This mode is the negative sign bit counterpart to the roundTowardPositive rounding mode. If the pre-rounded result is negative, and if either G or S is set, the mantissa is incremented. If the pre-rounded result is positive, no incrementing is done. In this mode the increment equation is

$$\text{Increment} = \text{sign} \ \& \ (G|S)$$

If we consider again Example 10.4, the S bit set and the sign positive (sign = 0) would dictate that the final significand would not be incremented. If the sign of the result were negative in the example, the roundTowardNegative rounding mode would dictate incrementing the final significand.

10.2.3.3 The roundTowardZero (RZ) Rounding Mode

The *roundTowardZero* (RZ) mode is also called *truncate*, and this rounding mode never increments an intermediate value, but simply drops any guard and sticky bits. Any bits computed beyond the L bit are ignored. This the mode commonly used in integer arithmetic in processors with divide operations. In this mode the increment equation is

$$\text{Increment} = 0$$

In other words, we never increment the result in the roundTowardZero rounding mode. In Example 10.4, even though the G and S bits are set, the final significand is not incremented. All the bits to the right of the L bit are simply truncated.

10.2.4 Rounding Mode Summary

The operation of these four rounding modes may be summarized by the diagram in Figure 10.3. The values n and $n + 1$ *ulp*[*] are two, contiguous, representable floating-point values. The value $n + 1/2$ *ulp* represents the point halfway between the two representable floating-point values but is not itself a representable floating-point value.

Consider first the upper two lines in Figure 10.3. These lines represent the behavior of the *RNE* rounding mode, and the handling of tie cases depends on the value of L, the least significant bit of the internal normalized significand (more on normalization in Section 10.5). In the top line, L is 1, indicating the normalized significand is *odd.* In this case the tie case rounds up to make the result significand even. In the second line the internal normalized significand is even before the rounding decision, and a tie case will not increment, leaving the result even. In both cases, if the infinitely precise internal value is greater than $n + 1/2$ *ulp,* the internal value is incremented, and if less, the internal value is not incremented. The third line indicates the behavior of *RP* for a positive result and *RM* for a negative result, while the last line indicates the behavior of *RP* for a negative result, *RM* for a positive result, and *RZ* always. In both *RP* and *RM,* the decision to increment is made on the sign of the result, the rounding mode, and whether the internal normalized significand is exactly a representable value. To complete Table 10.1, we add in the three directed rounding modes to form Table 10.3.

Let's consider a multiplication example to demonstrate the four rounding modes.

EXAMPLE 10.5

Multiply 0x3F800001 (1.00000011920928955) by 0xC4D00000 (–1664) in each rounding mode and compute the result.

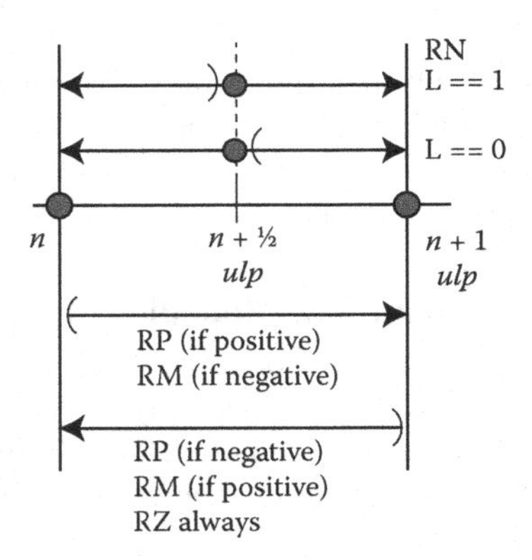

FIGURE 10.3 Rounding mode summary.

[*] A *ulp* is shorthand for a *unit-in-the-last-place*, or the bit with the smallest contribution to the final result. For single-precision, this is the least significant bit. This is a common term in floating-point error analysis.

TABLE 10.3
Rounding Mode Summary

Sign	Rounding Bits			Increment? Rounding Mode				
	L	**G**	**S**	**RNE**	**RP**	**RM**	**RZ**	**Data Characteristic**
0	0	0	0	No	No	No	No	Exact
1	0	0	0	No	No	No	No	Exact
0	0	0	1	No	Yes	No	No	Inexact—positive
1	0	0	1	No	No	Yes	No	Inexact—negative
0	0	1	0	No	Yes	No	No	Inexact—positive, tie case
1	0	1	0	No	No	Yes	No	Inexact—negative, tie case
0	0	1	1	Yes	Yes	No	No	Inexact—positive
1	0	1	1	Yes	No	Yes	No	Inexact—negative
0	1	0	0	No	No	No	No	Exact
1	1	0	0	No	No	No	No	Exact
0	1	0	1	No	Yes	No	No	Inexact—positive
1	1	0	1	No	No	Yes	No	Inexact—negative
0	1	1	0	Yes	Yes	No	No	Inexact—positive, tie case
1	1	1	0	Yes	No	Yes	No	Inexact—negative, tie case
0	1	1	1	Yes	Yes	No	No	Inexact—positive
1	1	1	1	Yes	No	Yes	No	Inexact—negative

Solution

Let OpA = 0x3F800001 and OpB = 0xC4D00000. The two operands are shown in their component parts in Table 10.4.

Recall that in integer multiplication, we take each digit of the multiplier, and if it is a 1 we include the multiplicand shifted to align bit [0] of the multiplicand with the multiplier bit. If the bit is a zero, we skip it and go on to the next bit. In the diagram below, there are two multiplier bits, OpA[0] and OpA[23], resulting in two partial product terms, called OpA[0] term and OpA[23] term. These two partial products are summed using binary addition to form the infinitely precise product. Since the two significands are 24 bits each, the product will be (24 + 24 −1), or 47 bits. If the summing operation produced a carry, the product would have 48 bits; however, in this case there is no carry, so only 47 bits are valid. The L bit and G bit, and the bits which will be ORed to make the S bit (identified with a lowercase s) are marked in the line pre-rounded product. In this example, L is 1, so our pre-rounded product is odd; G is 1, and S is 1. From Table 10.5, for the RNE and RM rounding modes the pre-rounded product will be incremented to form the result product, but for the RP and RZ rounding modes the pre-rounded product is not incremented. (Notice the product is negative.)

TABLE 10.4
Operands for Example 10.5

OpA:	0x3F800001	=	-1^0	x	2^0	x	1.00000000000000000000001
OpB:	0xC4D00000	=	-1^1	x	2^{10}	x	1.10100000000000000000000

TABLE 10.5
Example 10.5 Intermediate Values

			Operand Significand		
OpB Significand (Multiplicand)			1	101	00000000000000000000
OpA Significand (Multiplier)			1	000	00000000000000000001
			Internally aligned partial product terms		
OpA[0] term			1	101	00000000000000000000
OpA[23] term	1.101	0000000000000000	0000		
			Infinitely-precise internal significand		
Infinitely precise product	1.101	0000000000000000	0001	101	00000000000000000000
Pre-rounded product	1.101	0000000000000000	000L	Gss	ssssssssssssssssssss
Incremented product	1.101	0000000000000000	0010	101	00000000000000000000
		Result significand			
Result product - RNE	1.101	0000000000000000	0010		
Result product - RP	1.101	0000000000000000	0001		
Result product - RM	1.101	0000000000000000	0010		
Result product - RZ	1.101	0000000000000000	0001		

10.3 EXCEPTIONS

10.3.1 Introduction to Floating-Point Exceptions

An important difference between integer and floating-point operations is the intrinsic nature of exceptions. You may be familiar with one exception in the integer world, *division by zero*. If you have a processor with a hardware divider and attempt to divide an operand by zero, you will very likely see something like

#DIV/0!

and find your program has halted. The reason for this is that there is no value that would be appropriate to return as the result of this operation. With no suitable result, this operation in the program is flawed, and any results cannot be trusted. Even if the numerator was also a zero, this situation will be signaled with the same result. If the integer number set had a representation for infinity, this might not be a fatal situation, but all integer bit patterns represent real numbers, and no infinity representation exists.

In floating-point, we *do* have a representation for infinity, and division by zero is not fatal. When we use the term *exception* in the floating-point context, we do not mean a catastrophic failure, or even a situation that requires a programmer's or user's attention, but simply a case of which you, the programmer or user, *might* want to be made aware. We say might, because in many exceptional cases the program will continue execution with the exceptional condition and end successfully. Returning to our division by zero operation, we were taught in math class that a nonzero number divided by zero was not allowed. However, in a computational environment, we would expect the hardware to return a signed infinity according to the IEEE 754-2008 standard, and in *roundTiesToEven* rounding mode this is what we see. The program does not have to be halted. As we will see in Chapter 11, all floating-point operations have rational behaviors with infinities as operands.

The IEEE 754-2008 specification requires five exceptions, and each must deliver a default result to the destination, signal the exception by a corresponding flag, and not stop processing when encountered.* In our division by zero example, a signed infinity would be a proper return value. If this infinity were operated on further, the subsequent operations must respect the properties of an infinity, and this is what will happen. It's very difficult to change an infinity to a normal number (impossible, really) so the infinity will in most cases be the result of the computation, and the user will see the infinity as the output.

As mentioned above, the IEEE 754-2008 standard specifies five exceptions:

- Division by Zero
- Invalid Operation
- Overflow

* The IEEE 754-2008 standard specifies an *Alternate exception handling* mechanism in which a trap is taken to a trap handler, which allows the programmer to specify the behavior to be taken when the exception is detected. The Cortex-M4 does not allow for this option directly, but instead provides five pins on the boundary of the processor core that toggle with the exception flag. Each could be connected to an interrupt input and cause an interrupt routine to execute when the exception is signaled.

- Underflow
- Inexact

We will consider each one separately; however, be aware that Inexact may appear with Overflow and with Underflow, returning two exceptions on a single operation. Also, recall that all exception flags in the Cortex-M4 are cumulative, or "sticky," and once set by an exceptional condition remain set until written to a zero to clear it.

10.3.2 Exception Handling

The IEEE 754-2008 standard requires that a default result be written to the destination register, a corresponding flag be set, and the processing to continue uninterrupted. In the case of arithmetic exceptional conditions, such as overflow, underflow, division by zero, and inexactness, the default results may be the correct result and allow processing to continue without error. For example, in an overflow situation, a properly signed infinity is a correct result and may indicate that this computation is simply out of bounds. This may be a valid output, or it may signal that the data set operated on resulted in computations out of bounds for the algorithm, and a modification to the data set or the algorithm is required. The selection of default results shows the desire of the architects of the floating-point specification to have a system that will do mathematical processing with reasonable results, even with the limitations of the data types and operating on unknown data sets. It is possible to construct robust programs that can tolerate varying data sets and exceptions of the types we will discuss below and return reasonable and useful data. As you consider each of the exceptions below, see if you agree with the selection of the default result for each case.

10.3.3 Division by Zero

Division by zero occurs whenever a division operation is performed with a divisor of zero and the dividend is a normal or subnormal operand. When this occurs, the default result is a properly signed infinity. Properly signed here means that the sign rules learned in school apply, that is, if both signs are the same, a positive result is returned, and if the signs are different, a negative result is returned. When detected, a properly signed infinity is written to the destination register, the *division by zero* (DZC) status bit is set in the FPSCR, or remains set if it was set prior to this instruction, and processing continues with this result. Note that if the operation is a reciprocal operation, the dividend is assumed to be +1.0. The code below shows the behavior of the Cortex-M4 in a division-by-zero case.

EXAMPLE 10.6

```
; Example 10.6 - Divide by Zero
; In this example we load s0 with 5.0 and s1 with
; 0.0 and execute a divide. We expect to see +inf
; in the destination register (s2)
; Next we load s3 with -0.375 and perform the
; division, this time expecting -inf in the destination
```

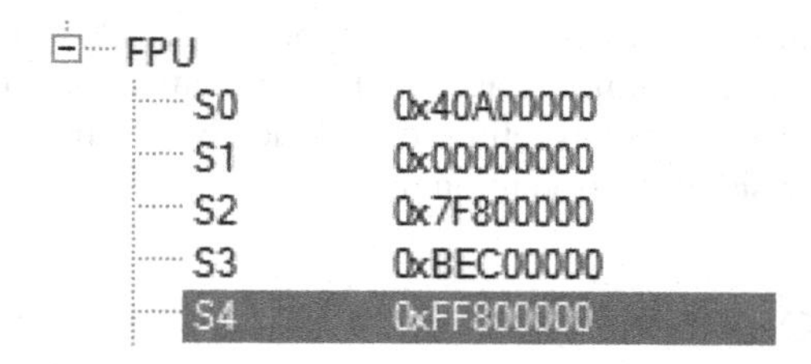

FIGURE 10.4 Output of Example 10.6.

```
; register s4

VMOV.F   s0, #5.0
LDR      r1, = 0x00000000 ; cannot load 0.0 using VMOV.F
VMOV.F   s1, r1
VDIV.F   s2, s0, s1       ; return positive infinity
VMOV.F   s3, #-0.375
VDIV.F   s4, s3, s1       ; return negative infinity
```

After running this code the floating-point registers contain the values shown in Figure 10.4. The contents of the FPSCR register show the DZC (Divide-by-zero Condition bit) is set as shown in Figure 10.5.

The result of the first division in register s2 is a positive infinity (convince yourself that this is the hexadecimal pattern for a positive infinity), and the result in register s4 is a negative infinity (again, make sure you are convinced that this is a negative infinity).

FPSCR	0x00000002
N	0
Z	0
C	0
V	0
AHP	0
DN	0
FZ	0
RMode	RN
IDC	0
IXC	0
UFC	0
OFC	0
DZC	1
IOC	0

FIGURE 10.5 FPSCR Contents after Example 10.6.

What if the dividend was not a normal or subnormal value? If it was a zero, NaN, or an infinity, there is no commonly accepted mathematical result for this situation. An *Invalid Operation* exception is returned, and the division by zero exception is not signaled. We will consider this exception next.

10.3.4 Invalid Operation

There are a host of conditions that will signal an *Invalid Operation* exception. Most are present to signal mathematical situations for which a commonly accepted result is not known, such as the division of zero by zero, as mentioned above. The conditions for *Invalid Operation* fall into three categories:

1. Operations with signaling NaNs (sNANs)—Of the two types of NaNs discussed in Section 9.6.5, the sNaN, or signaling NaN, will always signal the *Invalid Operation* exception when operated on by an arithmetic operation. Note that data moves will not trigger the exception. No other exceptions are signaled for an operation involving a sNaN, even if other exceptional conditions exist.
2. Arithmetic operations without a commonly accepted default result—Several operations simply don't have agreed upon results. Consider addition with unlike-signed infinities and multiplication of infinity by zero. These operations are defined as "undefined," or an "indefinite form." The complete list is given in Table 10.6 for floating-point operations; however, other operations that are not specified by the standard, such as transcendental functions, may also raise exception flags.
3. Conversion operations—When a conversion of a value in floating-point format to an integer or fixed-point format isn't possible because the value of the floating-point operand is too large for that destination format, the *Invalid Operation* exception is signaled. For example, if register s4 contains 0x60000000 ($\sim 3.7 \times 10^{15}$), conversion to a 32-bit integer would not be possible, since this value is much greater than is representable. In this case the largest integer value is returned, and the IOC and IXC bits are set in the FPSCR. Again, this is because there is no integer format that would indicate the error. All bit patterns in the integer formats represent valid numbers, and to return even the maximum value would not represent the input value or indicate the error condition. Why not use the *Overflow* exception for this case? The *Overflow* exception indicates the result of an arithmetic operation, and this is a format conversion issue and not an arithmetic operation.

When the *Invalid Operation* exception is detected for arithmetic and conversion operations, the default Quiet NaN (qNaN) is returned and the *Invalid Operation* (IOC) status bit is set in the FPSCR. The format of the default NaN is at the implementer's discretion; in Section 9.8.1.1 we saw what the developers of the Cortex-M4 chose as the default NaN. When an input is a sNaN, the sNaN is *quieted*, that is, the NaN type bit is set, making the sNaN into a qNaN. If more than one input operand is a NaN of either flavor, one of the NaNs will be returned, but always in a quiet form.

TABLE 10.6
Operations and Operands Signaling the Invalid Operation Exception

Instruction	Invalid Operation Exceptions
VADD	(+infinity) + (–infinity) or (–infinity) + (+infinity)
VSUB	(+infinity) – (+infinity) or (–infinity) – (–infinity)
VCMPE, VCMPEZ	Any NaN operand
VMUL, VNMUL	Zero × ±infinity or ±infinity × zero
VDIV	Zero/zero or infinity/infinity
VMAC, VNMAC	Any condition that can cause an Invalid Operation exception for VMUL or VADD can cause an Invalid Operation exception for VMAC and VNMAC. The product generated by the VMAC or VNMAC multiply operation is considered in the detection of the Invalid Operation exception for the subsequent sum operation
VMSC, VNMSC	Any of the conditions that can cause an Invalid Operation exception for VMUL or VSUB can cause an Invalid Operation exception for VMSC and VNMSC. The product generated by the VMSC or VNMSC multiply operation is considered in the detection of the Invalid Operation exception for the subsequent difference operation
VSQRT	Source is less than 0
VMLA/VMLS	Multiplier and multiplicand are zero and infinity or infinity and zero
VMLA/VMLS	The product overflows and the addend is an infinity, and the sign of the product is not the sign of the addend
Convert FP to Int	Source is NaN, Inf, or outside the range with RMode

10.3.5 Overflow

The *Overflow* exception is signaled when the result of an arithmetic operation cannot be represented because the absolute value of the result is too large for the destination format. In this way it is possible to overflow both with a positive and a negative result. In the default RNE rounding mode, a positive value too large for the single-precision format will return a positive infinity. Likewise, if the absolute value of the result is too large for the single-precision format, and the sign is negative, a negative infinity will be returned. In the general case, the default value returned depends on the signs of the operands and the rounding mode, as seen in Table 10.7. One thing to note here is that with an overflow exception both the overflow status bit (OFC) and the inexact status bit (IXC) are set.

TABLE 10.7
Default Values for the Overflow Exception

Rounding Mode	Positive Result	Negative Result
RNE	+ infinity	– infinity
RP	+ infinity	– maximum normal value
RM	+ maximum normal value	– infinity
RZ	+ maximum normal value	– maximum normal value

If you consider that the default rounding mode is RNE, returning positive and negative infinity values for overflows makes very good sense. Likewise, for the three directed rounding modes, returning the largest normal value indicates rounding in one direction, while infinity indicates rounding in the other direction. Overflow is possible in most arithmetic operations. Take note, overflow may be due to the operation resulting in a value outside the range *before* rounding, or it may be due to a pre-rounded result that is in the normal range but the *rounded result* overflows. The case of a pre-rounded result rounding to an overflow condition is left as an exercise.

EXAMPLE 10.7

Compute the factorial function for all integers from 1 to 35.

SOLUTION

The factorial computation is

$$n! = \prod_{k=1}^{n} k$$

or by the recurrence relation

$$n! = \begin{cases} 1 & \text{if } n=0, \\ (n-1)! \times n & \text{if } n>0. \end{cases}$$

We can construct a simple table of factorials, beginning with the factorial of 1 and continue until we have a factorial too large to fit in a single-precision value.

1 - 1
2 - 2
3 - 6
4 - 24
...
10 - 3,628,800, or 3.628×10^{6}
...
15 - 1,307,674,368,000, or 1.307×10^{12}
...
20 - 2,432,902,008,176,640,000, or 2.432×10^{18}
...
25 - 15,511,210,043,330,985,984,000,000, or 1.551×10^{25}
...
30 - 265,252,859,812,191,058,636,308,480,000,000, or 2.653×10^{32}
31 - 8,222,838,654,177,922,817,725,562,880,000,000, or 8.223×10^{33}
32 - 263,130,836,933,693,530,167,218,012,160,000,000, or 2.632×10^{35}
33 - 8,683,317,618,811,886,495,518,194,401,280,000,000, or 8.683×10^{36}
34 - 295,232,799,039,604,140,847,618,609,643,520,000,000, or 2.952×10^{38}
35 - 10,333,147,966,386,144,929,666,651,337,523,200,000,000, or 1.033×10^{40}

Recalling that the range of a single-precision value is up to 3.40×10^{38}, the factorial of 35 will result in a value too large for the single-precision format. According to Table 10.2, the result for the roundTiesToEven rounding mode and the roundTowardPositive rounding mode is a +infinity, while the roundTowardNegative and roundTowardZero would return the maximum normal value, or 3.40×10^{38} (0x7F7FFFFF). If we attempted to compute the factorial of 36 by multiplying the computed value of 35! by 36, we would get back the same value as computed for 35!. Why? In RNE and RP modes, we are multiplying +infinity by 36, which results in +infinity, and in RM and RP modes, the value of the maximum normal value multiplied by 36 again overflows, and the maximum normal value will be returned. Showing this in simulation is left as an exercise.

10.3.6 Underflow

Floating-point arithmetic operations can also *underflow* when the result of the operation is *too small* to fit in the destination format. You can imagine when two very small values, say 6.6261×10^{-34} (Planck's constant in J · s) and 1.602×10^{-19} (elementary charge in Coulombs), are multiplied, the product is 10.607×10^{-53}, but this value is outside the normal range of a single-precision value, and we have *underflowed.* In some systems the result would be a signed zero.* Underflow is unique among the exceptions in that it is at the discretion of the processor designer whether the determination of underflow is made *before rounding* or *after rounding*, but all underflow determinations must be made the same way. The Cortex-M4 chose to detect underflow before rounding. When the *Underflow* exception is detected, the default value returned is a subnormal value (if the result is within the subnormal range for the destination precision) or a signed zero. The Underflow status bit (UFC) and the Inexact status bit (IXC) are set in the FPSCR *if the result is not exact*; otherwise, neither status bit is set. For example, if the operation resulted in a subnormal value that was exact, neither the UFC or IXC bits will be set. The IEEE 754-2008 standard does not regard this as an underflow condition. However, if the result is subnormal and not exact, both the UFC and IXC bits will be set, since the result is below the normal range and inexact. In the same way, if the result is too small even to be represented as a subnormal, and a zero is returned, both the UFC and IXC bits will be set, since this is both an underflow and an inexact condition. Table 10.8 summarizes the several cases possible in underflow condition.

TABLE 10.8
Summary of the Flags and Results in Underflow Conditions

Result before Rounding	Returned Result	Flags Set
Subnormal range	Subnormal value	If exact, no flags. Otherwise, UFC and IXC
Below subnormal range	Signed zero	UFC and IXC

* We will see this option in the Cortex-M4 in the *Flush-to-zero* mode described in Section 11.5.1.

10.3.7 Inexact Result

As we saw in the section on rounding, not all floating-point computations are exact. More likely than not, most will result in an intermediate value which is between two representable values and requires rounding. When this occurs, the result is said to be *inexact*. That means simply that there was no representable value exactly matching the result, and another value was substituted for the computed result. When the *Inexact Exception* occurs, the *Inexact* flag (IXC) is set in the FPSCR and the computation continues. The programmer may check this flag at any time to see whether any of the operations returned an inexact result since the last time the flag was cleared.

10.4 ALGEBRAIC LAWS AND FLOATING-POINT

In school we were taught several laws of mathematics, and we're interested in three of these, namely the *commutative law*, the *associative law*, and the *distributive law*. Are they still useful in the world of floating-point? Let's take each one separately. The commutative law states that in addition and multiplication the operands may be swapped without affecting the answer. Such is not the case for subtraction and division. Does this law hold for floating-point addition and multiplication? Consider the following.

EXAMPLE 10.8

If register s7 contains 0x40200000 (2.5) and register s8 contains 0x42FD999A (126.8), will these two instructions produce the same result? You should try this for yourself.

```
VADD.F32 s10, s7, s8
VADD.F32 s11, s8, s7
```

Likewise, consider these instructions:

```
VMUL.F32 s12, s7, s8
VMUL.F32 s13, s8, s7
```

Is the value in register s10 the same as in register s11, and the value in register s12 the same as register s13? They do indeed have the same values, and we can expect that in all cases floating-point addition and multiplication abide by the commutative property. Note, however, that this applies only to a single addition or multiplication operation. When more than two operands are to be summed or multiplied, the IEEE 754-2008 standard requires the operations be performed in the order of the program code unless rearranging the operands would return the same result value and flags.

What about the associative law? If A, B, and C are single-precision floating-point values, is

$$(A + B) + C = A + (B + C)$$

as required by the associative property of addition? If we denote floating-point addition of single-precision values with a single-precision result as ⊕, is

$$(A \oplus B) \oplus C = A \oplus (B \oplus C)?$$

Consider the example below.

EXAMPLE 10.9

Let A be 0x50800000 (1.718×10^{10}) in register s13, let B be 0xD0800000 (-1.718×10^{10}) in register s14, and let C be 0x2FC00000 (3.492×10^{-10}) in register s15. What is the result of the following pair of instructions?

```
VADD.F32 s16, s13, s14 ; s13 = A, s14 = B, s16 = A ⊕ B
VADD.F32 s17, s16, s15 ; s15 = C, s17 = (A ⊕ B) ⊕ C
```

Is it different from this pair of instructions?

```
VADD.F32 s16, s14, s15 ; s16 = (B ⊕ C)
VADD.F32 s17, s13, s16 ; s17 = A ⊕ (B ⊕ C)
```

In this example A ⊕ B is zero, so the result of the first set of instructions is the value C in register s15, or 3.492×10^{-10}. However, when B and C are added, the result is B, since the contribution of C is too small and is lost in the rounding of the VADD operation. So the result of the second set of operations is zero! While it is not always the case that floating-point addition fails to satisfy the associative property of addition, it must be a consideration to a programmer that the *order* of addition operations may affect the final result.

Does the associative law hold for multiplication? If we again have 3 single-precision operands, A, B, and C, and floating-point multiplication is denoted by ⊗, is the following true in all cases?

$$(A \otimes B) \otimes C = A \otimes (B \otimes C)$$

EXAMPLE 10.10

Let A be 0x734C0000 (1.616×10^{31}) in register s20, let B be 0x5064E1C0 (1.536×10^{10}) in register s21, and let C be 0x2BF92000 (1.770×10^{-12}) in register s22. What will be the answer for each of the following pairs of instructions?

```
VMUL.F32 s23, s20, s21 ; s20 = A, s21 = B, s23 = A ⊗ B
VMUL.F32 s24, s23, s22 ; s22 = C, s24 = (A ⊗ B) ⊗ C
```

and

```
VMUL.F32 s25, s21, s22 ; s25 = B ⊗ C
VMUL.F32 s26, s20, s23 ; s26 = A ⊗ (B ⊗ C)
```

In the first pair, A multiplied by B returns positive infinity (0x7F800000), and the second multiplication with C results in a positive infinity. The first multiplication overflows, and the second multiplication of a normal value and an infinity returns an infinity. However, in the second pair of instructions, B and C are multiplied first and results in 2.719×10^{-2}. When this product is multiplied by A, the result is 4.395×10^{29}. All products in the second sequence of instructions are normal numbers; none of the products is an infinity. As with addition, the order of operands can play a critical role in the result of a series of additions or multiplications. When a result is not what is expected, as in the case of the infinity in the pair of multiplications, it is often a clue that the operand ordering played a part in the result.

This leaves the distributive law, which states that

$$A * (B + C) = (A * B) + (A * C)$$

From the above it should be clear that in floating-point operations this property can quite easily be shown to fail. This is left as an exercise for the reader.

10.5 NORMALIZATION AND CANCELATION

Often a floating-point computation will not be *normalized*, that is, it will not be in the correct form in the equation for a normal or subnormal value. It could be so because the computed significand is in the range [2.0, 4.0). To normalize the result, it must be shifted right one place, and the exponent must be incremented to be within the proper range. For example, if we multiply

$$1.7 \text{ (0x3FD9999A)} \times 1.4 \text{ (0x3FB33333)} = 2.38 \text{ (0x401851EC)}$$

you notice that both input operands have the exponent 0x3F8 (representation of 2^0); however, the result has the exponent 0x400 (representation of 2^1). Internally, the product of 1.7 and 1.4 results in a significand in the range [2.0, 4.0), specifically, 2.38. To form the final result value, the Cortex-M4 shifts the internal significand to the right 1 place to form a new significand of 1.19, and increments the exponent (then 0) to 1, and forms the result as

$$2.38 = -1^0 \times 2^1 \times 1.19$$

This is referred to as *post-normalization* and is all done internal to the processor—it's invisible to the user. Once the computed result is normalized, the guard and sticky bits can be generated. Similarly, in the case of an effective subtraction, it is possible for the upper bits to cancel out, leaving a string of zeros in the most significant bit positions. An effective subtraction is a subtraction operation on like signed operands, that is,

$$(+1.0) - (+0.45) \text{ or } (-5.3) - (-2.1),$$

or an addition of unlike signed operands, such as

$$(+1.0) + (-0.45) \text{ or } (+5.3) + (-2.1).$$

Any summation operation that produces a result *closer to 0* is an effective subtraction.

EXAMPLE 10.11

Consider in this decimal example

$$1.254675 \times 10^6 - 1.254533 \times 10^6 = 1.42 \times 10^2$$

Most of the upper digits cancel out, leaving only a value with an order of magnitude of 2 when the original operands were order of magnitude 6. This may occur when the exponents of the two operands in an effective subtraction are equal or differ by 1. The same situation occurs for floating-point, requiring the resulting significand to be left shifted until a 1 appears in the integer bit of the result, and the exponent must be decremented accordingly.

EXAMPLE 10.12

Add 0x3F9CE3BD and 0xBF9CD35B.

SOLUTION

Let OpA be 0x3F9CE3BD (1.2257) and OpB be 0xBF9CD35B (–1.2252). If the value in register s3 is OpA and register s4 contains OpB, and these are added in the FPU with the instruction

```
VADD.F32  s5, s3, s4
```

the result in register s5 will be 0x3A031000 (4.9996×10^{-4}, the closest representation in single-precision to 5.0×10^{-4}). Notice the exponent has been adjusted so the resulting significand is in the range [1.0, 2.0), as we saw with the multiplication and decimal examples above. Table 10.9 shows this process. The two operands are normalized (the leading binary bit is a 1). When subtracted, the upper bits cancel, leaving a string of zeros before the first 1. To normalize the significand, we shift it left until the most significant bit is a 1. The number of shift positions is 11, since the number of leading zeros is 11. To generate the final exponent, the initial exponent (0x7F) is decremented by 11, resulting in the final exponent of 0x74. The

TABLE 10.9
Internal Values for Example 10.12

	Exponent	Significand
OpA (Addend)	01111111	100111001110001110111100
OpA (Addend)	01111111	100011001101001101011011
Sum (pre-normalized)	01111111	000000000001000001100010
Post-normalized sum	01110100	100000110001000000000000
Result sum – RNE	01110100	100000110001000000000000
Result sum – RP	01110100	100000110001000000000000
Result sum – RM	01110100	100000110001000000000000
Result sum – RZ	01110100	100000110001000000000000

final result in single-precision format is 0x3A031000. Verify for yourself that the four results in binary in Table 10.9 are correct.

When a result is computed and the exponent is incremented or decremented, it may result in an overflow or underflow condition.

EXAMPLE 10.13

Multiply 0x3F800001 by 0x7F7FFFFE in each rounding mode.

Solution

In this example, the rounding of the internal infinitely precise product results in an overflow of the significand, and the exponent is incremented in the normalization of the final significand. When the exponent is incremented, it becomes too large for the single-precision format, and the final product overflows.

In Table 10.10, the OpA value is the multiplier and the OpB value is the multiplicand. Only two bits are set in the multiplier—OpA[0] and OpA[23] so there will only be two partial products to be summed. The infinitely precise product is all ones, except the final bit, and the L, G, and S rounding bits are each one. In the roundTiesToEven and roundTowardPositive rounding modes, this causes a rounding increment, which is done by adding a one to the bit in the L position. When the infinitely precise product is incremented, the resulting internal product is

```
10.0000000000000000000000011111111111111111111110
```

which is greater than 2.0, as shown in the line Incremented product in Table 10.10. To normalize this significand to the range [1.0, 2.0), it is shifted right one place and the exponent is incremented by one. Only the upper 24 bits are returned; the lower bits, which contribute to the rounding determination, are discarded. The exponent before the increment is 0xFE, the largest normal exponent. When incremented, the resulting exponent is 0xFF, which is too large to be represented in single-precision, and an infinity is returned. In the roundTowardNegative and roundTowardZero rounding modes the increment is not required, and the pre-rounded product is normalized and within the bounds of single-precision range.

It is also possible to round out of an underflow condition. Recall that the Cortex-M4 will signal underflow if the intermediate result is below the minimum normal range and imprecise, even if rounding would return the minimum normal value.

EXAMPLE 10.14

Multiply 0x3F000001 and 0x00FFFFFF in each rounding mode.

Solution

In this example, the two significands are exactly the same as in the previous example, resulting in the same infinitely precise product and rounding conditions. As we saw in the previous example, in the roundTiesToEven and roundToward-Positive rounding modes this causes a rounding increment. In this case the exponent before the rounding was 0x00, one less than is representable by a normal value. When the infinitely precise product is incremented, the resulting internal

TABLE 10.10
Internal Values for Example 10.13

			Operand Significand		
OpB Significand (Multiplicand)			1	111	11111111111111111110
OpA Significand (Multiplier)			1	000	00000000000000000001
		Infinitely-precise internal significand			
OpA[0] term			1	111	11111111111111111110
OpA[23] term	1.111	1111111111111111	1110		
		Infinitely-precise internal significand			
Infinitely precise product	1.111	1111111111111111	1111	111	11111111111111111110
Pre-rounded product	1.111	1111111111111111	111L	Gss	ssssssssssssssssssss
Incremented product	10.000	0000000000000000	0000	111	11111111111111111110
		Result significand			
Result product - RNE	1.000	0000000000000000	0000		
Result product - RP	1.000	0000000000000000	0000		
Result product - RM	1.111	1111111111111111	1111		
Result product - RZ	1.111	1111111111111111	1111		

product is again 2.0, and the normalization results in a new exponent of 0x01, the smallest exponent for normal numbers. The result for the roundTiesToEven and roundTowardPositive is the smallest normal number, 0x00800000, and underflow is signaled with IXC. However, for roundTowardNegative and roundTowardZero, no increment is required and the result is 0x007FFFFF, in the subnormal range, and UFC and IXC are signaled.

10.6 EXERCISES

1. For each rounding mode, show the resulting value given the sign, exponent, fraction, guard, and sticky bit in the cases below.

Rounding Mode	Sign	Exponent	Fraction	G bit	S bit
	0	011111111	11111111111111111111111	1	0
roundTiesToEven					
roundTowardPositive					
roundTowardNegative					
roundTowardZero					
	1	000000000	11111111111111111111111	0	1
roundTiesToEven					
roundTowardPositive					
roundTowardNegative					
roundTowardZero					
	0	11111110	11111111111111111111111	1	1
roundTiesToEven					
roundTowardPositive					
roundTowardNegative					
roundTowardZero					
	1	11111110	01111111111111111111110	1	0
roundTiesToEven					
roundTowardPositive					
roundTowardNegative					
roundTowardZero					

2. Rework Example 10.1 with the following values for OpB. For each, generate the product rounded for each of the four rounding modes:
   a. 0xc4900000
   b. 0xc4800000
   c. 0xc4b00000
   d. 0xc4c00000
   e. 0x34900000
   f. 0x34800000
   g. 0x34b00000
   h. 0x34c00000

3. Complete the following table for the given operands and operations, showing the result and the exception status bits resulting from each operation. Assume the multiply-accumulate operations are fused.

Operation	Operand A	Operand B	Operand C	Result	Exception Bit(s) set
A + B	0xFF800000	0x7F800000	–		
A * B	0x80000000	0x7F800000	–		
A – B	0xFF800000	0xFF800001	–		
A/B	0x7FC00011	0x00000000	–		
A/B	0xFF800000	0xFF800000	–		
A * B	0x10500000	0x02000000	–		
A * B	0x01800000	0x3E7FFFFF	–		
A * B	0x3E7F0000	0x02000000	–		
(A * B) + C	0x80000000	0x00800000	0x7FB60004		
(A * B) + C	0x3F800000	0x7F800000	0xFF800000		
(A * B) + C	0x6943FFFF	0x71000000	0xFF800000		

4. Write a program in a high-level language to input two single-precision values and add, subtract, multiply and divide the values. In this Exercise use the default rounding mode. Test your program with various floating-point values.

5. Using routines available in your high-level language, perform each of the computations in each of the four rounding modes. In C, you can use the *floating-point environment* by including <fenv.h> and changing the rounding mode by the function fsetround(RMODE), where RMODE is one of

   FE_DOWNWARD
   FE_TONEAREST
   FE_TOWARDZERO
   FE_UPWARD

   Test your program with input values, modifying the lower bits in the operands to see how the result differs for the four rounding modes. For example, use `0x3f800001` and `0xbfc00000`. What differences did you notice in the results for the four rounding modes?

6. Give 3 values that will hold to the distributive law and 3 which will not.

7. Is it is possible to have cancelation in an effective subtraction operation and have a guard and sticky bit? If so, show an example. If not, explain why.

8. Demonstrate that the distributive law can fail to hold for floating-point values.

9. Redo Example 10.4 using 0x3F800003 as the multiplier.

10. Show that 36! generates an overflow condition.

11. Demonstrate a case in which a multiply operation results in a normal value for the RZ and RN rounding modes, but overflows for the RNE and RP rounding modes.

11 Floating-Point Data-Processing Instructions

11.1 INTRODUCTION

Floating-point operations are not unlike their integer counterparts. The basic arithmetic operations are supported, such as add and subtract, multiply and multiply–accumulate, and divide. Three are unique, however, and they are negate, absolute value, and square root. You will also notice that the logic operations are missing. There are no floating-point Boolean instructions, and no bit manipulation instructions. For these operations, should you need them, integer instructions may be used once the floating-point operand is moved into an ARM register with a VMOV instruction.

Floating-point performance is measured in *flops*, or floating-point operations per second. Only arithmetic operations are included in the flops calculation, and this measurement has been a fundamental component in comparing floating-point units for decades. Even though the flops measurement is concerned only with arithmetic operations, real floating-point performance is a combination of data transfer capability and data processing. It's important to do the arithmetic fast, but if the data cannot be loaded and stored as fast as the arithmetic, the performance suffers. We have already considered the varied options for moving data between memory and the FPU, and in Chapter 13 another means, the load and store multiple instructions, will be introduced. In this chapter we look at the arithmetic, and non-arithmetic, instructions available in the Cortex-M4 for floating-point data. This chapter begins with a discussion of the status bits, and then considers the basic instructions in the ARM v7-M floating-point extension instructions.

11.2 FLOATING-POINT DATA-PROCESSING INSTRUCTION SYNTAX

Floating-point data-processing instructions have a consistent syntax that makes it easy to use them without having to consult a reference manual. The syntax is shown below.

V<operation>{cond}.F32 {<dest>}, <src1>, <src2>

All floating-point data processing instructions in the Cortex-M4 operate on single-precision data and write a single-precision result, so the only data format is F32 (which can be abbreviated .F). The src1, src2, and dest registers can be any of the single-precision registers, s0 to s31, in the register file. There are no restrictions on the use of registers. Also, the src1, src2, and dest registers can be the same register, different registers, or any two can be the same register. For example, to square a value in register s9 and place the result in register s0, the following multiply instruction could be used:

```
VMUL.F32 s0, s9, s9
```

If the value in register s9 is no longer necessary, it could be overwritten by replacing register s0 as the destination register with register s9.

11.3 INSTRUCTION SUMMARY

Table 11.1 shows the floating-point data-processing instructions available in the Cortex-M4.

TABLE 11.1
Cortex-M4 Floating-Point Instruction Summary

Operation	Format	Operation
Absolute value	VABS{cond}.F32 <Sd>, <Sm>	Sd = |Sm|
Negate	VNEG{cond}.F32 <Sd>, <Sm>	Sd = –1 * Sn
Addition	VADD{cond}.F32 <Sd>, <Sn>, <Sm>	Sd = Sn + Sm
Subtract	VSUB{cond}.F32 <Sd>, <Sn>, <Sm>	Sd = Sn – Sm
Multiply	VMUL{cond}.F32 <Sd>, <Sn>, <Sm>	Sd = Sn * Sm
Negate Multiply	VNMUL{cond}.F32 <Sd>, <Sn>, <Sm>	Sd = –1 * (Sn * Sm)
Chained Multiply–accumulate	VMLA{cond}.F32 <Sd>, <Sn>, <Sm>	Sd = Sd + (Sn * Sm)
Chained Multiply–Subtract	VMLS{cond}.F32 <Sd>, <Sn>, <Sm>	Sd = Sd + (–1 * (Sn * Sm))
Chained Negate Multiply–accumulate	VNMLA{cond}.F32 <Sd>, <Sn>, <Sm>	Sd = (–1 * Sd) + (–1 * (Sn * Sm))
Chained Negate Multiply–Subtract	VNMLS{cond}.F32 <Sd>, <Sn>, <Sm>	Sd = (–1 * Sd) + (Sn * Sm)
Fused Multiply–accumulate	VFMA{cond}.F32 <Sd>, <Sn>, <Sm>	Sd = Sd + (Sn * Sm)
Fused Multiply–Subtract	VFMS{cond}.F32 <Sd>, <Sn>, <Sm>	Sd = Sd + ((–1 * Sn) * Sm)
Fused Negate Multiply–accumulate	VFNMA{cond}.F32 <Sd>, <Sn>, <Sm>	Sd = (–1 * Sd) + (Sn * Sm)
Fused Negate Multiply–Subtract	VFNMS{cond}.F32 <Sd>, <Sn>, <Sm>	Sd = (–1 * Sd) + ((–1 * Sn) * Sm)
Comparison	VCMP{E}{cond}.F32 <Sd>, <Sm> VCMP{E}{cond}.F32 <Sd>, #0.0	Sets FPSCR flags based on comparison of Sd and Sm or Sd and 0.0
Division	VDIV{cond}.F32 <Sd>, <Sn>, <Sm>	Sd = Sn/Sm
Square root	VSQRT{cond} <Sd>, <Sm>	Sd = Sqrt(Sm)

11.4 FLAGS AND THEIR USE

As we saw in Chapter 2, the various Program Status Registers hold the flags and control fields for the integer instructions. Recall from Chapter 9 how the Floating-Point Status and Control Register, the FPSCR, performs the same function for the FPU. One difference between the integer handling of the flags and that of the FPU is in the operations that can set the flags. Only the two compare instructions, VCMP and VCMPE, can set the flags for the FPU. None of the arithmetic operations are capable of setting flags. In other words, there is no *S* variant for floating-point instructions as with integer instructions. As a result, you will see that the flags are much simpler in the FPU than their integer counterparts, however, the C and V flags are redefined to indicate one or both operands in the comparison is a NaN. The use of the V flag in integer operations to indicate a format overflow is not necessary in floating-point.

11.4.1 Comparison Instructions

The VCMP and VCMPE instructions perform a subtraction of the second operand from the first and record the flag information, but not the result. The two instructions differ in their handling of NaNs. The VCMPE instruction will set the Invalid Operation flag if either of the operands is a NaN, while the VCMP instruction does so only when one or more operands are sNaNs. The check for NaNs is done first, and if neither operand is a NaN, the comparison is made between the two operands. As we mentioned in Chapter 9, infinities are treated in an *affine* sense, that is,

$$-\text{infinity} < \textit{all finite numbers} < +\text{infinity}$$

which is what we would expect. If we compare a normal number and a positive infinity, we expect the comparison to show the infinity is greater than the normal number. Likewise, a comparison of a negative infinity with any value, other than a negative infinity or a NaN, will show the negative infinity is less than the other operand.

The VCMP and VCMPE instructions may be used to compare two values or compare one value with zero. The format of the instruction is

```
VCMP{E}{<cond>}.F32  <Sd>,  <Sm>
VCMP{E}{<cond>}.F32  <Sd>,  #0.0
```

The VCMP instruction will set the Invalid Operand status bit (IOC) if either operand is a sNaN. The VCMPE instruction sets the IOC if either operand is a NaN, whether the NaN is signaling or quiet. The flags are set according to Table 11.2.

11.4.2 The N Flag

The N flag is set only when the first operand is numerically smaller than the second operand. Since an overflow is recorded in the OFC status bit, there is no need for the N flag in detecting an overflow condition as in integer arithmetic.

TABLE 11.2
Floating-Point Status Flags

Comparison Result	N	Z	C	V
Less than	1	0	0	0
Equal	0	1	1	0
Greater than	0	0	1	0
Unordered	0	0	1	1

11.4.3 The Z Flag

The Z flag is set only when the first and second operands are not NaN and compare exactly. There is one exception to this rule, and that involves zeros. The positive zero and negative zero will compare equal. That is, when both operands are zero, the signs of the two zeros are ignored.

11.4.4 The C Flag

The C flag is set in two cases. The first is when the first operand is equal to or larger than the second operand, and the second is when either operand is NaN.

11.4.5 The V Flag

The V flag is set only when a comparison is *unordered*, that is, when a NaN is one or both of the comparison operands.

EXAMPLE 11.1

The comparisons in Table 11.3 show the operation of the Cortex-M4 compare instructions.

TABLE 11.3
Example Compare Operations and Status Flag Settings

Operands		Flags				
Sd	Sm	N	Z	C	V	Notes
0x3f800001	0x3f800000	0	0	1	0	Sd > Sm
0x3f800000	0x3f800000	0	1	1	0	Sd = = Sm
0x3f800000	0x3f800001	1	0	0	0	Sd < Sm
0xcfffffff	0x3f800000	1	0	0	0	Sd < Sm
0x7fc00000	0x3f800000	0	0	1	1	Sd is qNaN
0x40000000	0x7f800001	0	0	1	1	Sm is sNaN

11.4.6 Predicated Instructions, or the Use of the Flags

The flags in the FPU may be accessed by a read of the FPSCR and tested in an integer register. The most common use for these flags is to enable predicated operation, as was covered in Chapter 8. Recall that the flag bits used in the determination of whether the predicate is satisfied are the flag bits in the APSR. To use the FPU flags, a VMRS instruction must be executed to move the flags in the FPSCR to the APSR. The format of the VMRS is

VMRS{<cond>} <Rt>, FPSCR

The destination can be any ARM register, r0 to r14, but r13 and r14 are not reasonable choices. To replace the NZCV flag bits in the APSR the <Rt> field would contain "APSR_nzcv." This operation transfers the FPSCR flags to the APSR, and any predicated instruction will be executed or skipped based on the FPSCR flags until these flags are changed by any of the operations covered in Chapter 7. When using the flags, the predicates are the same as those for integer operations, as seen in Chapter 8 (see Table 8.1).

EXAMPLE 11.2

Transfer the flag bits in the FPSCR to the APSR.

Solution

The transfer is made with a VMRS instruction, with the destination APSR_nzcv:

```
VMRS.F32 APSR_nzcv, FPSCR
```

VMRS is what is known as a *serializing* instruction. It must wait until all other instructions have completed and the register file is updated to ensure any instruction that could alter the flag bits has completed. Other serializing instructions include the counterpart instruction, VMSR, which overwrites the FPSCR with the contents of an ARM register. This instruction is serializing to ensure changes to the FPSCR do not affect instructions that were issued before the VMSR but have not yet completed.

To modify the FPSCR, for example, to change the rounding mode, the new value must be read from memory or the new rounding mode inserted into the current FPSCR value. To change the current FPSCR value, first move it into an ARM register, modify the ARM register, and then use the VMSR instruction to move the new value back to the FPSCR. The modification is done using the integer Boolean operations. The format for the VMSR instruction is

VMSR{<cond>} FPSCR, <Rt>

EXAMPLE 11.3

Set the rounding mode to roundTowardZero.

Solution

The rounding mode bits are FPSCR[22:23], and the patterns for the rounding mode selection was shown in Chapter 9. To set the rounding mode to roundTowardZero, the bits [22:23] must be set to 0b11. Modifying the FPSCR is done using integer bit manipulation instructions, but the FPSCR must first be copied to an ARM register by the VMRS instruction. The bits can be ORed in using an ORR immediate instruction, and the new FPSCR written to the FPU with the VMSR instruction. The code sequence is below.

```
VMRS r2, FPSCR              ; copy the FPSCR to r2
ORR  r2, r2, #0x00c00000    ; force bits [22:23] to 0b11
VMSR FPSCR, r2              ; copy new FPSCR to FPU
```

After running this code, Figure 11.1 shows the register window in the Keil tools with the change in the FPSCR.

To set the rounding mode back to RN, the following code can be used:

```
VMRS r2, FPSCR              ; copy the FPSCR to r2
BIC  r2, r2, #0x00c00000    ; clear bits [22:23]
VMSR FPSCR, r2              ; copy new FPSCR to FPU
```

EXAMPLE 11.4

Find the largest value in four FPU registers.

Solution

Assume registers s4, s5, s6, and s7 contain four single-precision values. The VCMP.F32 instruction performs the compares and sets the flags in the FPSCR. These flags are moved to the APSR with the VMRS instruction targeting

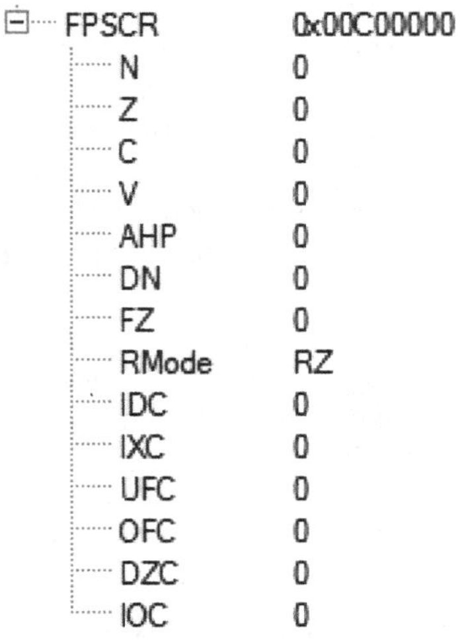

FPSCR	0x00C00000
N	0
Z	0
C	0
V	0
AHP	0
DN	0
FZ	0
RMode	RZ
IDC	0
IXC	0
UFC	0
OFC	0
DZC	0
IOC	0

FIGURE 11.1 FPSCR contents after the rounding mode change.

APSR_nzcv as the destination; the remaining bits in the APSR are unchanged. This allows for predicated operations to be performed based on the latest floating-point comparison.

```
    ; Find the largest value in four FPU registers
    ; s4-s7. Use register s8 as the largest value register
    ; First, compare register s4 to s5, and copy the largest to s8.
    ; Then compare s6 to s8, and copy s6 to s8 if it is
    ; larger. Finally, compare s7 to s8, copying s7 to s8 if
    ; it is the larger.

    ; Set up the contents of registers s4-s7 using VLDR
    ; pseudo-instruction
    VLDR.F32 s4, =45.78e5
    VLDR.F32 s5, =-0.034
    VLDR.F32 s6, =1.25e8
    VLDR.F32 s7, =-3.5e10

    ; The comparisons use the VCMP instruction, and the status
    ; bits copied to the APSR. Predicated operations perform
    ; the copies

    ; First, compare s4 and s5, and copy the largest
    ; to s8. The GT is true if the compare is signed >,
    ; and the LE is true if the compare is signed<=.
    VCMP.F32     s4, s5              ; compare s4 and s5
    VMRS         APSR_nzcv, FPSCR    ; copy only the flags to APSR
    VMOVGT.F32   s8, s4              ; copy s4 to s8 if larger than s5
    VMOVLE.F32   s8, s5              ; copy s5 if larger or equal to s4

    ; Next, compare s6 with the new largest. This time only
    ; move s6 if s6 is greater than s8.
    VCMP.F32     s6, s8              ; compare s6 and the new larger
    VMRS         APSR_nzcv, FPSCR    ; copy only the flags to APSR
    VMOVGT.F32   s8, s6              ; copy s6 to s8 if new largest

    ; Finally, compare s7 with the largest. As above, only
    ; move s7 if it is greater than s8.
    VCMP.F32     s7, s8              ; compare s6 and the new larger
    VMRS         APSR_nzcv, FPSCR    ; copy only the flags to APSR
    VMOVGT.F32   s8, s7              ; copy s7 to s8 if new largest

    ; The largest of the 4 registers is now in register s8.

Exit B Exit
```

11.4.7 A WORD ABOUT THE IT INSTRUCTION

The IT instruction was introduced in Chapter 8. Recall that ARM instructions are predicated, with the AL (Always) predicate the default case, used when an instruction is to be executed regardless of the status bits in the APSR. When execution is to be determined by the status bits, as in the example above, a field mnemonic is appended to the instruction, as in VMOVGT seen above. This is true in the ARM instruction set, but not in the Thumb-2 instruction set—this functionality is available through the IT instruction. In the disassembly file, the Keil assembler inserted an IT instruction before the VMOVGT and the VMOVLE instructions as shown below.

```
0x00000034 BFCC      ITE        GT
    63:       VMOVGT.F32 s8, s4    ; copy s4 to s8 if larger than s5
0x00000036 EEB04A42  VMOVGT.F32  s8,s4
    64:       VMOVLE.F32 s8, s5    ; copy s5 if larger or equal to s4
0x0000003A EEB04A62  VMOVLE.F32  s8,s5
```

Since the GT and LE conditions are opposites, that is, the pair covers all conditions, only a single IT block is needed.

The Keil tools allow for the programmer to write the assembly code as if the instructions are individually predicated, as in the example above. The assembler determines when an IT block is needed, and how many predicated instructions may be part of the IT block. Each IT block can predicate from one to four instructions. It is a very powerful tool and should be used when the result of a compare operation is used to select only a small number of operations.

11.5 TWO SPECIAL MODES

Early in the development of ARM FPUs, two modes were introduced which simplified the design of the FPU, enabling a faster and smaller design, but which were not fully IEEE 754-1985 compatible. These two modes are Flush-to-Zero and Default NaN. Both are enabled or disabled by bits in the FPSCR—the Flush-to-Zero mode is enabled by setting the FZ bit, bit [24], and the Default NaN mode is enabled by setting the DN bit, bit [25].

11.5.1 Flush-to-Zero Mode

When the Cortex-M4 is in Flush-to-Zero mode, all subnormal operands are treated as zeros with the sign bit retained, and any result in the subnormal range *before rounding* is returned as zero with the sign of the computed subnormal result. When an input subnormal operand is treated as a zero, the Input Denormal exception bit (IDC, bit [7] of the FPSCR) is set, but the Inexact status bit is not set. However, when a subnormal result is detected, the Underflow exception (UFC, bit [3] of the FPSCR) is set, but the Inexact status bit is not set.

Note that Flush-to-Zero mode is not compatible with the IEEE 754-2008 specification, which states that subnormal values must be computed faithfully. When would you consider using Flush-to-Zero mode? In early ARM FPUs, a subnormal input or a result in the subnormal range would cause a trap to library code to compute the operation, resulting in potentially thousands of cycles to process the operation faithfully. Unlike these older FPUs, the Cortex-M4 computes all operations with the same number of clock cycles, even when subnormal operands are involved or the result is in the subnormal range. It is unlikely you will ever need to enable the Flush-to-Zero mode.

11.5.2 Default NaN

When in Default NaN mode, the Cortex-M4 treats all NaNs as if they were the default NaN. Recall that the IEEE 754-2008 specification suggests that a NaN operand to an operation should be returned unchanged, that is, the payload, as we discussed in Chapter 9, should be returned as the result. In Default NaN mode this is not the case. Any input NaN results in the default NaN, shown in Table 9.1, regardless of the payload of any of the input NaNs. As with the Flush-to-Zero mode above, it was the case in the earlier FPUs that NaNs would cause a trap to library code to preserve the payload according to the recommended IEEE 754-2008 behavior. However, the Cortex-M4 handles NaNs according to the recommendations of the standard without library code, so it is unlikely you will ever need to enable the Default NaN mode.

11.6 NON-ARITHMETIC INSTRUCTIONS

Two instructions are referred to as "non-arithmetic" even though they perform arithmetic operations. They are Absolute Value (VABS) and Negate (VNEG). They differ from the other data-processing instructions in that they do not signal an Invalid Operation if the operand is a signaling NaN.

11.6.1 Absolute Value

As you recall, floating-point values are stored in *sign-magnitude* form, that is, a separate sign bit indicates the sign of the unsigned magnitude. So to make a floating-point value positive simply requires setting the sign bit to zero. This is true for normal and subnormal values, zeros, infinities and NaNs. Contrast this with changing the sign of a two's complement number and you will see how easy it is in floating-point. The format of the VABS instruction is shown below. While it is a two-operand instruction, it is not uncommon to overwrite the source if only the absolute value of the operand will be used.

VABS{cond}.F32 <Sd>, <Sm>

11.6.2 Negate

The VNEG operation simply flips the sign bit of the source operand and writes the modified result to the destination register. This is true of zero, infinity, and NaN results, as well as all normal and subnormal results. The format of the VNEG instruction is shown below. Like VABS above, it is a two-operand instruction, and it is not uncommon to overwrite the source if only the negative of the operand will be used.

VNEG{cond}.F32 <Sd>, <Sm>

EXAMPLE 11.5

The result of executing VABS and VNEG on the following values is shown in Table 11.4.

TABLE 11.4
Examples of VABS and VNEG

Input Value	VABS	VNEG	Note
0X5FEC43D1	0X5FEC43D1	0XDFEC43D1	Normal value, sign bit modified
0x80000000	0x00000000	0x00000000	Negative zero becomes positive zero for both operations
0x00000000	0x00000000	0x80000000	Positive zero becomes negative after VNEG
0xFF800055	0x7F800055	0x7F800055	Signaling NaN, only the sign bit is changed, and IOC is not set
0x800000FF	0x000000FF	0x000000FF	Subnormal sign changed, and UFC is not set

11.7 ARITHMETIC INSTRUCTIONS

Most of the data-processing instructions are arithmetic operations, that is, they will set exception status bits for signaling NaNs. At this point we'll spend some time looking at the operations in some detail.

11.7.1 Addition/Subtraction

The addition and subtraction instructions have the following format:

VADD{cond}.F32 <Sd>, <Sn>, <Sm>
VSUB{cond}.F32 <Sd>, <Sn>, <Sm>

Recall in Chapter 10 how addition and subtraction can result in unexpected values due to rounding and cancelation, and when operands are infinities and NaNs. For normal and subnormal values, the instructions are straightforward in their use, with any register available as a source or destination register. For example, to double a value in a register, the instruction

```
VADD.F32 s5, s5, s5
```

would do so for normal and subnormal values, and avoid having to store a factor of 2.0 in a register.

It is possible with the VADD and VSUB instructions to incur all of the exceptions except divide-by-zero. It is not difficult to see how these operations could overflow and underflow, and it is common to return an inexact result. A signaling NaN can cause an Invalid Operation exception, as can some operations with infinities. Table 11.5 shows how VADD and VSUB behave with the five classes of encodings. In this table, Operand A and Operand B may be either input operand. The notes are directed to the VADD instruction, but also apply to the VSUB when the signs of the operands are different.

TABLE 11.5
Floating-Point Addition and Subtraction Instructions Operand and Exception Table

Operand A	Operand B	Result	Possible Exceptions	Notes
Normal	Normal	Normal, Subnormal, Infinity	OFC, UFC, IXC	Normal + Normal can overflow, resulting in an infinity or max normal, and if opposite signs, can result in a subnormal value
Normal	Subnormal	Normal, Subnormal	OFC, UFC, IXC	If opposite signs a subnormal result is possible, otherwise, a normal would result
Normal	Infinity	Infinity	None	
Normal	Zero	Normal	None	
Subnormal	Subnormal	Normal, Subnormal, Zero	UFC, IXC	
Subnormal	Infinity	Infinity	None	
Subnormal	Zero	Subnormal	None	
Infinity	Infinity	Infinity	None	
Infinity	Zero	Infinity	None	
NaN	Anything	NaN	IOC	If a signaling NaN, IOC is set

EXAMPLE 11.6

Select three pairs of operands that when added using a VADD instruction result in a

1. Normal value
2. Subnormal value
3. Infinity

SOLUTION

1. The two operands could be 0x3F800000 (+1.0). The sum would be 0x40000000 (2.0).
2. To return a subnormal with two normal input operands, cancelation would have to take place. If the two input operands are 0x00800001 and 0x80800000, the result would be 0x00000001, the minimum subnormal value (1.401×10^{-45}).
3. To return an infinity, the rounding mode would have to be either RNE or RP, for two positive values, or RNE or RM, for two negative values. As an example, if the rounding mode was RNE, the operands 0x7F7FFFFF and 0x7F700001 would overflow and return an infinity. Likewise, the operands 0xFF7FFFFF and 0xFF700001, would return a negative infinity.

EXAMPLE 11.7

Select three pairs of operands that when subtracted using a VSUB instruction results in

1. Normal value
2. Subnormal value
3. Infinity

Solution

1. The two operands could be 0x3F800000 (+1.0) and 0xBF800000 (–1.0). The sum would be 0x40000000 (2.0).
2. To return a subnormal with 2 normal operands, cancelation would have to take place. If the two input operands are 0x00800001 and 0x00800000, the result would be 0x00000001, the minimum subnormal value (1.401×10^{-45}).
3. To return an infinity, if the rounding mode was RNE, the operands 0xFF7FFFFF and 0x7F700001 would overflow and return a negative infinity. Likewise, the operands 0x7F7FFFFF and 0xFF700001 would return a positive infinity.

11.7.2 Multiplication and Multiply–Accumulate

The Cortex-M4 has a rich variety of multiplication and multiply–accumulate operations, but some can be a bit confusing. Two varieties of multiply–accumulate instructions, chained and fused, are available with options to negate the addend and the product. In early ARM FPUs, only the chained operations were available. For example, the chained VNMLA instruction produces a result equivalent to a sequence of multiply and add operations. If the instruction is

```
VNMLA.F32     s1, s2, s3
```

an equivalent sequence of instructions producing the same result would be

```
VMUL.F32      s2, s2, s3
VNEG.F32      s2, s2
VADD.F32      s1, s1, s2
```

The advantage is in the single instruction and in compliance to the IEEE 754-1985 standard. Before the introduction of the IEEE 754-2008 standard, no multiply–accumulate could be compliant without rounding the product before the addition of the addend. The chained operations in the Cortex-M4 are a legacy of earlier ARM FPUs that were IEEE 754-1985 standard compliant by performing this rounding step on the product before adding in the addend. With the introduction of the IEEE 754-2008 standard a new set of instructions, referred to as fused multiply–accumulate instructions, were made part of the standard. These instructions compute the product equivalently to the infinitely precise product, and this value, unrounded, is added to the addend. The final sum is rounded to the destination precision. The fused operations are more accurate because they avoid the rounding of the intermediate product, and they are preferred to the chained operations. In some cases of legacy floating-point code, the chained operations may be used to exactly reproduce earlier IEEE 754-1985 standard results, while the fused operations may give different results.

We will first consider the multiply instructions, which include VMUL and VNMUL. Next we will consider the chained multiply–accumulate operations, and finally the fused multiply–accumulate operations.

11.7.2.1 Multiplication and Negate Multiplication

Two multiply instructions are available in the Cortex-M4. VMUL multiplies two operands, writing the result in a destination register. VNMUL first negates the second of the two operands before the multiplication. The formats of the two instructions are shown below.

VMUL{cond}.F32 <Sd>, <Sn>, <Sm>
VNMUL{cond}.F32 <Sd>, <Sn>, <Sm>

Any of the floating-point registers can be a source and destination register. As an example, the following instruction

```
VMUL.F32 s12, s12, s12
```

would square a normal value and leave the square in the register. Similarly, if the algorithm called for the negative of the square of a normal value, the instruction

```
VNMUL.F32 s12, s12, s12
```

would perform the operation with a single instruction.

The VMUL and VNMUL instructions can generate all of the exceptions apart from the divide-by-zero exception. Overflow and underflow are common, as multiplication can create values both too large and too small to fit in a finite data type. Inexact is also a common consequence of floating-point multiplication. Recall that the Inexact status bit is set whenever an overflow is detected.

Table 11.6 shows how VMUL and VNMUL instructions behave with the five classes of encodings. In this table, Operand A and Operand B may be either input operand.

11.7.2.2 Chained Multiply–Accumulate

There are four chained multiply–accumulate operations, providing options to subtract rather than add, and an option to negate the product. The formats of the instructions are shown below.

VMLA{cond}.F32 <Sd>, <Sn>, <Sm>
VMLS{cond}.F32 <Sd>, <Sn>, <Sm>
VNMLA{cond}.F32 <Sd>, <Sn>, <Sm>
VNMLS{cond}.F32 <Sd>, <Sn>, <Sm>

Each of these instructions can be represented by an equivalent set of operations, as shown in Table 11.7. The possible exceptions and default results are the same as those for the component operations discussed above. For example, in a VMLA operation in the RNE rounding mode, if the Sn and Sm operands are two very large normal values, which, when multiplied will overflow to a positive infinity, and the Sd operand is a normal value, the result of the VMLA will be the same as an addition of a normal value and a positive infinity, which is a positive infinity. In understanding

TABLE 11.6
Floating-Point Multiply Instructions Operand and Exception Table

Operand A	Operand B	Result	Possible Exceptions	Notes
Normal	Normal	Normal, Subnormal, Infinity, Zero	OFC, UFC, and IXC	Normal * Normal can overflow, resulting in an infinity or max normal, or result in a subnormal value or zero
Normal	Subnormal	Normal, Subnormal, Zero	UFC, IXC	The result may be in the normal range, subnormal range, or underflow to a zero
Normal	Zero	Zero	None	As expected in zero arithmetic
Normal	Infinity	Infinity	None	As expected in infinity arithmetic
Subnormal	Subnormal	Zero	UFC, IXC	This case will always result in underflow, with a zero result and UFC and IXC status bits set
Subnormal	Zero	Zero	None	As expected in zero arithmetic
Subnormal	Infinity	Infinity	None	As expected in infinity arithmetic
Zero	Zero	Zero	None	As expected in zero arithmetic
Infinity	Zero	NaN	IOC	Invalid operation
Infinity	Infinity	Infinity	None	As expected in infinity arithmetic
NaN	Anything	NaN	IOC	If a signaling NaN, IOC is set, and SNaN input is quieted

the behavior of chained operations, the individual component operations are considered in order separately of the others, and the final result of the chained instructions is the result of the last of the component operations (Table 11.7).

EXAMPLE 11.8

Execute each of VMLA, VMLS, VNMLA, and VNMLS instructions with the following operands:

Sn = −435.792
Sm = 10.0
Sd = 5832.553

SOLUTION

Using a tool such as the conversion tools at (http://babbage.cs.qc.cuny.edu/IEEE-754.old/Decimal.html), the single-precision values for Sn, Sm, and Sd are:

Sn: 0xC3D9E560
Sm: 0x41200000
Sd: 0x45B6446D

TABLE 11.7
Chained Multiply-Accumulate Operations

Instruction	Operation	Equivalent Operations
VMLA	Chained Multiply–accumulate Sd = Sd + (Sn * Sm)	Temp = Round(Sn * Sm) Sd = Round(Sd + Temp)
VMLS	Chained Multiply Subtract Sd + (–1 * ((Sn * Sm))	Temp = Round(Sn * Sm) Temp = Negate(Temp) Sd = Round(Sd + Temp)
VNMLA	Chained Negate Multiply–accumulate Sd = (–1 * Sd) + (–1 * (Sn * Sm))	Temp = Round(Sn * Sm) Temp = Negate(Temp) Temp2 = Negate(Sd) Sd = Round(Temp2 + Temp)
VNMLS	Chained Negate Multiply Subtract Sd = (–1 * Sd) + (Sn * Sm)	Temp = Round(Sn * Sm) Temp2 = Negate(Sd) Sd = Round(Temp2 + Temp)

The following code implements the solution.

```
    ADR r1, MulAddTestData

    VLDR.F32   s0, [r1]      ; Sn
    VLDR.F32   s1, [r1, #4]  ; Sm
    VLDR.F32   s2, [r1, #8]  ; Sd

    ; VMLA
    VMLA.F32   s2, s0, s1

    ; VMLS
    VLDR.F32   s2, [r1, #8]  ; Reload Sd
    VMLS.F32   s2, s0, s1

    ; VNMLA
    VLDR.F32   s2, [r1, #8]  ; Reload Sd
    VNMLA.F32  s2, s0, s1

    ; VNMLS
    VLDR.F32   s2, [r1, #8]  ; Reload Sd
    VNMLS.F32  s2, s0, s1

    B Exit

    ALIGN
MulAddTestData
    DCD 0xC3D9E560 ; -435.792
    DCD 0x41200000 ; 10.0
    DCD 0x45B6446D ; 5832.553
```

When this code is run, the destination register s2 contains the following for each of the four operations:

```
VMLA:  s2 = 0x44B85444, which is 1474.633.
VMLS:  s2 = 0x461F39E4, which is 10,190.473
VNMLA:  s2 = 0xC4B85444, which is -1474.633
VNMLS:  s2 = 0xC61F39E4, which is -10,190.473
```

Confirm for yourself that the answers are correct for each of the four operations.

11.7.2.3 Fused Multiply–Accumulate

The Cortex-M4 implements a second set of multiply–accumulate operations that are referred to as *fused*. Unlike the chained operations discussed above, the fused operations do not round the product, but maintain the product in an infinitely precise, unrounded form. The addend is then added to the product. By eliminating the rounding of the product, the result may have greater accuracy. In most cases this will never be an issue, but in algorithms such as those used to compute transcendental functions, the accuracy of the fused operations enables writing library functions with lower error bounds compared to discrete or chained instructions.

The fused multiply–accumulate instructions have the formats shown below.

```
VFMA{cond}.F32   <Sd>, <Sn>, <Sm>
VFMS{cond}.F32   <Sd>, <Sn>, <Sm>
VFNMA{cond}.F32  <Sd>, <Sn>, <Sm>
VFNMS{cond}.F32  <Sd>, <Sn>, <Sm>
```

It is useful to consider these instructions, as we did the chained instructions above, as implementing a series of operations. The first thing to notice is the function of the two negate instructions differs from the chained operations. The chained instruction VNMLA instruction is analogous to the fused VFNMS instruction, while the chained VNMLS instruction is analogous to the fused VFNMA instruction. Table 11.8 shows the instructions, the operations, and the equivalent operations.

EXAMPLE 11.9

Evaluate the accuracy of three random operands in the range (0,1.0) using both the VMLA and the VFMA instructions.

SOLUTION

Consider these three operands (each is computed to full precision):

Sn: 0x3F34FE23 (0.707002818584442138671875)
Sm: 0x3E78EE2A (0.2430960237979888916015625)
Sd: 0x3F7F3DCA (0.9970365762710571289 0625)

The computed result of Sd + (Sn * Sm) (to 24 digits, the precision of our input operands) is 1.16890615028290589805237. The results are shown in Table 11.9.

TABLE 11.8
Fused Multiply–Accumulate Instructions Equivalent Operations

Instruction	Operation	Equivalent Operations
VFMA	Fused Multiply–accumulate Sd = Sd + (Sn * Sm)	Temp = (Sn * Sm) Sd = Round(Sd + Temp)
VFMS	Fused Multiply Subtract Sd = Sd + ((–1 * Sn) * Sm)	Temp = Negate(Sn) Temp = (Temp * Sm) Sd = Round(Sd + Temp)
VFNMA	Fused Negate Multiply–accumulate Sd = (–1 * Sd) + (Sn * Sm)	Temp = (Sn * Sm) Temp2 = Negate(Sd) Sd = Round(Temp2 + Temp)
VFNMS	Fused Negate Multiply Subtract Sd = (–1 * Sd) + ((–1 * Sn) * Sm)	Temp = Negate(Sn) Temp = (Temp * Sm) Temp2 = Negate(Sd) Sd = Round(Temp2 + Temp)

As you can see, the difference is one ulp in the final result, or 2^{-23}. Not very much. In fact, the VMLA results differ from the full precision result by just over $5 \times 10^{-6}\%$, while the fused is only very slightly more accurate! Overall, the computations show very little difference between the two instructions; the computation for both instructions shows very high accuracy, and the use of fused over chained, in this example, has very little impact.

But this doesn't tell the whole story. Some pathological cases exist which can return very different results. Take the following inputs as an example:

s0 = 0x3F800001 ($1.0 + 2^{-23}$, or 1.0 + 1 ulp)
s1 = 0x3F800001
s2 = 0xBF800002 ($-(1.0 + 2^{-22})$, or –(1.0 + 2 ulps))

When input to the chained VMLA instruction the result will be zero. The square of 0x3F800001 will result in a fraction with 1's in the 2^0, 2^{-22}, and 2^{-46} bit positions internal to the hardware. When rounded in RNE, the 2^{-46} contribution is dropped, leaving only $2^0 + 2^{-22}$, the same fraction as the 0xBF800002 operand. The VFMA does not round, so the 2^{-46} contribution is retained, and the result is 0x2880000, or 2^{-46}. This case shows a greater error when the inputs are changed just a bit. Consider these operands input to both operations:

TABLE 11.9
Results of Example 11.9

Instruction	Cortex-M4 Result (Hex)	Cortex-M4 Result (Decimal)	Difference from the Computed Result (%)
VMLA	0x3F959EB8	1.16890621185302734375	0.00000527
VFMA	0x3F959EB7	1.16890609264373779296875	0.00000493

s0 = 0x3FC00001 (1.5 + 2^{-23}, or 1.5 + 1 ulp)
s1 = 0x3FC00001
s2 = 0xC0100002 (−(2.25 + 2^{-22}), or −(2.25 + 2 ulps))

The result of the VMLA will again be zero, but the VFMA returns 2^{-23} as the result. The answer to why this is so is left as an exercise.

One more situation is worth noting. When using the fused multiply–accumulate instructions, you don't have to worry about exceptions generated by the multiply operation, because these will be reflected in the final result *if they impact the final result.* For example, consider the following inputs to the VMLA instruction:

s0 = 0x7F000001 (just greater than ½ max normal)
s1 = 0x40000000 (2.0)
s2 = 0xFF7FFFFF (negative, and just under max normal)

If we execute the VMLA instruction:

```
VMLA.F32 s2, s0, s1
```

the result is a positive infinity, and the OFC and IXC status bits are set. Why? The product of 0x7F000001 and 0x40000000 overflowed, and an infinity was substituted for the product, and input to the final addition. The infinity plus the very large negative value resulted in an infinity.

If the same inputs are made to the VFMA instruction:

```
VFMA.F32 s2, s0, s1
```

the result is 0x74400000, and no exception status bits are set. Since the intermediate product is not evaluated for overflow but rather input with the extended range of the intermediate, infinitely precise value, it is not replaced with an infinity. The addition of the negative addend brings the sum back into the normal range.

11.7.3 Division and Square Root

Both division and square root instructions are available in the Cortex-M4. VDIV divides the first source operand by the second source operand, writing the result in a destination register. VSQRT performs the square root operation. The formats of the two instructions are shown below.

```
VDIV{cond}.F32   <Sd>, <Sn>, <Sm>
VSQRT{cond}.F32  <Sd>, <Sm>
```

Any of the floating-point registers can be a source and destination register. As an example, the following instruction

```
VDIV.F32 s21, s8, s1
```

would divide the value in register s8 by the value in register s1, and put the rounded quotient in register s21.

Division will result in a divide-by-zero exception, setting the DZC status bit if the divisor is a zero and the dividend is normal or subnormal. Overflow and underflow are possible when the result of the division would result in a value too large or too small, respectively, for representation in the single-precision format. Any division with normal or subnormal values can produce an inexact result and set the IXC status bit.

Table 11.10 shows how the VDIV instruction functions with the five classes of encodings. In this table, Operand A is the dividend and Operand B the divisor. If the result is not exact, the Inexact status bit, IXC, will be set. Recall that the Underflow status bit, UFC, is set when a subnormal result is generated and the result is either not exact or a zero due to a result smaller in magnitude than can be represented in the destination precision. If the computed result is too small, the UFC and IXC status bits will be set and a signed zero result returned.

TABLE 11.10
Floating-Point Divide Instruction Operand and Exception Table

Operand A	Operand B	Result	Possible Exceptions	Notes
Normal	Normal	Normal, Subnormal, Infinity, Zero	OFC, UFC, and IXC	Normal/Normal can overflow, resulting in an infinity or max normal, or result in a subnormal value or zero
Normal	Subnormal	Normal, Infinity	OFC, IXC	The result may be in the normal range, subnormal range, or underflow to a zero.
Normal	Zero	Infinity	DZC	Divide by zero
Normal	Infinity	Zero	None	As expected for an infinity divisor
Subnormal	Normal	Normal, Subnormal, Zero	UFC, IXC	Subnormal/Normal may be normal or subnormal, or zero. UFC and IXC are set if subnormal and inexact, or if zero.
Subnormal	Subnormal	Normal	IXC	If exact, IXC is not set
Subnormal	Infinity	Zero	None	As expected for an infinity divisor
Subnormal	Zero	Infinity	DZC	Divide by zero
Zero	Normal	Zero	None	As expected for a zero dividend
Zero	Subnormal	Zero	None	
Zero	Infinity	Zero	None	
Zero	Zero	NaN	IOC	Invalid operation
Infinity	Normal	Infinity	None	As expected for an infinity dividend
Infinity	Subnormal	Infinity	None	
Infinity	Infinity	NaN	IOC	Invalid operation
Infinity	Zero	Infinity	None	Odd, perhaps, but the infinity governs the result, which is infinity.
Anything	NaN	NaN	IOC	If a signaling NaN, IOC is set and NaN is quieted.

TABLE 11.11
Floating-Point Square Root Instruction Operand and Exception Table

Operand A	Result	Possible Exceptions	Notes
+Normal	+Normal	IXC	
–Normal	Default NaN	IOC	Any input below zero results in an invalid operation
+Subnormal	+Normal	IXC	
–Subnormal	Default NaN	IOC	Any input below zero results in an invalid operation
+Infinity	+Infinity	None	
–Infinity	Default NaN	IOC	
+Zero	+Zero	None	
–Zero	–Zero	None	
NaN	NaN	IOC	If a signaling NaN, IOC is set. NaN should be input NaN

Only the Invalid Operation and Inexact exceptions are possible with square root. Any positive normal or subnormal operand may produce an inexact result and set the IXC status bit. When an operand is a negative normal or subnormal value, and not a negative NaN, the operation is invalid, the default NaN is returned and the IOC status bit is set.

Table 11.11 shows how VSQRT instruction functions with the five classes of encodings and signed values.

11.8 PUTTING IT ALL TOGETHER: A CODING EXAMPLE

In this section, we'll tie everything together by coding a routine for the bisection algorithm, which is a simple method of finding a root (a zero crossing) of a continuous function. The algorithm begins with two points on the function that have opposite signs and computes a third point halfway between the two points. The new third point replaces one of the original points that has the same sign as the function evaluated at that new third point, and the algorithm repeats. In Figure 11.2, the original points are labeled a and b, and we see that $f(a)$ and $f(b)$ have opposite signs. The computed third point, c, is the result of one iteration. Further iteration will result in a computed new point closer to the true crossing. The algorithm is ended when the computed point is exactly on the zero crossing ($f(c) = 0$) or the difference between the input point with the same sign as the computed point is below a threshold.

The algorithm is written in pseudo-code as shown below.*

```
INPUT: Function f, endpoint values a, b, tolerance TOL, maximum iterations NMAX
CONDITIONS: a<b, either f(a) < 0 and f(b) > 0 or f(a) > 0 and f(b) < 0
OUTPUT: value which differs from a root of f(x) = 0 by less than TOL

N← 1
While N≤NMAX {limit iterations to prevent infinite loop
c← (a+b)/2 new midpoint
```

* The code is taken from the Wikipedia entry on "Bisection method," taken from http://en.wikipedia.org/wiki/Bisection_method#CITEREFBurdenFaires1985.

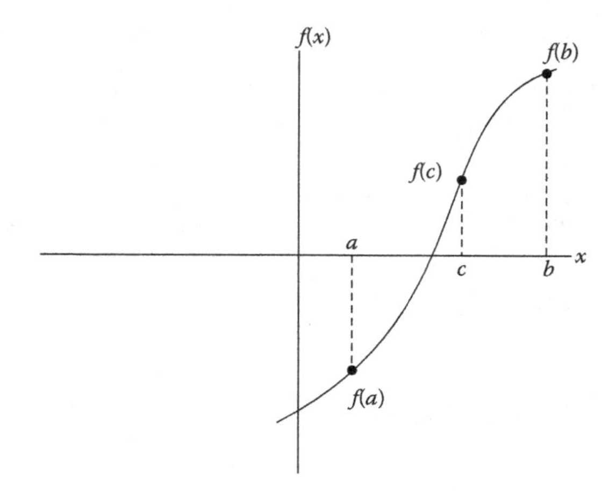

FIGURE 11.2 Bisection method for two initial points and a computed third point.

```
If (f(c) = 0 or (b - a)/2 < TOL then {solution found
   Output(c)
   Stop
}
N← N+1 increment step counter
If sign(f(c)) = sign(f(a)) then a← c else b← c new interval
}
Output("Method failed.") max number of steps exceeded
```

```
        ; Bisection code
        ; The algorithm requires the first two points,
        ; a, and b, to be one below and one above the
        ; root; that is, f(a) will have an opposite
        ; sign to f(b). The algorithm computes the midway
        ; point between a and b, called c, and computes f(c).
        ; This new point replaces the point with the same
        ; function result sign. If f(a) is positive and f(c)
        ; is positive, c replaces a, and the algorithm reruns
        ; with b and c as points. The algorithm exits when f(c)
        ; is zero, or (a-b) is less than a threshold value.

        ; FPU registers
        ; s6 - the threshold value, 0.0002
        ; s7 - 2.0, to divide the sum for the new operand
        ; s8 - operand a
        ; s9 - operand b
        ; s10 - the new point, operand c
        ; s11 - f(a)
        ; s12 - f(b)
        ; s13 - f(c)
        ; ARM registers
        ; r1 - sign of f(a)
        ; r2 - sign of f(b)
        ; r3 - sign of f(c)
        ; r4 - iteration count
```

```
        ; Choose 0 and 4 as initial samples

        ; Initialize the divisor register and threshold
        VMOV.F32 s7, #2.0
        VLDR.F32 s6, =0.0002

        ; Initialize the operand registers
        VSUB.F32 s8, s7, s7           ; a lazy way to create 0.0
        VMOV.F32 s9, #4.0

        ; Initialize our loop counter
        MOV     r4, #0

Loop
        ; Increment our loop counter
        ADD     r4, r4, #1

        ; Test a and b for (a-b)<threshold
        ; Use s11 for the difference, we will overwrite
        ; it in the eval of operand a
        VSUB.F32 s11, s8, s9          ; compute the difference
        VABS.F32 s11, s11             ; make sure diff is positive
        VCMP.F32 s11, s6              ; test diff>threshold?
        VMRS.F32 APSR_nzcv, FPSCR     ; copy status bits to APSR
        BLS      Exit                 ; if neg or eq, exit

        ; Evaluate the function for operand a
        VMOV.F32 s1, s8
        BL       func
        VMOV.F32 s11, s0

        ; Evaluate the function for operand b
        VMOV.F32 s1, s9
        BL       func
        VMOV.F32 s12, s0

        ; Compute the midpoint for operand c,
        ; the point halfway between operands
        ; a and b
        VADD.F32 s10, s8, s9
        VDIV.F32 s10, s10, s7

        ; Evaluate the function for operand c
        VMOV.F32 s1, s10
        BL       func
        VMOV.F32 s13, s0

        ; Test the signs of the three operands
        VCMP.F32 s11, #0              ; set status bits only on operand a
        VMRS.F32 r1, FPSCR
        AND      r1, r1, #0x80000000  ; isolate the N status bit
        VCMP.F32 s12, #0              ; set status bits only on operand b
        VMRS.F32 r2, FPSCR
        AND      r2, r2, #0x80000000  ; isolate the N status bit
        VCMP.F32 s13, #0              ; set status bits only on operand c
        VMRS.F32 r3, FPSCR
```

```
        TST        r3, #0x4000000          ; test for zero
        BEQ        Exit                    ; the value in s10 is exactly
                                           ; the root
        AND        r3, r3, #0x80000000     ; isolate the N status bit

        ; If sign(a) !=sign(c), copy s10 into s9;
        ; else sign(b) !=sign(c), copy s10 into s8;
        EORS       r1, r3                  ; test if sign(a)=sign(c)
        VMOVEQ.F32 s8, s10                 ; if 0, copy c to a
        BLEQ       Loop                    ; run it again with a new a
        VMOV.F32   s9, s10                 ; if not a, then copy c into b
        BL         Loop                    ; run it again with a new b
Exit
        B          Exit

        ; Test functions
        ; Assumes ATPCS - regs s0-s15 parameters and/or scratch
        ; Register usage:
        ;  s0 - return result
        ;  s1 - input operand
        ;  s2 - scratch
        ;  s3 - scratch
func
        ; Function - x^3+2x - 8
        VMOV.F32 s0, #2.0              ; use s0 to hold 2.0 temporarily
        VMUL.F32 s2, s1, s1            ; initial squaring of input
        VMUL.F32 s3, s1, s0            ; multiply input by 2
        VMOV.F32 s0, #8.0              ; use s0 to hold 8.0 temporarily
        VMUL.F32 s2, s2, s1            ; finish cubing of input
        VSUB.F32 s3, s3, s0            ; subtract off 8.0 from 2x
        VADD.F32 s0, s2, s3            ; add in x^3 to return reg
        BX       lr                    ; return
```

11.9 EXERCISES

1. Complete the table for a VCMP instruction and the following operands.

Operand A	Operand B	N	Z	C	V
0xFF800000	0x3F800000				
0x00000000	0x80000000				
0x7FC00005	0x00000000				
0x7F80000F	0x7F80000F				
0x40000000	0xBF000000				

2. Give the instructions to implement the following algorithm. Assume y is in register s0, and return x in register s0.

$$x = 8y^2 - 7y + 12$$

3. Expand the code above to create a subroutine that will resolve an order 2 polynomial with the constant for the square term in register s0, the order 1 term in register s1, and the constant factor in register s2. Return the result in register s0.

4. Give the instructions to perform the following loop over an one-dimensional array X of 20 data values in memory at address 0x40000100 and an one-dimensional array Y of data values in memory at address 0x40000200. Use a constant value of A between 2.0 and 10.0. The new y value should overwrite the original value.

$$y = Ax + y$$

5. Modify the program in Example 11.4 to order the four values in registers s8 to s11 in order from smallest to largest.

6. In the third case in Example 11.9 (Section 11.7.2.3) the error of the VMLA was 2^{-23}. Show why this is the case.

7. Write a division routine that checks for a divisor of zero, and if it is, returns a correctly signed infinity without setting the DZC bit. If the divisor is not zero, the division is performed.

8. Add to the program of Exercise 7 a check on a divisor that is 2.0. If it is, perform a multiplication of 0.5 rather than do the division.

9. Write a subroutine that would perform a reciprocal operation on an input in register s0, returning the result in register s0.

12 Tables

12.1 INTRODUCTION

In the last few chapters, we've dealt primarily with manipulating numbers and performing logical operations. Another common task that microprocessors usually perform is searching for data in memory from a list of elements, where an element could be sampled data stored from sensors or analog-to-digital converters (ADCs), or even data in a buffer that is to be transmitted to an external device. In the last 10 years, research into sorting and search techniques have, in part, been driven by the ubiquity of the Internet, and while the theory behind new algorithms could easily fill several books, we can still examine some simple and very practical methods of searching. Lookup tables are sometimes efficient replacements for more elaborate routines when functions like log(x) and tan(x) are needed; the disadvantage is that you often trade memory usage and precision for speed. Before we examine subroutines, it's worth taking a short look at some of the basic uses of tables and lists, for both integer and floating-point algorithms, as this will ease us into the topic of queues and stacks.

12.2 INTEGER LOOKUP TABLES

Consider a list of elements ordered in memory starting at a given address. Suppose that each element in the list is a word in length, as shown in Figure 12.1. Addressing a particular element in the list becomes quite easy, since the ARM addressing modes allow pre-indexed addressing with an offset. More precisely, if the starting address were held in register r5, then a given element could either be addressed by putting an offset in another register, or the element number can be used to generate an offset by scaling. The third element in the list could be accessed using either

```
LDR  r6, [r5, r4]
```

or

```
LDR  r6, [r5, r4, LSL #2]
```

where register r4 would contain the value 8, the actual offset, in the first case, or 2, one less than the element number, in the second case (for our discussion, the first element is number zero). The latter addressing mode accounts for the size of the data by scaling the element number by 4.

Certainly the same concepts apply if the elements are halfwords, only now the load instructions would be

```
LDRH  r6, [r4, r5]
```

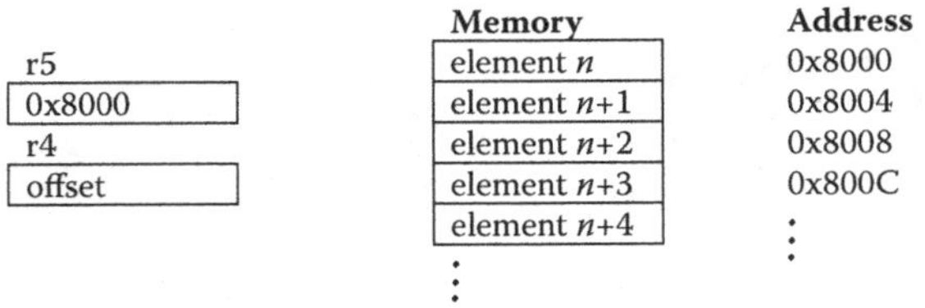

r5	Memory	Address
0x8000	element *n*	0x8000
r4	element *n*+1	0x8004
offset	element *n*+2	0x8008
	element *n*+3	0x800C
	element *n*+4	⋮
	⋮	

FIGURE 12.1 A simple list in memory.

and

```
LDRH  r6, [r4, r5, LSL #1]
```

EXAMPLE 12.1

Many control and audio applications require computing transcendental functions, such as log(x), tan(x), and sin(x). An easy way to compute the sine of an angle is to use a lookup table. There are obvious limits to the precision available with such a method, and if greater precision is required, there are very good routines for computing these types of functions, such as those by Symes (Sloss, Symes, and Wright 2004). However, a lookup table can return a value in Q31 notation for integer values of the angle between 0 and 360 degrees, and the implementation is not at all difficult.

To begin, it's necessary to create a table of sine values for angles between 0 and 90 degrees using Q notation. A short C program can generate these values very quickly, and if you throw in a little formatting at the end, it will save you the time of having to add assembler directives. The C code* below will do the trick:

```
#include <stdio.h>
#include <string.h>
#include <math.h>
main()
{
   int i;
   int index=0;
   signed int j[92];
   float sin_val;

   FILE *fp;

   if ((fp=fopen("sindata.txt","w"))==NULL)
   {
       printf("File could not be opened for writing\n");
       exit(1);
   }
   for (i=0; i<=90; i++){
       /* convert to radians */
       sin_val=sin(M_PI*i/180.0);
       /* convert to Q31 notation */
```

* Depending on how you compile your C code, your table may be slightly different. A #DEFINE statement may also be necessary for pi.

```
        j[i] = sin_val * (2147483648);
        }
    for (i = 1; i <=23; i++){
        fprintf(fp,"DCD ");
        fprintf(fp,"0x%x,",j[index]);
        fprintf(fp,"0x%x,",j[index + 1]);
        fprintf(fp,"0x%x,",j[index + 2]);
        fprintf(fp,"0x%x",j[index + 3]);
        fprintf(fp,"\n");
        index += 4;
        }
    fclose(fp);
}
```

It's important to note that while generic C code like this will produce accurate values for angles between 0 and 89 degrees, it's still necessary to manually change the value for 90 degrees to 0x7FFFFFFF, since you cannot represent the number 1 in Q31 notation (convince yourself of this). Therefore, we will just use the largest value possible in a fractional notation like this. The next step is to take the table generated and put this into an assembly program, such as the ones shown below for the ARM7TDMI and the Cortex-M4. While this is clearly not optimized code, it serves to illustrate several points.

```
; Example for the ARM7TDMI
     AREA SINETABLE, CODE
     ENTRY
; Registers used:
; r0 = return value in Q31 notation
; r1 = sin argument (in degrees, from 0 to 360)
; r2 = temp
; r4 = starting address of sine table
; r7 = copy of argument

main
     MOV    r7,r1          ; make a copy of the argument
     LDR    r2, = 270      ; constant won't fit into rotation scheme
     ADR    r4, sin_data   ; load address of sin table
     CMP    r1, #90        ; determine quadrant
     BLE    retvalue       ; first quadrant?
     CMP    r1, #180
     RSBLE  r1,r1,#180     ; second quadrant?
     BLE    retvalue
     CMP    r1, r2
     SUBLE  r1, r1, #180   ; third quadrant?
     BLE    retvalue
     RSB    r1, r1, #360   ; otherwise, fourth
retvalue
     ; get sin value from table
     LDR    r0, [r4, r1, LSL #2]
     CMP    r7, #180       ; do we return a neg value?
     RSBGT  r0, r0, #0     ; negate the value if so
done B      done
ALIGN
```

```
sin_data
        DCD 0x00000000,0x023BE164,0x04779630,0x06B2F1D8
        DCD 0x08EDC7B0,0x0B27EB50,0x0D613050,0x0F996A30
        DCD 0x11D06CA0,0x14060B80,0x163A1A80,0x186C6DE0
        DCD 0x1A9CD9C0,0x1CCB3220,0x1EF74C00,0x2120FB80
        DCD 0x234815C0,0x256C6F80,0x278DDE80,0x29AC3780
        DCD 0x2BC750C0,0x2DDF0040,0x2FF31BC0,0x32037A40
        DCD 0x340FF240,0x36185B00,0x381C8BC0,0x3A1C5C80
        DCD 0x3C17A500,0x3E0E3DC0,0x40000000,0x41ECC480
        DCD 0x43D46500,0x45B6BB80,0x4793A200,0x496AF400
        DCD 0x4B3C8C00,0x4D084600,0x4ECDFF00,0x508D9200
        DCD 0x5246DD00,0x53F9BE00,0x55A61280,0x574BB900
        DCD 0x58EA9100,0x5A827980,0x5C135380,0x5D9CFF80
        DCD 0x5F1F5F00,0x609A5280,0x620DBE80,0x63798500
        DCD 0x64DD8900,0x6639B080,0x678DDE80,0x68D9F980
        DCD 0x6A1DE700,0x6B598F00,0x6C8CD700,0x6DB7A880
        DCD 0x6ED9EC00,0x6FF38A00,0x71046D00,0x720C8080
        DCD 0x730BAF00,0x7401E500,0x74EF0F00,0x75D31A80
        DCD 0x76ADF600,0x777F9000,0x7847D900,0x7906C080
        DCD 0x79BC3880,0x7A683200,0x7B0A9F80,0x7BA37500
        DCD 0x7C32A680,0x7CB82880,0x7D33F100,0x7DA5F580
        DCD 0x7E0E2E00,0x7E6C9280,0x7EC11A80,0x7F0BC080
        DCD 0x7F4C7E80,0x7F834F00,0x7FB02E00,0x7FD31780
        DCD 0x7FEC0A00,0x7FFB0280,0x7FFFFFFF
        END


; Program for the Cortex-M4
        MOV    r7,r1          ; make a copy of the argument
        LDR    r2,=270        ; constant won't fit into rotation scheme
        ADR    r4, sin_data   ; load address of sin table
        CMP    r1, #90        ; determine quadrant
        BLE    retvalue       ; first quadrant?
        CMP    r1, #180
        ITT    LE
        RSBLE  r1,r1,#180     ; second quadrant?
        BLE    retvalue
        CMP    r1, r2
        ITT    LE
        SUBLE  r1, r1, #180   ; third quadrant?
        BLE    retvalue
        RSB    r1, r1, #360   ; otherwise, fourth
retvalue
        ; get sin value from table
        LDR    r0, [r4, r1, LSL #2]
        CMP    r7, #180       ; do we return a neg value?
        IT     GT
        RSBGT  r0, r0, #0     ; negate the value if so
done    B      done
        ALIGN

sin_data
           DCD 0x00000000,0x023BE164,0x04779630,0x06B2F1D8
           DCD 0x08EDC7B0,0x0B27EB50,0x0D613050,0x0F996A30
           DCD 0x11D06CA0,0x14060B80,0x163A1A80,0x186C6DE0
```

```
DCD 0x1A9CD9C0,0x1CCB3220,0x1EF74C00,0x2120FB80
DCD 0x234815C0,0x256C6F80,0x278DDE80,0x29AC3780
DCD 0x2BC750C0,0x2DDF0040,0x2FF31BC0,0x32037A40
DCD 0x340FF240,0x36185B00,0x381C8BC0,0x3A1C5C80
DCD 0x3C17A500,0x3E0E3DC0,0x40000000,0x41ECC480
DCD 0x43D46500,0x45B6BB80,0x4793A200,0x496AF400
DCD 0x4B3C8C00,0x4D084600,0x4ECDFF00,0x508D9200
DCD 0x5246DD00,0x53F9BE00,0x55A61280,0x574BB900
DCD 0x58EA9100,0x5A827980,0x5C135380,0x5D9CFF80
DCD 0x5F1F5F00,0x609A5280,0x620DBE80,0x63798500
DCD 0x64DD8900,0x6639B080,0x678DDE80,0x68D9F980
DCD 0x6A1DE700,0x6B598F00,0x6C8CD700,0x6DB7A880
DCD 0x6ED9EC00,0x6FF38A00,0x71046D00,0x720C8080
DCD 0x730BAF00,0x7401E500,0x74EF0F00,0x75D31A80
DCD 0x76ADF600,0x777F9000,0x7847D900,0x7906C080
DCD 0x79BC3880,0x7A683200,0x7B0A9F80,0x7BA37500
DCD 0x7C32A680,0x7CB82880,0x7D33F100,0x7DA5F580
DCD 0x7E0E2E00,0x7E6C9280,0x7EC11A80,0x7F0BC080
DCD 0x7F4C7E80,0x7F834F00,0x7FB02E00,0x7FD31780
DCD 0x7FEC0A00,0x7FFB0280,0x7FFFFFFF
```

The first task of the program is to determine in which quadrant the argument lies. Since

$$\sin(x) = \sin(180° - x)$$

and

$$\sin(x - 180°) = -\sin(x)$$
$$\sin(360° - x) = -\sin(x)$$

we can simply compute the value of sine for the argument's reference angle in the first quadrant and then negate the result as necessary. The first part of the assembly program compares the angle to 90 degrees, then 180 degrees, then 270 degrees. If it's over 270 degrees, then by default it must be in the fourth quadrant. The reference angle is also calculated to use as a part of the index into the table, using SUB or RSB as necessary. For values that lie in either the third or fourth quadrant, the final result will need to be negated.

The main task, obtaining the value of sine, is actually just one line of code:

```
LDR  r0, [r4, r1, LSL #2] ; get sin value from table
```

Since the starting address of our lookup table is placed in register r4, we index the entry in the table with an offset. Here, we're using pre-indexed addressing, with the offset calculated by multiplying the value of the argument (a number between 0 and 90) by four. For example, if the starting address of the table was 0x4000, and the angle was 50, then we know we have to skip 50 words of data to get to the entry in the table that we need.

The reverse subtract at the end of the routine negates our final value if the argument was in either quadrant three or four. The exact same technique could be used to generate a cosine table, which is left as an exercise, or a logarithm table.

12.3 FLOATING-POINT LOOKUP TABLES

Analogous to the integer lookup tables in Section 12.2, floating-point lookup tables are addressed with load instructions that have offsets, only the values for most cases are single-precision floating-point numbers. Instead of an LDR instruction, we use a VLDR instruction to move data into a register, or something like

```
VLDR.F   s2, [r1, #20]  ; offset is a multiple of 4
```

EXAMPLE 12.2

In this example, we set up a constant table with the label ConstantTable and load this address into register r1, which will serve as an index register for the VLDR instruction. The offset may be computed as the index of the value in the table entry less one, then multiplied by 4, since each data item is 4 bytes in length.

```
     ADR            r1, ConstantTable ; Load address of
                                      ; the constant table

     ; load s2 with pi, s3 with 10.0,
     ; and multiply them to s4

     VLDR.F         s2, [r1, #20] ; load pi to s2
     VLDR.F         s3, [r1, #12] ; load 10.0 to s3
     VMUL.F         s4, s2, s3

loop B              loop

     ALIGN

ConstantTable
     DCD            0x3F800000 ; 1.0
     DCD            0x40000000 ; 2.0
     DCD            0x80000000 ; -0.0
     DCD            0x41200000 ; 10.0
     DCD            0x42C80000 ; 100.0
     DCD            0x40490FDB ; pi
     DCD            0x402DF854 ; e
```

A common use of the index-with-offset addressing mode is with literal pools, which we encountered in Chapter 6. Literal pools are very useful in floating-point code since many floating-point data items are not candidates for the immediate constant load, which we will discuss in a moment. When the assembler creates a literal pool, it uses the PC as the index register. The Keil assembler allows for constants to be named with labels and used with their label.

EXAMPLE 12.3

The following modification to the example above shows how labels can be used in constant tables, should your assembler support this.

```
      ; load s2 with pi, s3 with 10.0,
      ; and multiply them to s4
      VLDR.F          s5, C_Pi
      VLDR.F          s6, C_Ten
      VMUL.F          s7, s5, s6
loop B                loop

      ALIGN

C_One       DCD       0x3F800000 ; 1.0
C_Two       DCD       0x40000000 ; 2.0
C_NZero     DCD       0x80000000 ; -0.0
C_Ten       DCD       0x41200000 ; 10.0
C_Hun       DCD       0x42C80000 ; 100.0
C_Pi        DCD       0x40490FDB ; pi
C_e         DCD       0x402DF854 ; e
```

Since the labels C_Pi and C_Ten translate to addresses, the distances between the current value of the Program Counter and the constants are calculated, then used in a PC-relative VLDR instruction. This technique allows you to place floating-point values in any order, since the tools calculate offsets for you.

EXAMPLE 12.4 RECIPROCAL SQUARE ROOT ESTIMATION CODE

In graphics algorithms, the reciprocal square root is a common operation, used frequently in computing the normal of a vector for use in lighting and a host of other operations. The cost of doing the full-precision, floating-point calculation of a square root followed by a division can be expensive. On the Cortex-M4 with floating-point hardware, these operations take 28 cycles, which is a relatively small amount for division and square root. So this example, while not necessarily an optimal choice in all cases, demonstrates the use of a table of half-precision constants and the use of the conversion instruction. The reciprocal square root is calculated by using a conversion table for the significand and adjusting the exponent as needed.

The algorithm proceeds as follows. If we first consider the calculation of a reciprocal square root, the equation is

$$\frac{1}{\sqrt{x}} = \frac{1}{\sqrt{1.f \cdot 2^n}}$$

where $x = 1.f \cdot 2^n$

And we know that

$$\frac{1}{\sqrt{1.f \cdot 2^n}} = \frac{1}{\sqrt{1.f} \cdot \sqrt{2^n}}$$

Resulting in

$$\frac{1}{\sqrt{1.f} \cdot \sqrt{2^n}} = 1.g \cdot 2^{-n/2}$$

where 1.g is the table estimate.

The sequence of operations then becomes:

1. Load pointers to a table for even exponent input operands and a table for odd exponent operands. The tables take a small number of the most significant fraction bits as the index into the table.
2. The oddness of the exponent is determined by ANDing all bits except the LSB to a zero, and testing this with the TEQ instruction. If odd, the exponent is incremented by 1.
3. Divide the exponent by 2 and negate the result. A single-precision scale factor is generated from the computed exponent.
4. Extract the upper 4 bits of the fraction, and if the exponent is odd, use them to index into table RecipSQRTTableOdd for the estimate (the table estimate 1.g), and if the exponent is even, use the table RecipSQRTTableEven for the estimate.
5. Convert the table constant to a single-precision value using the VCVTB instruction, then multiply by the scale factor to get the result.

Note that this code does not check for negative values for x, or whether x is infinity or a NaN. Adding these checks is left as an exercise for the reader.

```
Reset_Handler

      ; Enable the FPU
      ; Code taken from ARM website
      ; CPACR is located at address 0xE000ED88
      LDR.W      r0, =0xE000ED88     ; Read CPACR
      LDR        r1, [r0]
      ; Set bits 20-23 to enable CP10 and CP11 coprocessors
      ORR        r1, r1, #(0xF << 20)
      ; Write back the modified value to the CPACR
      STR        r1, [r0]            ; wait for store to complete
      DSB

      ; Reciprocal Square Root Estimate code
      ; r1 holds the address to the odd table
      ADR        r0, RecipSQRTTableOdd
      ; r2 holds the address to the even table
      ADR        r1, RecipSQRTTableEven

      ; Compute the reciprocal square root estimate for a
      ; single precision value X x 2^n as
      ; 1/(X)^-1/2. The estimate table is stored in two
      ; halves, the first for odd exponents
      ; RecipSqrtTableOdd) and the second for
      ; even exponents (RecipSqrtTableEven).

      VLDR.F     s0, InputValue
      VMOV.F     r2, s0
      ; Process the exponent first - we assume positive input
      MOV        r3, r2              ; exp in r2, frac in r3
      LSR        r2, #23             ; shift the exponent for subtraction
      SUB        r2, #127            ; subtract out the bias
      AND        r4, r2, #1          ; capture the lsb to r4
      TEQ        r4, #1              ; check for odd exponent
```

```
        ; Odd Exponent - add 1 before the negate and shift
        ; right operations
        ADDEQ           r2, #1          ; increment to make even
        ; All exponents
        LSR             r2, r2, #1      ; shift right by 1 to divide by 2
        NEG             r2, r2          ; negate
        ADD             r2, #127        ; add in the bias
        LSL             r2, #23         ; return the new exponent - the
        ; Extract the upper 4 fraction bits for the table lookup
        AND             r3, #0x00780000
        LSR             r3, #18         ; shift so they are *2
        ; Select the table and the table entry based on
        ; the upper fraction bits
        LDRHEQ          r4, [r3, r0]    ; index into the odd table
        LDRHNE          r4, [r3, r1]    ; index into the even table
        VMOV.F          s3, r4          ; copy the selected half-precision
        VCVTB.F32.F16   s4, s3          ; convert the estimate to sp
        VMOV.F          s5, r2          ; move the exp multiplier to s5
        VMUL.F          s6, s5, s4      ; compute the recip estimate

loop B              loop

        ALIGN

InputValue
; Test values. Uncomment the value to convert
;       DCD     0x42333333    ; 44.8, recip sqrt is 0.1494, odd exp
;       DCD     0x41CA3D71    ; 25.28, recip sqrt is 0.19889, even exp

        ALIGN

RecipSQRTTableEven
        DCW     0x3C00                  ; 1.0000 -> 1.0000
        DCW     0x3BC3                  ; 1.0625 -> 0.9701
        DCW     0x3B8B                  ; 1.1250 -> 0.9428
        DCW     0x3A57                  ; 1.1875 -> 0.9177
        DCW     0x3B28                  ; 1.2500 -> 0.8944
        DCW     0x3AFC                  ; 1.3125 -> 0.8729
        DCW     0x3AD3                  ; 1.3750 -> 0.8528
        DCW     0x3AAC                  ; 1.4375 -> 0.8340
        DCW     0x3A88                  ; 1.5000 -> 0.8165
        DCW     0x3A66                  ; 1.5625 -> 0.8000
        DCW     0x3A47                  ; 1.6250 -> 0.7845
        DCW     0x3A29                  ; 1.6875 -> 0.7698
        DCW     0x3A0C                  ; 1.7500 -> 0.7559
        DCW     0x39F1                  ; 1.8125 -> 0.7428
        DCW     0x39D8                  ; 1.8750 -> 0.7303
        DCW     0x39BF                  ; 1.9375 -> 0.7184

        ALIGN

RecipSQRTTableOdd
        DCW     0x3DA8                  ; 0.5000 -> 1.4142
        DCW     0x3D7C                  ; 0.5322 -> 1.3707
        DCW     0x3D55                  ; 0.5625 -> 1.3333
        DCW     0x3D31                  ; 0.5938 -> 1.2978
        DCW     0x3D0F                  ; 0.6250 -> 1.2649
        DCW     0x3CF0                  ; 0.6563 -> 1.2344
        DCW     0x3CD3                  ; 0.6875 -> 1.2060
        DCW     0x3CB8                  ; 0.7186 -> 1.1795
        DCW     0x3C9E                  ; 0.7500 -> 1.1547
```

```
        DCW     0x3C87                  ; 0.7813 -> 1.1313
        DCW     0x3C70                  ; 0.8125 -> 1.1094
        DCW     0x3C5B                  ; 0.8438 -> 1.0886
        DCW     0x3C47                  ; 0.8750 -> 1.0690
        DCW     0x3C34                  ; 0.9063 -> 1.0504
        DCW     0x3C22                  ; 0.9375 -> 1.0328
        DCW     0x3C10                  ; 0.9688 -> 1.0160
```

12.4 BINARY SEARCHES

Searching through lists or tables of information is considered to be something of a standard problem in computer science. Tables are usually organized to hold data in a regular structure so that they can be searched quickly, using an identifier at the beginning of an entry. A *key* is defined as a tag of some sort that identifies an entry in the table. Sometimes it's just important to know whether or not a key exists in a table. Sometimes you need the data associated with that key. Either way, the techniques used for gathering this information date back almost as far as the computer itself, and while volumes have been written on the subject (Knuth 1973), we'll start by examining a basic search technique called a binary search.

If you have a list of entries in a table, as shown in Figure 12.2, where each entry consists of a numerical key and some type of data to go along with that key, e.g., character data such as an address or numerical data such as a phone number, you could try to find a key by sequentially comparing each key in the table to your value. Obviously this would take the longest amount of time, especially if the key of interest happened to be at the end of the table. If the keys are sorted already, say in increasing order, then you can significantly reduce your search efforts by starting at the middle of the table. This can immediately halve your search effort. Again referring to Figure 12.2, you can see that if our key is less than the middle key, we don't even have to look in the latter half of the table. We know it's somewhere between the first and middle keys, if it's there at all. Next, we further refine the search by making the last key in our search the key just before the middle one. The new middle key is defined as the average of the first and last keys, and the procedure is repeated until the key is either found or we confirm that it's not in the table. If the middle key happens to match our key, then the algorithm is finished. In a like manner, if the key we're looking for is between the middle and last keys in

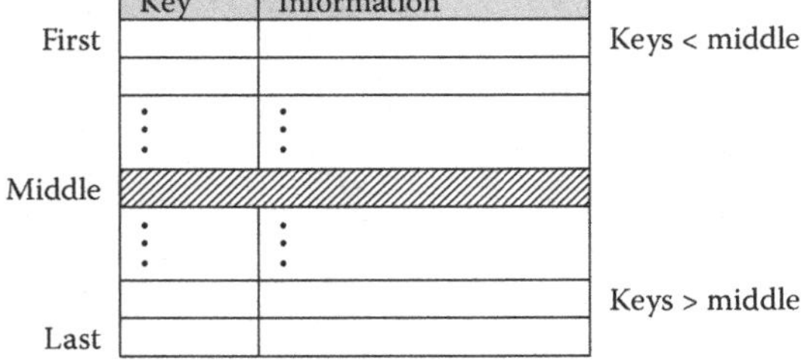

	Key	Information	
First			Keys < middle
	⋮	⋮	
Middle			
	⋮	⋮	
			Keys > middle
Last			

FIGURE 12.2 Binary search table.

the table, we move the start of our search to the entry just after the middle key and compute a new middle key for comparisons. The C equivalent of this technique could be described as

```
first=0;
last=num_entries - 1;
index=0;
while ((index == 0) & (first <= last)){
      middle=(first+last)/2;
      if (key == table[middle]) index=middle;
      else if (key<table[middle]) last=middle - 1;
      else first=middle+1;
      }
```

where `num _ entries` is the number of entries in the table.

Figure 12.3 shows an example of how this works. Suppose we have a table with nine entries in it, and the key of interest is 992. On the first pass of the search, we compute the middle of the table to be index number 4, since this is the average of the first and last entry numbers, 0 and 8, respectively. A comparison is then made against the table entry with this index. Since our number is greater than the middle number, the search focuses on the half of the table where the keys are even larger. The new starting position is entry number 5, while the last entry remains the same. A new middle index is found by averaging 5 and 8, which is 6 (remember they have to be integers). The comparison against the table entry with this index happens to match, so we've actually found the entry and the algorithm terminates.

In coding the binary search, we should examine a few aspects of the algorithm and of the data first, since most of the work can be done with just a few instructions. The rest of them are used to control the loop. Consider a table starting at address 0x4000, where each entry is 16 bytes, and say that 4 bytes, or a word, is used as a key. This leaves the remaining 12 bytes for character data, as shown in Figure 12.4. Examining the address of the tag, we can see that if the index i ranges from 0 to some number n–1, and the starting address of the table is table_addr, the address of the ith entry would be

$$\text{address} = \text{table\_addr} + i * \text{size\_of\_entry}.$$ 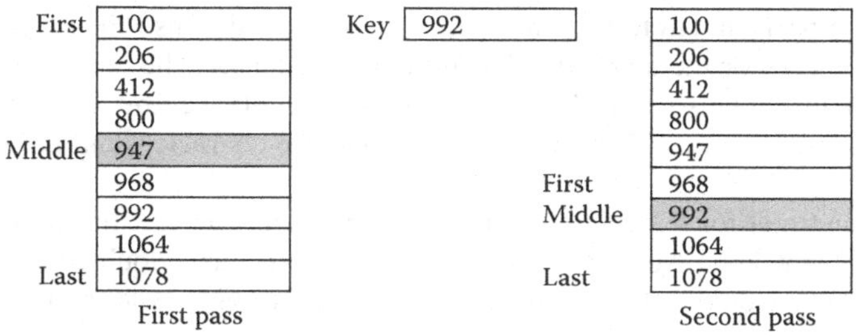

Key: 992

First pass		Second pass	
First	100		100
	206		206
	412		412
	800		800
Middle	947		947
	968	First	968
	992	Middle	992
	1064		1064
Last	1078	Last	1078

FIGURE 12.3 Two passes through a binary search.

Address	4 Byte key	12 Bytes of data
0x4000	0x00000034	Vacuums
0x4010	0x00000243	Clothes
0x4020	0x00003403	Candy
0x4030	0x0010382C	Telephones
.		
.		
.		

Table base address
↓
LDR r7, [r6, r2, LSL #4]
↑
Index

FIGURE 12.4 Structure of an example table.

For our table, the second entry would start at address 0x4000 + 2 × 16 = 0x4020.

Using this approach, we can simply use an LDR instruction with pre-indexed addressing, offsetting the table's base address with the scaled index, which is held in a register. The scaling (which is based on the entry size in bytes) can actually be specified with a constant—ESIZE—so that if we change this later, we don't have to recode the instructions. With this approach comes a word of caution, since we assume that the entry size is a power of two. If this is not the case, all hope is not lost. You can implement a two-level table, where an entry now consists of a key and an address pointing to data in memory, and the data can be any size you like. The size of each entry is again fixed, and it can be set to a power of two. However, for our example, the entry size is a power of two.

We can load our table entries with a single pre-indexed instruction:

```
LDR r7, [r6, r2, LSL #ESIZE]
```

This just made short work of coding the remaining algorithm. The first four instructions in Figure 12.5 are just initialization—the base address of the table is loaded into a register, the *first* index is set to 0, and the *last* index is set to the last entry, in this case, the number of entries we have, called NUM, minus one. The instructions inside of the loop test to see whether the *first* index is still smaller than or equal to the *last* index. If it is, a new *middle* index is generated and used to load the table entry into a register. The data is loaded from the table and tested against our key. We can effectively use conditional execution to change the *first* and *last* indices, since the comparison will test for mutually exclusive conditions. The loop terminates with either a zero or the key index loaded into register r3. The data that is used in the example might be someone's address on a street, followed by his or her favorite pizza toppings. Note that each key is 4 bytes and the character data is 12 bytes for each entry.

Execution times for search algorithms are important. Consider that a linear search through a table would take twice as long to execute if the table doubled in size, where a binary search would only require one more pass through its loop. The execution time increases logarithmically.

```
NUM     EQU     14      ; insert # of entries here
ESIZE   EQU     4       ; log 2 of the entry size (16 bytes)
                        ; NB: This assumes entry size is a power of 2
        AREA    BINARY, CODE
        ENTRY
; Registers used:
; R0 - first
; R1 - last
; R2 - middle
; R3 - index
; R4 - size of the entries (log 2)
; R5 - the key (what you're searching for)
; R6 - address of the list
; R7 - temp
        LDR     r5, =0x200              ; let's look for PINEAPPLE

        ADR     r6, table_start         ; load address of the table
        MOV     r0, #0                  ; first = 0
        MOV     r1, #NUM-1              ; last = number of entries in the list - 1

loop    CMP     r0, r1                  ; compare first and last
        MOVGT   r2, #0                  ; first > last, no key found, middle = 0
        BGT     done

        ADD     r2, r0, r1              ; first + last
        MOV     r2, r2, ASR #1          ; first + last /2

        LDR     r7, [r6, r2, LSL #ESIZE] ; load the entry
        CMP     r5, r7                  ; compare key to value loaded
        ADDGT   r0, r2, #1              ; first = middle + 1
        SUBLT   r1, r2, #1              ; last = middle - 1
        BNE     loop                    ; go again

done    MOV     r3, r2                  ; move middle to 'index'
stop    B       stop

table_start
        DCD     0x004
        DCB     "PEPPERONI    "
        DCD     0x005
        DCB     "ANCHOVIES    "
        DCD     0x010
        DCB     "OLIVES       "
        DCD     0x012
        DCB     "GREEN PEPPER"
        DCD     0x018
        DCB     "BLACK OLIVES"
        DCD     0x022
        DCB     "CHEESE       "
        DCD     0x024
        DCB     "EXTRA SAUCE "
        DCD     0x026
        DCB     "CHICKEN      "
        DCD     0x030
        DCB     "CANADIAN BAC"
        DCD     0x035
        DCB     "GREEN OLIVES"
        DCD     0x038
        DCB     "MUSHROOMS    "
        DCD     0x100
        DCB     "TOMATOES     "
        DCD     0x200
        DCB     "PINEAPPLE    "
        DCD     0x300
        DCB     "PINE NUTS    "
        END
```

FIGURE 12.5 Assembly code for the binary search.

12.5 EXERCISES

1. Using the sine table as a guide, construct a cosine table that produces the value for cos(x), where $0 < x < 360$. Test your code for values of 84 degrees and 105 degrees.

2. It was mentioned in Section 12.4 that a binary search only works if the entries in a list are sorted first. A bubble sort is a simple way to sort entries. The basic idea is to compare two adjacent entries in a list—call them entry[j] and entry[j + 1]. If entry[j] is larger, then swap the entries. If this is repeated until the last two entries are compared, the largest element in the list will now be last. The smallest entry will ultimately get swapped, or "bubbled," to the top. This algorithm could be described in C as

```
last = num;
while (last > 0){
pairs = last - 1;
for (j = 0; j <= pairs; j++) {
   if(entry[j] > entry[j + 1]) {
                  temp = entry[j];
                  entry[j] = entry[j + 1];
                  entry[j + 1] = temp;
                  last = j;
   }
}
}
```

   where num is the number of entries in the list. Write an assembly language program to implement a bubble sort algorithm, and test it using a list of 20 elements. Each element should be a word in length.

3. Using the bubble sort algorithm written in Exercise 2, write an assembly program that sorts entries in a list and then uses a binary search to find a particular key. Remember that your sorting routine must sort both the key and the data associated with each entry. Create a list with 30 entries or so, and data for each key should be at least 12 bytes of information.

4. Create a queue of 32-bit data values in memory. Write a function to remove the first item in the queue.

5. Using the sine table as a guide, construct a tangent table that produces the value for tan(x), where $0 \leq x \leq 45$. Test your code for values of 12 degrees and 43 degrees. You may return a value of 0x7FFFFFFF for the case where the angle is equal to 45 degrees, since 1 cannot be represented in Q31 notation.

6. Implement a sine table that holds values ranging from 0 to 180 degrees. The implementation contains fewer instructions than the routine in Section 12.2, but to generate the value, it uses more memory to hold the sine values themselves. Compare the total code size for both cases.

7. Implement a cosine table that holds values ranging from 0 to 180 degrees. The implementation contains fewer instructions than the routine in Section 12.2, but to generate the value, it uses more memory to hold the cosine values themselves. Compare the total code size for both cases.

8. Rewrite the binary search routine for the Cortex-M4.

13 Subroutines and Stacks

13.1 INTRODUCTION

Subroutines, which are routines dedicated to focused or shorter tasks, occur in nearly all embedded code, often to perform initialization routines or to handle algorithms that require handcrafted assembly. You're very likely to write subroutines when you come across a large problem to solve. The notion of divide-and-conquer aptly applies to writing assembly code—it's just easier to get your head around a problem by describing it as a sequence of events, worrying about the low-level details when you write the event itself. For example, you could break a large program that controls a motor into something that looks like

```
main    BL      ConfigurePWM
        BL      GetPosition
        BL      CalcOffset
        BL      MoveMotor
        ...
```

and then write each of the smaller tasks as subroutines. Even if you're coding at a high level, say in C or C++, it's natural to break down a large task into smaller blocks or functions, each function being a routine that can be called from a main routine.

To write a proper subroutine, we also have to look at ways of saving and restoring data, passing information to and from subroutines, and building stacks. This chapter will cover some instructions that we skipped in Chapter 5, the load and store multiple operations LDM and STM, and their synonymous mnemonics PUSH and POP, as they're used frequently in stack operations. We'll briefly look at ARM standards that govern stack creation and define a standard way to call subroutines, so that a common protocol exists for all developers. Without such standards, programmers could easily write code that is incompatible with code created using third-party tools, or even tools from ARM. Before writing any code, though, we have to look at those new instructions and define what we mean by a stack.

13.2 THE STACK

Stacks are conceptually Last In-First Out (LIFO) queues that can be used to describe systems from the architectural level down to the hardware level. Stacks can be used for software operations, too. More abstract descriptions of stacks can be used as data types by languages such as Java or Python, and there are even stack-based computer systems. When referring to hardware, generally these are areas in memory that have

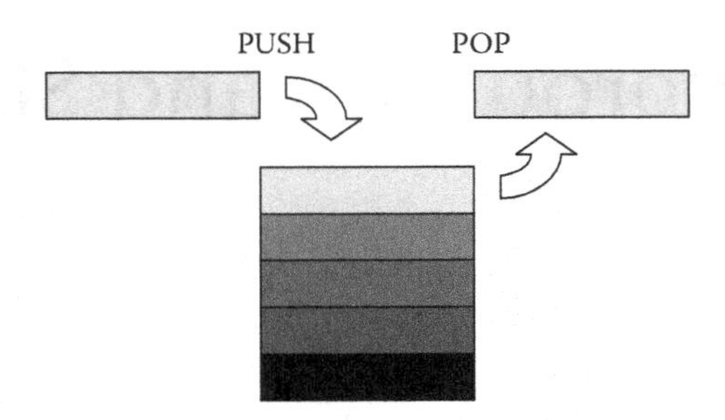

FIGURE 13.1 A hardware stack in memory.

a variable length and a fixed starting address. Figure 13.1 shows a description of a hardware stack, with each entry being a fixed number of bytes in memory (for our case, generally these are word-length values). Data is written, or pushed, onto the top of the stack, and also read, or popped, from the top of the stack, where the processor adjusts the stack pointer before or after each operation. ARM processors have a stack pointer, register r13, which holds the address of either the next empty entry or the last filled entry in the queue, depending on what type of stack you have. We'll see the different types shortly.

13.2.1 LDM/STM Instructions

Back in Chapter 5, we covered all of the basic load and store instructions, leaving off one pair until now—the load and store multiple instructions. Where LDR and STR instructions transfer words, halfwords, and bytes, the LDM and STM instructions always transfer one or more words using registers and a pointer to memory, known as the base register. This type of load/store appears most often in the context of stacks and exception handling, since the processor only has a limited number of registers, and at times, you just have to make some room for new data *somewhere*. By saving off the contents of the registers before handling an exception or going to a subroutine, you free up registers for different uses. Obviously, these must be restored when you're finished, and there are instructions for doing exactly that. There are also advantages in using a multiple register transfer instruction instead of a series of single data transfer instructions, to wit, the code size decreases. A single LDM instruction can load multiple registers from memory using only a single instruction, rather than individual LDR instructions. Execution time also shortens, since only one instruction must be fetched from memory.

On the ARM7TDMI, the syntax of the LDM instruction is

```
LDM <address-mode> {<cond>}    <Rn> {!}, <reg-list> {^}
```

where {<cond>} is an optional condition code; <address-mode> specifies the addressing mode of the instruction, which tells us how and when we change the base register; <Rn> is the base register for the load operation; and <reg-list> is a comma-delimited list of symbolic register names and register ranges enclosed in braces. We'll talk more about the "!" and "^" symbols in a moment.

On the Cortex-M3/M4, the syntax of the LDM instruction is

LDM <address-mode> {<cond>} <Rn> {!}, <reg-list>

where {<cond>} is an optional condition code; <address-mode> specifies the addressing mode of the instruction (although as we'll see in the next section, there are only two); <Rn> is the base register for the load operation; and <reg-list> is a comma-delimited list of symbolic register names and register ranges enclosed in braces.

EXAMPLE 13.1

Suppose you wanted to load a subset of all registers, for example, registers r0 to r3, from memory, where the data starts at address 0xBEEF0000 and continues upward in memory. The instruction would simply be

```
LDMIA r9, {r0-r3}
```

where the base register r9 holds the address 0xBEEF0000. The addressing mode used here is called Increment After, or IA. This says to increment the address *after* each value has been loaded from memory, which we'll see shortly. This has the same effect as four separate LDR instructions, or

```
LDR    r0, [r9]
LDR    r1, [r9, #4]
LDR    r2, [r9, #8]
LDR    r3, [r9, #12]
```

Notice in the example above that at the end of the load sequence, register r9 has not been changed and still holds the value 0xBEEF0000. If you wanted to load data into registers r0 through r3 and r12, you could simply add it to the end of the list, i.e.,

```
LDMIA r9, {r0-r3, r12}
```

Obviously, there must be at least one register in the list, but it doesn't actually matter in what order you list the registers. The lowest register will always be loaded from the lowest address in memory, and the highest register will be loaded from the highest address. For example, you could say

```
LDMIA r9,  {r5, r3, r0-r2, r14}
```

and register r0 will be loaded first, followed by registers r1, r2, r3, r5, and r14.

Analogous to the load multiple instruction, the store multiple instruction (STM) transfers register data *to* memory, and for the ARM7TDMI, its syntax is

STM <address-mode> {<cond>} <Rn> {!}, <reg-list> {^}

where the options are identical to those for the LDM instruction. The syntax for the Cortex-M3/M4 is

STM <address-mode> {<cond>} <Rn> {!}, <reg-list>

The options on LDM and STM instructions are used sparingly, but they're worth mentioning here. Starting from the value in the base register, the address accessed is decremented or incremented by one word for each register in the register list. Since the base register is not modified after an LDM or STM instruction completes, you can force the address to be updated by using the "!" option with the mnemonic. If you happen to have the base register in the register list, then you must not use the writeback option. The caret (^) option is discussed in Chapter 14, since it relates more to the procedures of handling exceptions.

The addressing modes go by different names, as we'll see in a moment, but basically there are four:

IA—Increment After
IB—Increment Before
DA—Decrement After
DB—Decrement Before

The suffix on the mnemonic indicates how the processor modifies the base register during the instruction. For example, if register r10 contained 0x4000,

```
LDMIA  r10, {r0, r1, r4}
```

would begin by loading register r0 with data from address 0x4000. The value in the base register is incremented by one word *after* the first load is complete. The second register, r1, is loaded with data from 0x4004, and register r4 is loaded with data from 0x4008. Note here that the base register is *not* updated after the instruction completes. The other three suffixes indicate whether the base register is changed before or after the load or store, as well as whether it is incremented or decremented, as shown in Figure 13.2. In the following sections, we'll examine stacks and the other addressing mode suffixes that are easier to use for stack operations.

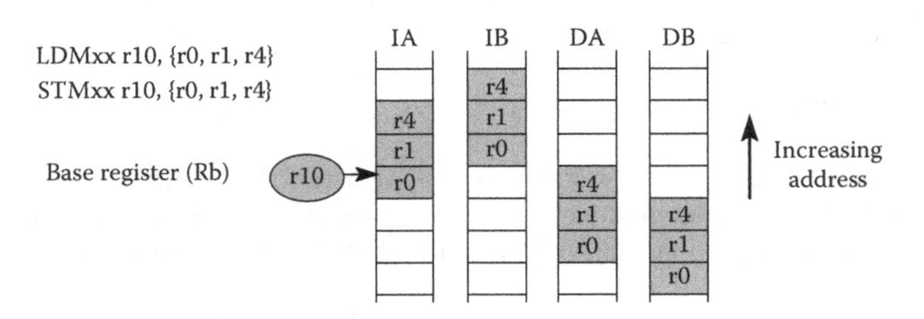

FIGURE 13.2 LDM/STM operations.

While the ARM7TDMI supports all four addressing modes, version 7-M processors like the Cortex-M3/M4 have more restrictive options and a few cautions to mind. There are only two addressing modes from which to choose:

IA—Increment After
DB—Decrement Before

since the Cortex-M3/M4 supports only one type of stack, which we'll examine more closely in the next few sections. If an LDM instruction is used to load the Program Counter, ensure that bit 0 of the loaded value is set to a 1; otherwise a fault exception occurs. For both the STM and LDM instructions, the stack pointer should not be included in the register list. Also be aware that if you have an LDM instruction that has the Link Register in the register list, you cannot include the Program Counter in the list. Consult the *ARM v7-M Architectural Reference Manual* (ARM 2010a) for other restrictions when using LDM and STM instructions.

13.2.2 PUSH AND POP

There are two instructions that are synonymous with STMDB and LDMIA, namely PUSH and POP, respectively. PUSH can be used in place of a STMDB instruction with both the ARM7TDMI and the Cortex-M4, as it falls in line with the new preferred UAL mnemonics. The syntax for the two instructions is

```
PUSH{<cond>}   <reglist>
POP{<cond>}    <reglist>
```

where {<cond>} is an optional condition code and <reg-list> is a comma-delimited list of symbolic register names and register ranges enclosed in braces. PUSH has similar restrictions to the STM instruction, e.g., the register list must not contain the PC. POP has similar restrictions to the LDM instruction, e.g., the register list must not contain the PC if it contains the LR.

EXAMPLE 13.2

PUSH and POP make it very easy to conceptually deal with stacks, since the instruction implicitly contains the addressing mode. Suppose we have a stack that starts at address 0x20000200 on the Tiva TM4C123GH6ZRB, grows downward in memory (a full descending stack), and has two words pushed onto it with the following code:

```
            AREA    Example3, CODE, READONLY
            ENTRY
SRAM_BASE   EQU     0x20000200
            LDR     sp, =SRAM_BASE

            LDR     r3, =0xBABEFACE
            LDR     r4, =0xDEADBEEF
            PUSH    {r3}
            PUSH    {r4}
```

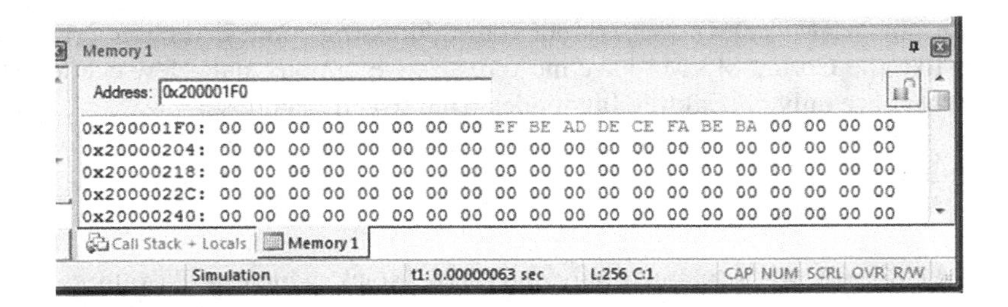

FIGURE 13.3

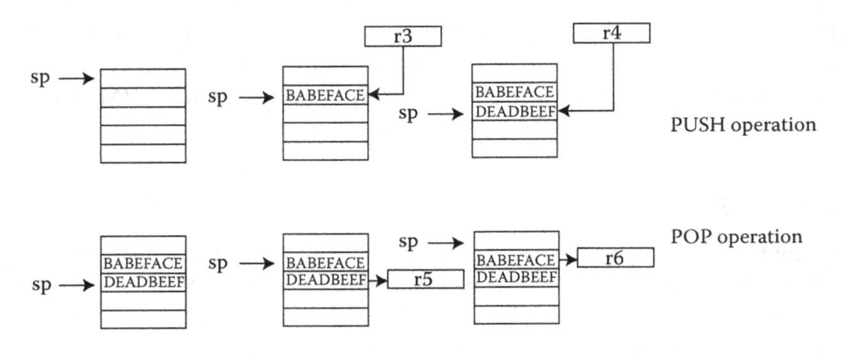

FIGURE 13.4

```
              POP     {r5}
              POP     {r6}

    stop B    stop    ; stop program
```

As we'll see in the next section, a full descending stack implies that the stack pointer is pointing to the last (full) item stored in the stack (at address 0x20000200) and that the stack items are stored at addresses that decrease with each new entry. Therefore, our stack pointer must be decremented before anything new is placed on the stack. The first word in our program would be stored in memory at address 0x200001FC. The second word would be stored at address 0x200001F8. If you run the code above in a simulator and view a memory window, such as the one in Figure 13.3, you will see the two words stored at successively decreasing addresses. The POP instructions will read the data into whichever registers we choose, so the value 0xDEADBEEF is popped off the top of the stack into register r5. The stack pointer is incremented afterward. The second POP instruction moves the value 0xBABEFACE into register r6, shown in Figure 13.4, returning the stack pointer value to 0x20000200.

13.2.3 Full/Empty Ascending/Descending Stacks

Stack operations are easy to implement using LDM and STM instructions, since the base register is now just the stack pointer, register r13. Several types of stacks can be built, depending on personal preferences, programming, or hardware requirements.

TABLE 13.1
Stack-Oriented Suffixes

Stack Type	PUSH	POP
Full descending	STMFD (STMDB)	LDMFD (LDMIA)
Full ascending	STMFA (STMIB)	LDMFA (LDMDA)
Empty descending	STMED (STMDA)	LDMED (LDMIB)
Empty ascending	STMEA (STMIA)	LDMEA (LDMDB)

Your software tools will probably build a particular type of stack by default. Fortunately, they all use the same instructions—the differences lie with suffixes on those instructions. The options are

Descending or ascending—The stack grows downward, starting with a high address and progressing to a lower one (a descending stack), or upward, starting from a low address and progressing to a higher address (an ascending stack).

Full or empty—The stack pointer can either point to the last item in the stack (a full stack), or the next free space on the stack (an empty stack).

To make it easier for the programmer, stack-oriented suffixes can be used instead of the increment/decrement and before/after suffixes. For example, you can just use the FD suffix to indicate that you're building a full descending stack; the assembler will translate that into the appropriate instructions. Pushing data onto a full descending stack is done with an STMDB instruction. The stack starts off at a high address and works its way toward a lower address, and full refers to the stack pointer pointing at the last item in the stack, so before moving new data onto it, the instruction has to decrement the pointer beforehand. Popping data off this type of stack requires the opposite operation—an LDMIA instruction. Because the address always points to the last item on the stack, the processor reads the data first, *then* the address is incremented. Refer to Table 13.1 for a list of stack-oriented suffixes.

EXAMPLE 13.3

Let's build a full descending stack in memory, using register r13 as the pointer to the stack. Further suppose that this code is part of a routine that will require the Link Register and registers r4 through r7 to be saved on the stack. Assume that the SRAM starts at address 0x20000200 for the Tiva TM4C123GH6ZRB microcontroller. Our code might start something like this:

```
          AREA    Test, CODE, READONLY
SRAM_BASE EQU     0x20000200
          ENTRY
          ; set up environment
          LDR     sp, =SRAM_BASE        ;r13 = ptr to stack memory
          ;
          ; your main code is here
          ;
```

```
            ; call your routine with a branch and link instruction
            BL      Myroutine
            ;
Myroutine   ; Routine code goes here. First, create space in the register
            ; file by saving r4-r7, then save the Link Register for the return,
            ; all with a single store multiple to the stack
            STMDB     sp!, {r4-r7,lr}    ;Save some working registers
            ;
            ; Routine code
            ;
            ; Restore saved registers and move the Link Register contents
            ; into the Program Counter, again with one instruction
            LDMIA        sp!, {r4-r7,pc}  ;restore registers and return
            END
```

Recall that full descending stacks can be created by using the STMDB/LDMIA combination, identical to the PUSH/POP combination. Notice that the LDM instruction pops the value of the Link Register into the Program Counter, so if we were to call our stacking routine as part of another function, the return address is moved into the Program Counter automatically and fetching begins from there. This is exactly how subroutines are called, which brings us to our next section.

13.3 SUBROUTINES

Most large programs consist of many smaller blocks of code, or subroutines, where functions can be called at will, such as a print routine or a complicated arithmetic function like a logarithm. A large task can be described more easily this way. Subroutines also allow programmers to write and test small blocks of code first, building on the knowledge that they've been proven to work. Subroutines should follow certain guidelines, and software should be able to interrupt them without causing any errors. A routine that can be interrupted and then called by the interrupting program, or more generally, can be called recursively without any problems, is called *reentrant*. By following a few simple procedures at the start of your code, you should be able to write reentrant subroutines with few headaches.

We've already seen a few examples of subroutines in Chapter 6, where the subroutine is called with the BL (branch and link) instruction. This instruction transfers the branch target (the starting address of your subroutine) into the Program Counter and also transfers the return address into the Link Register, r14, so that the subroutine can return back to the calling program.

Subroutines can also call other subroutines, but caution must be taken to ensure information is not lost in the process. For example, in Figure 13.5, a main routine calls a subroutine called func1. The subroutine should immediately push the values of any registers it might corrupt, as well as the Link Register, onto the stack. At some point in the code, another subroutine, func2, is called. This subroutine should begin the exact same way, pushing the values of any used registers and the Link Register onto the stack. At the end of func2, a single LDM instruction will restore any corrupted registers to their original values and move the address in the Link Register (that we pushed onto the stack) into the Program Counter. This brings us back to point where we left off in subroutine func1. At the end of func1, registers are restored and the saved Link Register value coming off the stack is moved into the Program Counter, taking us back to the main routine. If func1 doesn't save the Link Register

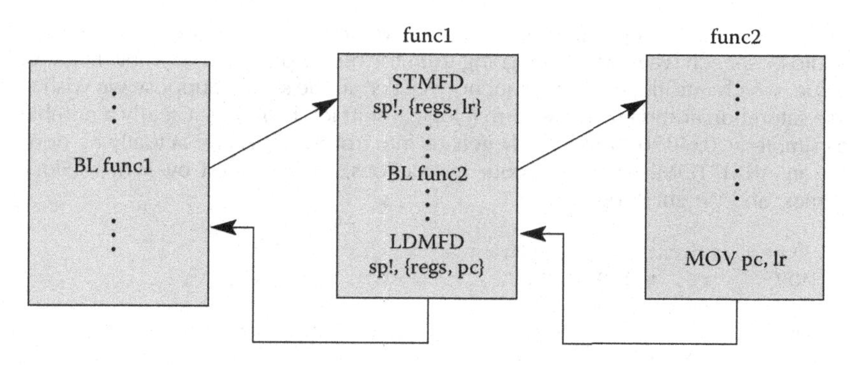

FIGURE 13.5 Stacking the Link Register during entry to a subroutine.

at the start of its routine, when func2 is called, the Link Register will get overwritten with a new return address, so when func1 finishes and tries to return to the main routine, it has the wrong return address.

13.4 PASSING PARAMETERS TO SUBROUTINES

Subroutines are often written to be as general as possible. A subroutine that computes the arctangent of 8 is of little use, but one that computes the arctangent of any given number can potentially be used throughout your code. Therefore, data or addresses need to be able to move into or out of a subroutine, and these values are called *parameters*. These can be passed to a subroutine through a predefined set of registers, a specified block of memory, or on the stack. We'll see what the trade-offs and requirements are for the different approaches, starting with the use of registers.

13.4.1 PASSING PARAMETERS IN REGISTERS

Passing parameters in registers is a fast way of transferring data between the calling program and a subroutine, but the subroutine must expect the data to be in specific registers.

EXAMPLE 13.4

Saturation arithmetic is used frequently in signal processing applications. For situations where the output of a digital filter or digital controller cannot exceed a certain value, saturation arithmetic can be used to effectively drive the result to either the most positive or most negative value that can be represented with a given number of bits. For example, a 32-bit value such as 0x7FFFFFFC can be seen as a large, positive value in a two's complement representation. However, if you were to add another small, positive number to it, such as 9, then the value becomes 0x80000005, which is now a very large negative number. If we were to use saturation arithmetic, the value returned from the addition would be 0x7FFFFFFF, which is the largest positive value you can represent in 32 bits using two's complement. Similarly, a large negative movement because of subtraction, e.g., 0x80000001 minus 2, would produce 0x80000000, the largest negative number, using saturation arithmetic. Imposing these two limits could be used to prevent an audio or speech waveform from going from the most positive to the most negative value, which introduces high-frequency "clicks" in the signal. Suppose we wish to use saturation arithmetic to perform a logical shift left by *m* bits. Clearly a number as simple as 0x40000000 already gets us into trouble. This can actually be done on an ARM7TDMI using only four instructions, as described by Symes (Sloss, Symes, and Wright 2004):

```
; r4 = saturate32(r5 << m)
MOV     r6, #0x7FFFFFFF
MOV     r4, r5, LSL m
TEQ     r5, r4, ASR m            ; if (r5! = (r4 >> m))
EORNE   r4, r6, r5, ASR #31      ; r4 = 0x7FFFFFFF^sign(r5)
```

Let's convert this algorithm into a subroutine, and then pass the parameters through registers. For our example, the parameters are the value to be shifted, the shift count, and the return value. Our target microcontroller can again be the LPC2132 from NXP. The code might look something like the following:

```
SRAM_BASE   EQU    0x40000000
      AREA  Passbyreg, CODE, READONLY
      ENTRY

      LDR   sp, =SRAM_BASE
      ; try out a positive case (this should saturate)
      MOV   r0, #0x40000000
      MOV   r1, #2
      BL    saturate

      ; try out a negative case (should not saturate)
      MOV   r0, #0xFFFFFFFE
      MOV   r1, #8
      BL    saturate

stop
      B     stop
saturate
      ; Subroutine saturate32
      ; Performs r2 = saturate32(r0 << r1)
      ; Registers used:
      ; r0 - operand to be shifted
      ; r1 - shift amount (m)
      ; r2 = result
      ; r6 - scratch register

      STMIA sp!,{r6,lr}
      MOV   r6, #0x7FFFFFFF
      MOV   r2, r0, LSL r1
      TEQ   r0, r2, ASR r1         ; if (r0! = (r2 >> m))
      EORNE r2, r6, r0, ASR #31    ; r2 = 0x7FFFFFFF^sign(r0)
      LDMDB sp!,{r6,pc}            ; return

      END
```

There are a few things to note in this example. The first is that we have three parameters to pass between the calling routine and the subroutine: the operand to be shifted, the shift amount, and the result. We can use registers r0, r1, and r2 for these parameters. Note that the subroutine also expects the parameters to be in these specific registers. The second point is that one register, r6, is corrupted in our subroutine, and we should, therefore, stack it to preserve its original value. Our stack pointer, register r13, is loaded with the base address of SRAM on the LPC2132. Our stack starts at this address and goes upward in memory. The Link Register is also stacked so that we ensure our subroutine can be interrupted, if necessary. We exit the subroutine by using only a single LDM instruction, since the last register to be updated is the PC, and this is loaded with the LR value, returning us to the calling routine.

13.4.2 Passing Parameters by Reference

A better approach to passing parameters is to send a subroutine *information* to locate the arguments to a function. Memory, such as a block of RAM, could hold the parameters, and then the calling program could pass just the address of the data to the subroutine, known as calling by reference. This allows for changing values, and in fact, is more efficient in terms of register usage for some types of data, e.g., a long string of characters. Rather than trying to pass large blocks of data through registers, the starting address of the data is the only parameter needed.

EXAMPLE 13.5

The same shift routine we wrote earlier could be written as shown below, now passing the address of our parameters in SRAM to the subroutine through register r3. Again, the target is the LPC2132.

```
SRAM_BASE      EQU       0x40000000
   AREA        Passbymem, CODE, READONLY
   ENTRY

   LDR    sp, =SRAM_BASE       ; stack pointer initialized
   LDR    r3, =SRAM_BASE + 100 ; writable memory for parameters

   ; try out a positive case (this should saturate)
   MOV    r1, #0x40000000
   MOV    r2, #2
   STMIA  r3, {r1,r2}          ; save off parameters
   BL     saturate

   ; try out a negative case (should not saturate)
   MOV    r1, #0xFFFFFFFE
   MOV    r2, #8
   STMIA  r3, {r1,r2}
   BL     saturate

stop
   B      stop
```

```
saturate
    ; Subroutine saturate32
    ; Parameters are read from memory, and the
    ; starting address is in register r3. The result
    ; is placed at the start of parameter memory.
    ; Registers used:
    ; r3 - holds address of parameters in memory
    ; r4 - result
    ; r5 - operand to be shifted
    ; r6 - scratch register
    ; r7 - shift amount (m)

    ; r4 = saturate32 (r5 << m)
    STMIA  sp!,{r4-r7,lr}          ; save off used registers
    LDMIA  r3, {r5,r7}             ; get parameters
    MOV    r6, #0x7FFFFFFF
    MOV    r4, r5, LSL r7
    TEQ    r5, r4, ASR r7          ; if (r5! = (r4 >>m))
    EORNE  r4, r6, r5, ASR #31     ; r4 = 0x7FFFFFFF^sign(r5)
    STR    r4, [r3]                ; move result to memory
    LDMDB  sp!,{r4-r7,pc}          ; return

    END
```

The operand to be shifted and the shift count are stored in memory starting at address SRAM_BASE + 100, where they are read by the subroutine. The entry to the subroutine does some housekeeping by saving off the registers about to be corrupted to the stack, including the Link Register. This is required by the ARM Application Procedure Call Standard (AAPCS), which is covered shortly.

There are two options for returning a value from this subroutine. The first is to just store it back in memory for later reading by some other code. The second option is to return the value in a register, say register r3. In our example, the value is stored back to memory. If you were doing string comparisons, you might call a subroutine and send the addresses of the two strings to the subroutine, expecting either a one (they matched) or a zero (they did not match) to be stored in a register as the result.

13.4.3 Passing Parameters on the Stack

One of the most straightforward ways to pass parameters to a subroutine is to use the stack. This is very similar to passing parameters in memory, only now the subroutine uses a dedicated register for a pointer into memory—the stack pointer, register r13. Data is pushed onto the stack before the subroutine call; the subroutine grabs the data off the stack to be used; and results are then stored back onto the stack to be retrieved by the calling routine.

At this point it's worth mentioning that a programmer should be mindful of *which* stack pointer he or she is using. Recall from Chapter 2 that the ARM7TDMI has different stack pointers for Supervisor mode, the exception modes, and for User mode, allowing different stacks to be built for the different modes if the programmer wishes to do so. The Cortex-M4 has two stack pointers, a main stack pointer (MSP), which is the default stack pointer, and a process stack pointer (PSP). The choice of stack pointers is controlled through the CONTROL Register, which was mentioned briefly in Chapter 2. We'll examine these more when dealing with exceptions in Chapter 15.

EXAMPLE 13.6

Rewriting the same saturated shift routine using the stack would look something like the code that follows:

```
SRAM_BASE       EQU       0x40000000
      AREA      Passbystack, CODE, READONLY
      ENTRY

      LDR       sp, =SRAM_BASE  ; stack pointer initialized

      ; try out a positive case (this should saturate)
      MOV       r1, #0x40000000
      MOV       r2, #2
      STMIA     sp!, {r1,r2}     ; push parameters on the stack
      BL        saturate
      ; pop results off the stack
      ; now r1 = result of shift
      LDMDB     sp!, {r1,r2}

      ; try out a negative case (should not saturate)
      MOV       r1, #0xFFFFFFFE
      MOV       r2, #8
      STMIA     sp!, {r1,r2}
      BL        saturate
      LDMDB     sp!, {r1,r2}
stop
      B         stop
saturate
      ; Subroutine saturate32
      ; Parameters are read from the stack, and
      ; registers r4 through r7 are saved on the stack.
      ; The result is placed at the bottom of the stack.
      ; Registers used:
      ; r4 - result
      ; r5 - operand to be shifted
      ; r6 - scratch register
      ; r7 - shift amount (m)

      ; r4 = saturate32 (r5 << m)
      STMIA     sp!,{r4-r7,lr}          ; save off used registers
      LDR       r5, [sp, #-0x20]        ; get first parameter off stack
      LDR       r7, [sp, #-0x1C]        ; get second parameter off stack
      MOV       r6, #0x7FFFFFFF
      MOV       r4, r5, LSL r7
      TEQ       r5, r4, ASR r7          ; if (r5! = (r4 >>m))
      EORNE     r4, r6, r5, ASR #31     ; r4 = 0x7FFFFFFF^sign(r5)
      STR       r4, [sp, #-0x20]        ; move result to bottom of stack
      LDMDB     sp!,{r4-r7,pc}          ; return

      END
```

The stack structure is drawn in Figure 13.6. The two parameters are pushed to the bottom of the stack, and then the saved registers are stacked on top of them,

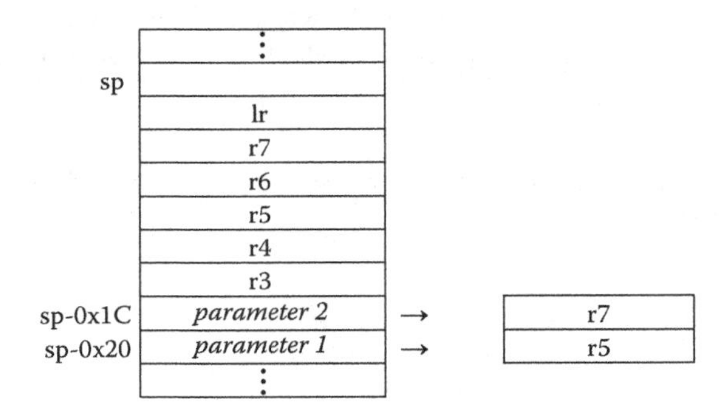

FIGURE 13.6 Stack configuration.

ending with the Link Register at the top. Since register r13 now points to the top of the stack, it's necessary to use the stack pointer with a negative offset to address the parameters. When the result is computed, it must be stored back to the bottom of the stack, again using the stack pointer with a negative offset.

If the example above used full descending stacks, then the PUSH and POP instructions could be used just as easily. To see how this might look using a Cortex-M4, let's examine the same algorithm that uses full descending stacks.

EXAMPLE 13.7

Since the object of the example is to compare the stack structures rather than the algorithm itself, the following code shows how to push two values onto the stack, call the subroutine, and then pop two values off the stack. Careful readers will have noticed that if the shift value were fixed, rather than variable as it is in our subroutine, you could save quite a bit of coding by just using the SSAT instruction that we saw in Chapter 7. For this example, the SRAM block begins at address 0x20000000 on the Tiva TM4C123GH6ZRB.

```
SRAM_BASE   EQU     0x20000200

    LDR     sp, =SRAM_BASE          ; stack pointer initialized

    ; try out a positive case (this should saturate)
    MOV     r1, #0x40000000
    MOV     r2, #2
    PUSH    {r1,r2}                 ; push parameters on the stack
    BL      saturate
    ; pop results off the stack
    ; now r1 = result of shift
    POP     {r1,r2}

    ; try out a negative case (should not saturate)
    MOV     r1, #0xFFFFFFFE
    MOV     r2, #8
    PUSH    {r1,r2}
```

```
        BL      saturate
        POP     {r1,r2}
stop
        B       stop

saturate
        ; Subroutine saturate32
        ; Parameters are read from the stack, and
        ; registers r4 through r7 are saved on the stack.
        ; The result is placed at the bottom of the stack.
        ; Registers used:
        ; r4 - result
        ; r5 - operand to be shifted
        ; r6 - scratch register
        ; r7 - shift amount (m)

        ; r4 = saturate32(r5 << m)
        PUSH    {r4-r7,lr}             ; save off used registers
        LDR     r5, [sp, #0x14]        ; get first parameter off stack
        LDR     r7, [sp, #0x18]        ; get second parameter off stack
        MOV     r6, #0x7FFFFFFF
        MOV     r4, r5, LSL r7
        ASR     r4, r7
        TEQ     r5, r4                 ; if (r5! = (r4 >>m))
        IT      NE
        EORNE   r4, r6, r5, ASR #31    ; r4 = 0x7FFFFFFF^sign(r5)
        STR     r4, [sp, #0x14]        ; move result to bottom of stack
        POP     {r4-r7,pc}             ; return
```

Notice at the end of the subroutine that the Link Register value that was pushed onto the stack is now loaded into the Program Counter using the POP instruction, similar to the method used by the ARM7TDMI.

13.5 THE ARM APCS

It turns out that there's a standard called the ARM Application Procedure Call Standard (AAPCS) for the ARM architecture, which is part of the Application Binary Interface (ABI) (ARM 2007b). The AAPCS defines how subroutines can be separately written, separately compiled, and separately assembled to work together. It describes a contract between a calling routine and a called routine that defines:

- Obligations on the caller to create a program state in which the called routine may start to execute
- Obligations on the called routine to preserve the program state of the caller across the call
- The rights of the called routine to alter the program state of its caller

The standard is also designed to combine the ease, speed, and efficiency of passing parameters through registers with the flexibility and extensibility of passing in the stack. While the document describes procedures for writing code, it also defines

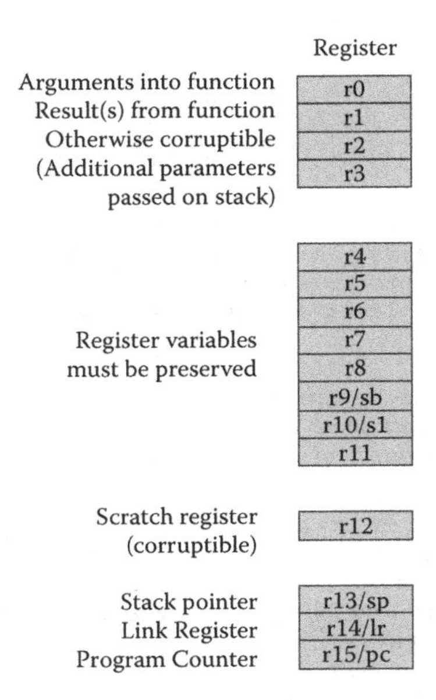

FIGURE 13.7 The ARM APCS specification for register usage.

the use of registers, shown in Figure 13.7. Some parts of the specification include the following:

The first four registers r0-r3 (also called a1-a4, for argument) are used to pass argument values into a subroutine and to return a result value from a function. They may also be used to hold intermediate values within a routine (but in general, only *between* subroutine calls).

Register r12 (IP) may be used by a linker as a scratch register between a routine and any subroutine it calls. It can also be used within a routine to hold intermediate values *between* subroutine calls.

Typically, the registers r4-r8, r10, and r11 are used to hold the values of a routine's local variables. Of these, only r4-r7 can be used uniformly by the whole Thumb instruction set, but the AAPCS does not require that Thumb code only use those registers.

A subroutine must preserve the contents of the registers r4-r8, r10, r11, and SP (and r9 in some Procedure Call Standard variants).

Stacks must be eight-byte aligned, and the ARM and Thumb C and C++ compilers always use a full descending stack.

For floating-point operations, similar rules apply. According to the standard, registers s16-s31 must be preserved across subroutine calls; registers s0-s15 do not need to be preserved, so you could use these for passing arguments to a subroutine. The only status register that may be accessed by conforming code is the FPSCR, and within this register, certain bits must be left unchanged. While it's important to write code that conforms to the specification, beginning programmers would do well to practice with the specification in mind, and as time permits, rework your code to follow the standard.

EXAMPLE 13.8

Let's look at a short floating-point routine for the Cortex-M4 that uses a Taylor series expansion to compute the value of sin(x). The subroutine uses registers s0-s10, so no floating-point registers need to be stacked. The input to the routine and the final result are stored in register s0. As we saw in Chapter 7, sometimes divisions can be avoided entirely by calculating constants beforehand and using multiplication operations instead.

```
;*************************************************************
;
; This is the code that gets called when the processor first starts execution
; following a reset event.
;
;*************************************************************
   EXPORT Reset_Handler
      ENTRY

Reset_Handler

      ; Enable the FPU
      ; Code taken from ARM website
      ; CPACR is located at address 0xE000ED88
      LDR.W    r0, =0xE000ED88

      LDR      r1, [r0]                  ; Read CPACR
      ; Set bits 20-23 to enable CP10 and CP11 coprocessors
      ORR      r1, r1, #(0xF << 20)
      ; Write back the modified value to the CPACR
      STR      r1, [r0]                  ; wait for store to complete
      DSB

      ; Reset pipeline now that the FPU is enabled
      ISB

      ;
      ; The calculation of the sin(x) will be done in the
      ; subroutine SinCalc. The AAPCS dictates the first
      ; 16 FPU registers (s0-s15) are not preserved, so we will
      ; use them in the calling routine to pass the operand and
      ; return the result. Registers s16-s31 must be preserved in
      ; a subroutine, so they are used in the calling routine.

      ; FPU registers
      ; s0 - Passed operand and returned result

      ; Evaluate the function for operand the test operand
      VLDR.F32         s0, =1.04719
      BL       SinCalc
Exit  B        Exit

      ; Sine code
      ; The algorithm is a Taylor series with
      ; 4 terms (x=x - x^3/3!+x^5/5! - x^7/7!)
      ; Optimized, we have 9 multiplications and 3 adds.
      ; We can avoid the divisions by computing 1/3!, 1/5!, etc. and
      ; using the constant in a multiplication.
```

```
            ;
            ; This formula works for all x in the range [0, pi/2]
            ; [0, pi/2]
            ;
            ; This routine assumes AAPCS -
            ; regs s0-s15 parameters and/or scratch
            ; Register usage:
            ; s0 - input operand and return result
            ; s1 - 1/3! (invfact3)
            ; s2 - 1/5! (invfact5)
            ; s3 - 1/7! (invfact7)
            ; s4 - x * s1 (xdiv3), then s4 * s7 (x^2 * xdiv3) (x3div3)
            ; s5 - x * s2 (xdiv5), then s5 * s8 (x^4 * xdiv5) (x5div5)
            ; s6 - x * s3 (xdiv7), then s6 * s9 (x^6 * xdiv7) (x7div7)
            ; s7 - x^2
            ; s8 - x^4
            ; s9 - x^6
            ; s10 - scratch
SinCalc
            ; set up the three inverse factorial constants
            VLDR.F32   s1, invfact3
            VLDR.F32   s2, invfact5
            VLDR.F32   s3, invfact7

            ;
            VMUL.F32   s4, s0, s1 ; compute xdiv3
            VMUL.F32   s7, s0, s0 ; compute x^2
            VMUL.F32   s5, s0, s2 ; compute xdiv5
            VMUL.F32   s4, s4, s7 ; compute x3div3
            VMUL.F32   s8, s7, s7 ; compute x^4
            VMUL.F32   s6, s0, s3 ; compute xdiv7
            VSUB.F32   s10, s0, s4 ; compute terms12, x-x^3/3!
            VMUL.F32   s9, s7, s8 ; compute x^6
            VMUL.F32   s5, s5, s8 ; compute x5div5
            VMUL.F32   s6, s6, s9 ; compute x7div7
            VADD.F32   s10, s10, s5 ; compute terms123, x-x^3/3!+x^5/5!
            VSUB.F32   s0, s10, s6 ; compute result

            BX         lr ; return

invfact3    DCD        0x3E2AAAAB ; 1/3!
invfact5    DCD        0x3C088888 ; 1/5!
invfact7    DCD        0x39500CD1 ; 1/7!
```

13.6 EXERCISES

1. What's wrong with the following ARM7TDMI instructions?
   a. `STMIA    r5!, {r5, r4, r9}`
   b. `LDMDA    r2, {}`
   c. `STMDB    r15!, {r0-r3, r4, lr}`

2. On the ARM7TDMI, if register r6 holds the address 0x8000 and you executed the instruction

   ```
   STMIA  r6, {r7, r4, r0, lr}
   ```

   what address now holds the value in register r0? Register r4? Register r7? The Link Register?

3. Assume that memory and ARM7TDMI registers r0 through r3 appear as follows:

Address			Register
0x8010	0x00000001	0x13	r0
0x800C	0xFEEDDEAF	0xFFFFFFFF	r1
0x8008	0x00008888	0xEEEEEEEE	r2
0x8004	0x12340000	0x8000	r3
0x8000	0xBABE0000		

Describe the memory and register contents after executing the instruction

```
LDMIA  r3!, {r0, r1, r2}
```

4. Suppose that a stack appears as shown in the first diagram below. Give the instruction or instructions that would push or pop data so that memory appears in the order shown. In other words, what instruction would be necessary to go from the original state to that shown in (a), and then (b), and then (c)?

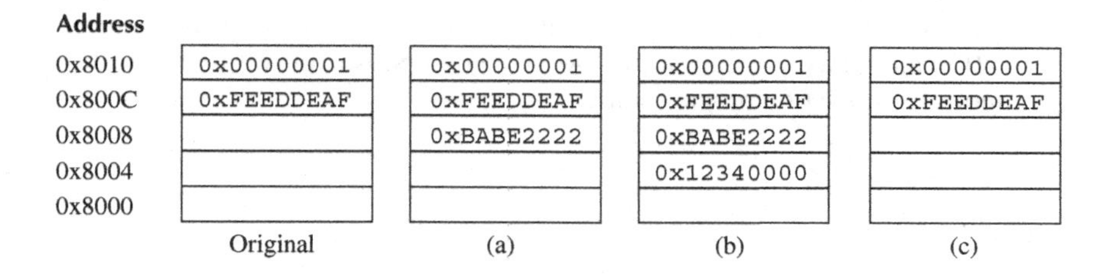

Address	Original	(a)	(b)	(c)
0x8010	0x00000001	0x00000001	0x00000001	0x00000001
0x800C	0xFEEDDEAF	0xFEEDDEAF	0xFEEDDEAF	0xFEEDDEAF
0x8008		0xBABE2222	0xBABE2222	
0x8004			0x12340000	
0x8000				

5. Convert the cosine table from Problem 1 in Chapter 12 into a subroutine, using a full descending stack.

6. Rewrite Example 13.4 using full descending stacks.

7. Rewrite Example 13.5 using full descending stacks.

8. Rewrite Example 13.6 using full descending stacks.

9. Convert the factorial program written in Chapter 3 into a subroutine, using full descending stacks. Pass arguments to the subroutine using both pass-by-register and pass-by-stack techniques.

10. Write the ARM7TDMI division routine from Chapter 7 as a subroutine that uses empty ascending stacks. Pass the subroutine arguments using registers, and test the code by dividing 4000 by 32.

11. Match the following terms with their definitions:

 a. Recursive 1. Subroutine can be interrupted and called by the interrupting routine

b. Relocatable	2. A subroutine that calls itself
c. Position independent	3. The subroutine can be placed anywhere in memory
d. Reentrant	4. All program addresses are calculated relative to the Program Counter

12. Write the ARM7TDMI division routine from Chapter 7 as a subroutine that uses full descending stacks. Pass the subroutine arguments using the stack, and test the code by dividing 142 by 7.

13. Write ARM assembly to implement a PUSH operation without using LDM or STM instructions. The routine should handle three data types, where register r0 contains 1 for byte values, 2 for halfword values, and 4 for word values. Register rl should contain the data to be stored on the stack. The stack pointer should be updated at the end of the operation.

14. Write ARM assembly to check whether an N × N matrix is a magic square. A magic square is an N × N matrix in which the sums of all rows, columns, and the two diagonals add up to $N(N^2 + 1)/2$. All matrix entries are unique numbers from 1 to N^2. Register rl will hold N. The matrix starts at location 0x4000 and ends at location $(0x4000 + N^2)$. Suppose you wanted to test a famous example of a magic square:

16	3	2	13
5	10	11	8
9	6	7	12
4	15	14	1

The numbers 16, 3, 2, and 13 would be stored at addresses 0x4000 to 0x4003, respectively. The numbers 5, 10, 11, and 8 would be stored at addresses 0x4004 to 0x4007, etc. Assume all numbers are bytes. If the matrix is a magic square, register r9 will be set upon completion; otherwise it will be cleared. You can find other magic square examples, such as Ben Franklin's own 8 × 8 magic square, on the Internet to test your program.

15. Another common operation in signal processing and control applications is to compute a dot product, given as

$$a = \sum_{m=0}^{N-1} c_m x_m$$

where the dot product a is a sum of products. The coefficients c_m and the input samples x_m are stored as arrays in memory. Assume sample data and coefficients are 16-bit, unsigned integers. Write the assembly code to compute a dot product for 20 samples. This will allow you to use the LDM instruction to load registers with coefficients and data efficiently. You probably want to bring in four or five values at a time, looping as needed to exhaust all values. Leave the dot product in a register, and give the register the name DPROD using the RN directive in your code. If you use the newer v7-M SIMD instructions, note that you can perform two multiples on 16-bit values at the same time.

16. Suppose your stack was defined to be between addresses 0x40000000 and 0x40000200, with program variables located at address 0x40000204 and higher in memory, and your stack pointer contains the address 0x400001FC. What do you think would happen if you attempt to store 8 words of data on an ascending stack?

14 Exception Handling
ARM7TDMI

14.1 INTRODUCTION

Large applications, including operating systems, often have to deal with inputs from various sources, such as keyboards, mice, USB ports, and even power management blocks telling the processor its battery is about to run dry. Sometimes an embedded microcontroller has only one or two external input sources (e.g., from sensors in an engine), but it may still have peripheral devices that may need attention from time to time, such as a watchdog timer. Universal asynchronous receiver/transmitters (UARTs), wake-up alerts, analog-to-digital converters (ADCs), and I^2C devices can all demand the processor's time. In the next two chapters, we're going to examine the different types of exceptions a processor can face in light of the fact that they are not isolated, only running code and talking to no one. In our definition of an exception, the events that can cause one must not be immediately thought of as bad or unwanted. Exceptions include benign events like an interrupt, and this can be any kind of interrupt, like someone moving a mouse or pushing a button. Technically, anything that breaks a program's normal flow could be considered an exception, but it's worth detailing the different types, since some can be readily handled and others are unexpected and can cause problems. At this end of the spectrum, catastrophic faults, such as a bus error when trying to fetch an instruction, may have no solution in software and the best outcome may be to alert the user before halting the entire system. Certain events can lead to a serious system failure, and while they are rare, they should be anticipated to help find the cause of the problem during application development or to plan for a graceful shutdown. For example, a rogue instruction in the processor's pipeline or a memory access to an address that doesn't exist should not occur once the software is finished and tested. Version 4T cores and version 7-M cores handle exceptions differently, and we'll therefore examine the exception model for the Cortex-M4 in Chapter 15. In this chapter, we'll start with the exception model for the ARM7TDMI, and we'll examine exceptions in two large classes—interrupts and error conditions.

14.2 INTERRUPTS

Interrupts are very common in microprocessor systems. They provide the ability for a device such as a timer or a USB interface to poke the processor in the ribs and loudly announce that it wants attention. Historically, large computers only took a set of instructions and some data, calculated an answer, and then stopped. These machines had no worries about dozens of interfaces and devices all vying for part of the CPU's time. Once microprocessors became ubiquitous in electronic devices, they had to deal with supporting an operating system and application software in addition to calculating and moving data for other parts of a system. Microcontrollers are, in effect, smaller versions of complete systems, where motor controllers, timers, real-time clocks, and serial interfaces all demand some face time from the processor. So what's the best way to let the processor do its main tasks while allowing other peripherals to ask for assistance every so often?

Say you had a UART, which is a type of serial interface, attached to a processor that received a character from another device, say a wireless keyboard. When a character shows up in the UART, it's basically sitting at a memory location assigned to the UART waiting for the processor to get the data. There are roughly three ways the processor can handle this situation. The first, and by far the least efficient, is for the processor to sit in a loop doing absolutely nothing except waiting for the character to show up. Given the speed at which processors run, where billions of instructions can now be processed in a single second, waiting even 1/100th of a second for a device to transmit the data wastes millions of cycles of bandwidth. The second option is for the processor to occasionally check the memory location to see if there is some new data there, known as polling. While the processor can do other things while it waits, it still has to take time to examine the (possibly empty) memory location. The third option, and clearly the best one, is to have the device tell the processor when there is new data waiting for it. This way, the processor can spend its time performing other functions, such as updating a display or converting MP3 data to an analog waveform, while it waits for a slower device to complete its task. An interrupt is therefore an efficient method for telling the processor that something (usually a device or peripheral) needs attention. If you refer back to the diagram of the ARM7TDMI in Chapter 1 (Figure 1.4), you will notice two external lines coming into the part—nIRQ and nFIQ, where the "n" denotes an active low signal. These are the two interrupt lines going into the processor, with a low priority interrupt called IRQ and a high priority interrupt called FIQ. In addition to hardware interrupts, software has one as well, called aptly enough, Software Interrupt or SWI in the older notation, and SVC in the newer notation. We will look at all of these in detail to see how they work.

14.3 ERROR CONDITIONS

While you hope not to have these exceptions in a system, they do occur often enough that software needs to be sufficiently robust to handle them. The ARM cores recognize a few error conditions, some of which are easy to handle, some of which are not. An undefined instruction in the program can cause an error, but this may or may not be intentional. In a completely tested system where no new code is introduced (e.g., an embedded processor in an MP3 player that only handles the display), one would not expect to see a strange instruction suddenly show up in the application code. However, if you know that you have a design that requires floating-point operations, but the processor does not support floating-point in hardware, you could decide to use floating-point instructions and emulate them in software. Once the processor sees floating-point instructions (which aren't listed in this book but can be found in the *ARM Architectural Reference Manual* (ARM 2007c)), it will take an undefined instruction exception since there is no hardware to perform the operations. The processor can then take the necessary actions to perform the operations anyway, using only software to emulate the floating-point operation, appearing to the user as if floating-point hardware were present.

Data and prefetch aborts are the exception types that often cause programmers the most angst. A prefetch abort occurs when the processor attempts to grab an instruction in memory but something goes wrong—if memory doesn't exist or the address is outside of a defined memory area, the processor should normally be programmed to recover from this. If the address is not "expected" but still permitted, then the processor may have additional hardware (known as a memory management unit or MMU) to help it out, but this topic is outside the scope of this book. A data abort occurs when the processor attempts to grab data in memory and something goes wrong (e.g., the processor is in an unprivileged mode and the memory is marked as being readable only in a privileged mode). Certain memory regions may be configured as being readable but not writable, and an attempt to write to such a region can cause a data abort. As with prefetch aborts, the processor usually needs to be able to recover from some situations and often has hardware to assist in the recovery. We will see more about aborts in Section 14.8.4.

14.4 PROCESSOR EXCEPTION SEQUENCE

When an exception occurs, the ARM7TDMI processor has a defined sequence of events to start the handling and recovery of the exception. In all cases except a reset exception, the current instruction is allowed to complete. Afterward, the following sequence begins automatically:

- The CPSR is copied into SPSR_<mode>, where <mode> is the new mode into which the processor is about to change. Recall from Chapter 2 that the register file contains SPSR registers for exceptional modes, shown in Figure 14.1.
- The appropriate CPSR bits are set. The core will switch to ARM state if it was in Thumb state, as certain instructions do not exist in Thumb that are needed to access the status registers. The core will also change to the new exception mode, setting the least significant 5 bits in the CPSR register. IRQ interrupts are also disabled automatically on entry to all exceptions. FIQ interrupts are disabled on entry to reset and FIQ exceptions.
- The return address is stored in LR_<mode>, where <mode> is the new exception mode.
- The Program Counter changes to the appropriate vector address in memory.

Note that the processor is responsible for the above actions—no code needs to be written. At this point, the processor begins executing code from an exception handler, which is a block of code written specifically to deal with the various exceptions. We'll look at how handlers are written and what's done in them shortly. Once

Mode					
User/System	Supervisor	Abort	Undefined	Interrupt	Fast Interrupt
R0	R0	R0	R0	R0	R0
R1	R1	R1	R1	R1	R1
R2	R2	R2	R2	R2	R2
R3	R3	R3	R3	R3	R3
R4	R4	R4	R4	R4	R4
R5	R5	R5	R5	R5	R5
R6	R6	R6	R6	R6	R6
R7	R7	R7	R7	R7	R7
R8	R8	R8	R8	R8	R8_FIQ
R9	R9	R9	R9	R9	R9_FIQ
R10	R10	R10	R10	R10	R10_FIQ
R11	R11	R11	R11	R11	R11_FIQ
R12	R12	R12	R12	R12	R12_FIQ
R13	R13_SVC	R13_ABORT	R13_UNDEF	R13_IRQ	R13_FIQ
R14	R14_SVC	R14_ABORT	R14_UNDEF	R14_IRQ	R14_FIQ
PC	PC	PC	PC	PC	PC

CPSR	CPSR	CPSR	CPSR	CPSR	CPSR
	SPSR_SVC	SPSR_ABORT	SPSR_UNDEF	SPSR_IRQ	SPSR_FIQ

= *banked register*

FIGURE 14.1 Register organization.

the handler completes, the processor should then return to the main code—whether or not it returns to the instruction that caused the exception depends on the type of exception. The handler may restore the Program Counter to the address of the instruction after the one that caused the exception. Either way, the last two things that remain to be done, and must be done by the software handler, are

- The CPSR must be restored from SPSR_<mode>, where <mode> is the exception mode in which the processor currently operates.
- The PC must be restored from LR_<mode>.

These actions can only be done in ARM state, and fortunately, the software can usually do these two operations with a single instruction at the end of the handler.

It is worth noting at this point that while most ARM cores have similar exception handling sequences, there are some differences in the newest cores (e.g., the Cortex-M3/M4 has a different programmer's model, and the Cortex-A15 has even more exception types and modes). The Technical Reference Manuals for the individual cores contain complete descriptions of exception sequences, so future projects using version 7 and version 8 processors might require a little reading first.

14.5 THE VECTOR TABLE

Earlier in Section 2.3.3, we saw the exception vector table for the first time, but we didn't do much with it. At this point, we can start using these addresses to handle the various types of exceptions covered in this chapter. Figure 14.2 shows the table again, with the vectors listed as they would be seen in memory. Recall that while some processors, e.g., the 6502 and Freescale's 680x0 families, put addresses in their vector tables, ARM uses actual instructions, so the reset exception vector (at address 0x0) would have a change-of-flow instruction of some type sitting there. It may not be the actual instruction B, as we'll see in a moment.

Having covered literal pools, we can now begin to examine the way that real ARM code would be written and stored in memory with regard to exceptions. Figure 14.3 shows a memory map from address 0x0 to 0xFFFFFFFF and an example layout for the exception handlers. Note that this is only an example, and may not be applicable to your application, so these are just options. For each type of exception, there is usually a dedicated block of code, called an exception handler, that is responsible for acknowledging an exceptional condition and, more often than not, fixing it. Afterward, the code should return the processor back to the point from where it left, now able to continue without the exception. Not all exceptions need handlers, and in some deeply embedded systems, the processor may not be able to recover. Consider the hypothetical situation where the processor tries to read a memory location that is not physically present. Further suppose that an address in the memory map does not point to a memory chip or memory block but rather points to nothing. The machine may be programmed to reset itself if something like that ever happens. Larger applications, such as a cell phone, will have to deal with all exceptions and provide robust methods to recover from them, especially in light of having hardware that can change, e.g., if a memory card can be added or removed.

Since exception vectors contain instructions at their respective addresses in memory, an exception such as an IRQ, which is a low-priority interrupt, would have some kind of change-of-flow instruction in its exception vector to force the processor to

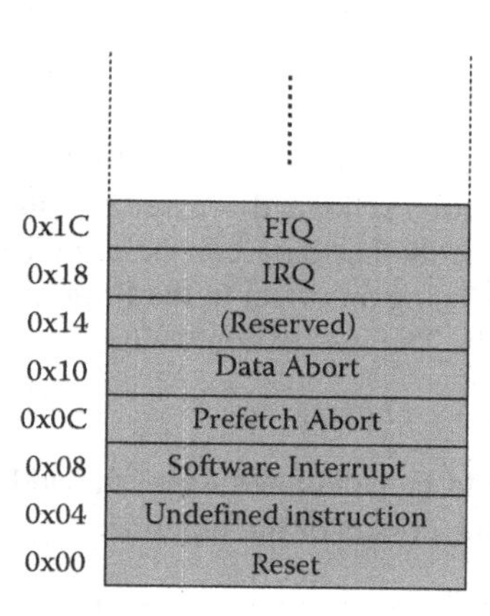

FIGURE 14.2 Exception vector table.

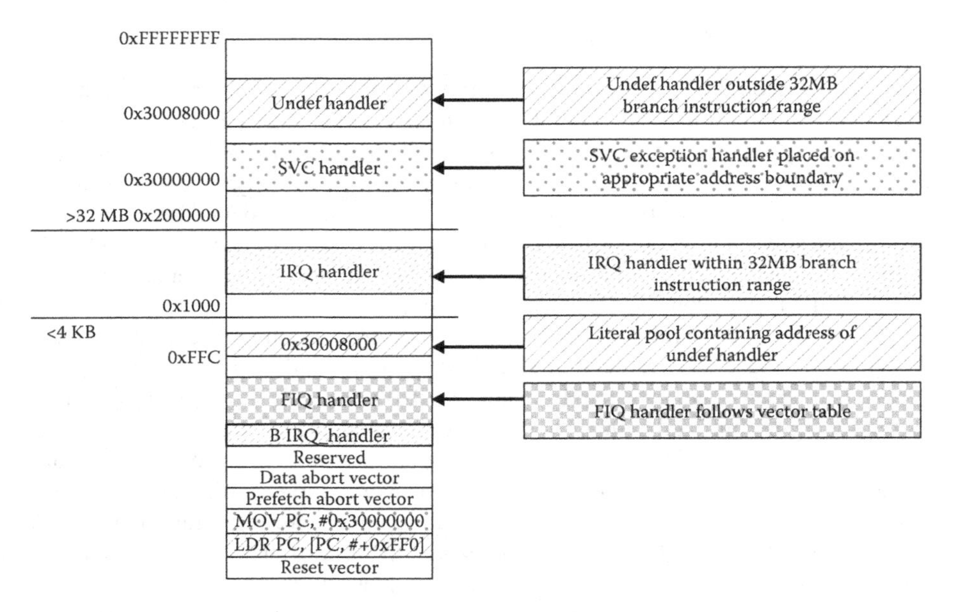

FIGURE 14.3 Example memory map with exception handlers.

begin fetching instructions from its handler. These change-of-flow instructions are one of the following:

- Branch instruction—A direct branch can be used to jump to your exception handler; however, the range of a B instruction is only 32 MB, and this may not always work with every memory organization. If your exception handler is more than 32 MB away, you must use another type of instruction.
- MOV instruction—A MOV instruction can change the PC simply by loading the register with a value. The value loaded could be created from a byte rotated by an even number of bits, so that it fits within a 32-bit instruction, for example,

```
MOV    PC, #0xEF000000
```

  Notice also that this instruction contains a 32-bit address, but it can be constructed using the rotation scheme discussed in Chapter 6.
- LDR instruction—Recall that data can be stored in instruction memory and then accessed by using an offset to the Program Counter, as we saw in Chapter 6 with literals. The instruction would have the form

```
LDR    PC, [PC+offset]
```

  where the offset would be calculated using the address of the handler, the vector address, and the effects of the pipeline.

Looking at our example memory map in Figure 14.3, we see that the reset exception can use a simple branch (B) instruction, provided that we place the reset handler within a 32 MB range. The next exception, the undefined instruction exception, uses a load instruction with an offset to access the value sitting at address 0xFFC in memory. When the processor is executing the load instruction at address 0x4, the Program Counter contains the value 0xC, since it is the address of the instruction being fetched. The offset would then be 0xFF0. When the value at address 0xFFC, 0x30008000, is loaded into the Program Counter, it has the same effect as jumping there with a branch instruction. One other thing to note is the size of the offset used—there are only 12 bits to create an offset with a load instruction of this type; hence, the value 0xFFC is the last word that could be accessed within that 4 KB range. The next word has the address 0x1000 and is too far away.

Continuing up the table, we come to the SWI (or SVC, as it is now known) exception. This example shows that an address like 0x30000000 can be generated using the byte rotation scheme and therefore can be reached using a simple MOV instruction. The SVC handler is then placed at that location in memory. Skipping the two abort exceptions and the reserved vector, we continue to address 0x18, where the IRQ exception vector contains a simple branch instruction, and the IRQ handler starts at an address that is located within 32 MB of the branch.

The last exception vector, to which we alluded back in Chapter 2, sits at the top of the vector table for a reason. FIQ interrupts are fast interrupts, meaning that if you have a critical event that must be serviced immediately, and it holds the highest priority among interrupts, then you want to spend as little time as possible getting to the code to service it. We also know from Chapter 8 that branches cause a pipelined architecture to throw away instructions, so rather than cause the processor to take some type of branch to get to the handler, the FIQ vector is the first instruction of the handler! There is no need to branch, since the Program Counter will be set to the address 0x1C when the processor acknowledges and begins to service the interrupt. The FIQ handler is then executed as the Program Counter increments through memory.

14.6 EXCEPTION HANDLERS

Exceptions require some housekeeping. Normally, the processor is busy moving data or crunching numbers, but when an exception occurs, processors have to prepare to save the status of the machine, since at some point they must return to crunching numbers or moving data. When an exceptional condition is seen by the processor, the first thing it must do is copy the Current Program Status Register into a Saved Program Status Register, and in particular the SPSR belonging to the new mode associated with the exception. Recall that for five of the seven modes, there are unique SPSRs (e.g., Abort Mode has an SPSR_abort). The Current Program Status Register must then be changed to reflect what happened—the mode bits will be changed, further interrupts may be disabled, and the state will change from Thumb state to ARM state if the processor was executing Thumb instructions (there's more on Thumb in Chapter 17). Since exceptions cause the code to jump to a new location in memory, it's imperative to save off a return address, akin to what was done for subroutines, so that the processor can return later. This return address is stored in the Link Register associated with the exception type (e.g., R14_FIQ if the processor took an FIQ exception). The Program Counter is then changed to the appropriate vector address. All of this work is done by the processor, so the focus for the programmer is to write the appropriate exception handler.

Once the mode of the processor has been changed, the exception handler will have to access its own stack pointer (R13_FIQ, for example), its own Link Register, and its own Saved Program Status Register. There are general-purpose registers that can be used by the handler, but normally some of these are saved off to a stack before using them, since any registers that are corrupted by a handler must be restored before returning (refer back to the ARM Application Procedure Call Standard that we saw in Chapter 13). This whole process of storing registers to external memory takes time, and again, depending on the type of exception, may cause an unacceptable delay in processing an exception. Going back to the idea of a fast interrupt, we've already seen that the FIQ vector sits at the top of the vector table, saving a branch instruction. To prevent having to store data on the stack before handling the FIQ interrupts, there are also five additional general-purpose registers (R8_FIQ to R12_FIQ) that the handler can access.

Exception handlers can be intentionally short, or long, robust routines depending on how much needs to be done for any given exception. In Section 14.8.2, we'll examine an undefined exception handler of only a few lines. At the end of all handlers, the programmer is responsible for restoring the state of the machine and returning back to the original instruction stream before the exception. This can be done as an atomic operation, moving the contents of the SPSR back into the CPSR while moving the Link Register into the Program Counter. The instructions to do these operations only exist in the ARM instruction set, which was why the processor had to switch from Thumb to ARM if it was executing Thumb code. The various return methods are discussed more in the next few sections.

14.7 EXCEPTION PRIORITIES

Exceptions must be prioritized in the event that multiple exceptions happen at the same time. Consider the case where a peripheral on a microcontroller has generated a low-priority interrupt, say an A/D converter has finished sampling some data and alerts the processor by pulling on the IRQ line. At the exact same time, the processor tries to access a memory location that is undefined while another high-priority interrupt tries to tell the processor that we're about to lose power in two minutes. The processor must now decide which exception type gets handled first. Table 14.1 shows the exception types in order of their priority.

For complicated reasons, data aborts are given the highest priority apart from the reset exception, since if they weren't, there would be cases where if two or more exceptions occurred simultaneously, an abort could go undetected. If an FIQ and an IRQ interrupt occur at the same time, the FIQ interrupt handler goes first, and afterward, the IRQ will still be pending, so the processor should still be able to service it. An SVC and an Undefined Instruction exception are mutually exclusive, since an SVC instruction is defined and cannot generate an Undefined Instruction exception. To settle the contention described earlier, the Data Abort exception would be handled first, followed by the FIQ interrupt alerting the system to a power failure, and then the A/D converter will have its turn.

TABLE 14.1
ARM7TDMI Exception Priorities

Priority	Exception	Comment
Highest	Reset	Handler usually branches straight to the main routine.
	Data Abort	Can sometimes be helped with hardware (MMU).
	FIQ	Current instruction completes, then the interrupt is acknowledged.
	IRQ	Current instruction completes, then the interrupt is acknowledged. Used more often than FIQ.
	Prefetch Abort	Can sometimes be helped with hardware (MMU).
	SVC	Execution of the instruction causes the exception.
Lowest	Undefined Instruction	SVC and Undef are actually mutually exclusive, so they have the same priority.

A situation could present itself where the processor is already handling an exception and another exception occurs. For example, suppose the processor is working on an FIQ interrupt and has already begun executing the handler for it. During the course of executing this code, a data abort occurs—one that could be helped by additional MMU hardware. The processor would begin exception processing again, storing the CPSR into the register SPSR_abort, changing the mode, etc., and then jump to the new exception handler. Once the second exception completes and the Program Counter points back to the first handler, the original FIQ exception can finish up. If another FIQ interrupt tried to interrupt instead of a data abort, it would be blocked, because FIQ interrupts are automatically disabled by the processor upon entry to FIQ mode. The software could enable them again, but this is not typical practice.

14.8 PROCEDURES FOR HANDLING EXCEPTIONS

As mentioned before, sometimes handlers can be very short and sometimes they can be quite complicated—it all depends on what the handler is responsible for doing. In this next section, we'll examine the basic requirements for the different exception types, along with some detailed code examples using the STR910FM32 and LPC2132 microcontrollers included in the Keil simulation tools.

14.8.1 Reset Exceptions

When the processor first receives power, it will put the value 0x00000000 on the 32-bit address bus going to memory and receive its first instruction, usually a branch. This branch then takes it to the first instruction of the reset handler, where initialization of the processor or microcontroller is started. Depending on what's needed, a reset handler can be very simple, or it may need to perform tasks such as:

- Set up exception vectors
- Initialize the memory system (e.g., if a memory management unit [MMU] or memory protection unit [MPU] is present)

- Initialize all required processor mode stacks and registers
- Initialize any critical I/O devices
- Initialize any peripheral registers, control registers, or clocks, such as a phase-locked loop (PLL)
- Enable interrupts
- Change processor mode and/or state

We'll see in the example shortly how registers are configured and how interrupts are handled. Once the handler sets up needed registers and peripherals, it will jump to the main routine in memory. Reset handlers do not have a return sequence at the end of the code.

14.8.2 Undefined Instructions

We saw in Chapter 3 that the ARM7TDMI has about 50 instructions in its instruction set, plus all of the combinations of addressing modes and registers. With exactly 2^{32} possible instruction bit patterns, that leaves quite a few combinations of ones and zeros that are classified as an undefined instruction! An exception can occur if the processor doesn't recognize a bit pattern in memory, but it can also take an Undefined exception in two other cases. The first is if the processor encounters an instruction that is intended for a coprocessor (such as a floating-point unit or other special bit of hardware that was attached to the ARM7TDMI's coprocessor interface), but the coprocessor either doesn't exist or it doesn't respond. This first case was mentioned in Section 14.3, where the processor can emulate floating-point instructions by building a very smart exception handler that goes into the instruction memory and examines the offending instruction. If it turns out to be one of the instructions that the software wishes to support (e.g., a floating-point addition), then it begins to decode it. Software determines the operation that is needed, which would have to use integer registers and an integer datapath to perform the operation, and then calculates the result using a floating-point format. We then return to the main routine again. Theoretically, it allows software to be written only once using real floating-point instructions, and this could save money and power if speed isn't critical. Should a hardware floating-point unit be present (maybe a silicon vendor makes two slightly different models of microcontroller or SoC), the code will execute more quickly in hardware.

The second case that can generate an undefined instruction exception involves a coprocessor not responding to an instruction. As an example, Vector Floating-Point (VFP) coprocessors appear on some of the more advanced ARM cores, such as the Cortex-A8 and ARM1136JF-S. They have unique instructions, such as FDIVS and FSQRTS, and the ability to generate errors just like the main integer processor. However, if the VFP coprocessor generates an exception while processing one of its own instructions (suppose it tried to divide by zero), it will simply not respond when the integer processor tries to give it another instruction. The exception handler will then have to determine that the VFP coprocessor generated an exception on the last instruction that it accepted.

The last case that will generate an exception is when the processor sees a legitimate coprocessor instruction but is not in a privileged mode. For example, on most advanced applications processors, such as the ARM926EJ-S, ARM1136JF-S, or Cortex-A15, caches are included to improve performance (think of a cache as a small block of memory used to hold instructions and data so that the processor doesn't have to go to external memory as often). Caches always have a cache control register to set things up, and ARM uses Coprocessor 15, or CP15, to do this. While there isn't a real coprocessor in hardware, the instructions can be used anyway—have a look at the STC (Store Coprocessor) instruction and notice that bits 8 through 11 designate a coprocessor number. Coprocessor 15 is reserved for cache and MMU control registers. Meddling with these registers is only allowed if you're in a privileged mode, so a user's code would not be allowed to change the hardware configurations. The processor would reject the offending instruction by taking an Undefined Instruction exception.

EXAMPLE 14.1

Let's examine a simple bit of code, running on an LPC2132 microcontroller from NXP, that forces an Undefined Instruction exception. In order to demonstrate how the processor behaves during such an exception, we'll use a contrived situation where we wish to allow an instruction, normally undefined, to be emulated in software. This is analogous to floating-point emulation mentioned earlier, except our handler will be very short and very clumsy. Suppose that we call our instruction ADDSHFT. It takes one argument—the contents of register Rm, which can range from r0 to r7—and adds the contents of register r0 to it, shifting the result left by 5 bits. The assembler certainly wouldn't recognize the mnemonic, so we will call the instruction manually using DCD statements. When the processor fetches the word of data in memory, it proceeds through the pipeline as an instruction. Once the processor tries to execute our new bit pattern, it will take an Undefined Instruction exception, where our handler will decode the instruction and perform the operation.

There are a few things to observe in the example. The first is which operations the processor does for us, and which operations must be done by a programmer. Recall the switching the mode is normally done by the processor during an exceptional condition; however, as we'll see shortly, the programmer can also manually change the mode to set up a stack pointer. On the ARM7TDMI, saving registers and state information to the stack must be done by the programmer. The second thing to observe is the register file. Since the machine will change to Undef mode, we will be using a new register r13 and r14, so when you simulate this program, be sure to note the values in all of the registers in the processor, since we will now be working with more than just the traditional r0 through r15 in a single mode.

Below, you can see the complete code listing:

```
; Area Definition and Entry Point

SRAM_BASE    EQU     0x40000000      ; start of RAM on LPC2132
Mode_UND     EQU     0x1B
Mode_SVC     EQU     0x13
I_Bit        EQU     0x80
F_Bit        EQU     0x40

             AREA    Reset, CODE, READONLY
             ARM
             ENTRY
```

```
; Exception Vectors
; Dummy Handlers are implemented as infinite loops which can be modified.

Vectors        LDR     PC, Reset_Addr
               LDR     PC, Undef_Addr
               LDR     PC, SVC_Addr
               LDR     PC, PAbt_Addr
               LDR     PC, DAbt_Addr
               NOP                          ; Reserved Vector
               LDR     PC, IRQ_Addr
               LDR     PC, FIQ_Addr

Reset_Addr     DCD     Reset_Handler
Undef_Addr     DCD     UndefHandler
SVC_Addr       DCD     SVCHandler
PAbt_Addr      DCD     PAbtHandler
DAbt_Addr      DCD     DAbtHandler
               DCD     0                    ; Reserved Address
IRQ_Addr       DCD     IRQHandler
FIQ_Addr       DCD     FIQHandler

SVCHandler     B       SVCHandler
PAbtHandler    B       PAbtHandler
DAbtHandler    B       DAbtHandler
IRQHandler     B       IRQHandler
FIQHandler     B       FIQHandler

; Reset Handler

; Undefined Instruction test
; 31 30 29 28 27 26 25 24 23 22 21 20 19 18 17 16 15 14 13 12 11 10 9 8 7 6 5 4  3 2 1 0
;|0  1  1  1 |0  1  1  1 |1  1  1  1 |0  0  0  0 |0  0  0  0 |1  1  1 1 1 1 1 1| Rm  |
;|CC = AL    |          OP           |  Rn = 0   |  Rd = 0   |        Rm             |

Reset_Handler
; The first order of business is to set up a stack pointer in
; UNDEF mode, since we know our simulation will hit an undefined
; instruction.
                MSR     CPSR_c, #Mode_UND:OR:I_Bit:OR:F_Bit
                LDR     sp, =SRAM_BASE+80; initialize stack pointer
                ; switch back to Supervisor mode
                MSR     CPSR_c, #Mode_SVC:OR:I_Bit:OR:F_Bit

                MOV     r0, #124; put some test data into r0
                MOV     r4, #0x8B; put some test data into r4
ADDSHFTr0r0r4   DCD     0x77F00FF4; r0 = (r0+r4) LSL #5
                NOP

Loop            B       Loop
                NOP

;/*****************************************************************/
;/* Undefined Handler                                             */
;/*****************************************************************/
; Note that this handler is NOT AAPCS compliant. See the
; RealView Compilation Tools Developer Guide for examples of
; AAPCS-compliant handlers, specifically for maintaining 8-byte
; alignment and stack requirements. We're taking some shortcuts
; here just so we can concentrate on a simple mechanism to deal
; with an undefined instruction.
```

```
UndefHandler
    STMFD  sp!, {r0-r12, LR}     ; Save Workspace & LR to Stack
    MRS    r0, SPSR              ; Copy SPSR to r0
    STR    r0, [sp, #-4]!        ; Save SPSR to Stack

    LDR    r0, [lr,#-4]          ; r0 = undefined instruction
    BIC    r2, r0, #0xF00FFFFF   ; clear out all but opcode bits
    TEQ    r2, #0x07F00000       ; r1 = opcode for ADDSHFT
    BLEQ   ADDSHFTInstruction    ; if a valid opcode, handle it

    ; insert tests for other undefined instructions here

    LDR    r1, [sp], #4          ; Restore SPSR to R1
    MSR    SPSR_cxsf, r1         ; Restore SPSR
    LDMFD  sp!, {r0-r12, PC}^    ; Return to program after
                                 ; Undefined Instruction

; ADDSHFT instruction adds r0 + Rm (where Rm can only be between r0 and r7),
; shifts the result left 5 bits, and stores result in r0. It also does not
; decode immediates, CC, S-bit, etc.)

ADDSHFTInstruction

    BIC    r3, r0, #0xFFFFFFF0   ; mask out all bits except Rm
    ADD    r3, r3, #1            ; bump past the SPSR on the stack
    LDR    r0, [sp, #4]          ; grab r0 from the stack
    LDR    r3, [sp, r3, LSL #2]  ; use the Rm field as an offset
    ADD    r0, r0, r3            ; calculate r0 + Rm
    MOV    r0, r0, LSL #5        ; r0 = (r0 + Rm) << 5
    STR    r0, [sp, #4]          ; store r0 back on the stack
    BX     lr

    END
```

Figure 14.4 shows the basic flow of the program. The first few instructions of the program form the vector table, using PC-relative load instructions at each exception vector. Notice that the reset handler's address is referenced in the DCD statement

```
Reset_Addr    DCD    Reset_Handler
```

so that when the processor comes out of reset, the first instruction it executes is LDR, which will load the Program Counter with a constant it fetches from memory. Examine the assembler listing and you will notice the PC-relative load instruction and the offset calculated by the linker. The constant in memory is the address of the reset handler, by design. Inside the reset handler, the machine is forced into Undef mode so that we can set up a stack pointer, since we know in advance we are going to switch modes after hitting our strange instruction. The machine is switched back into Supervisor mode afterward. When the processor then tries to execute the bit pattern we deliberately put into the pipeline,

```
ADDSHFTr0r0r4      DCD     0x77F00FF4 ; r0 = (r0 + r4) LSL #5
```

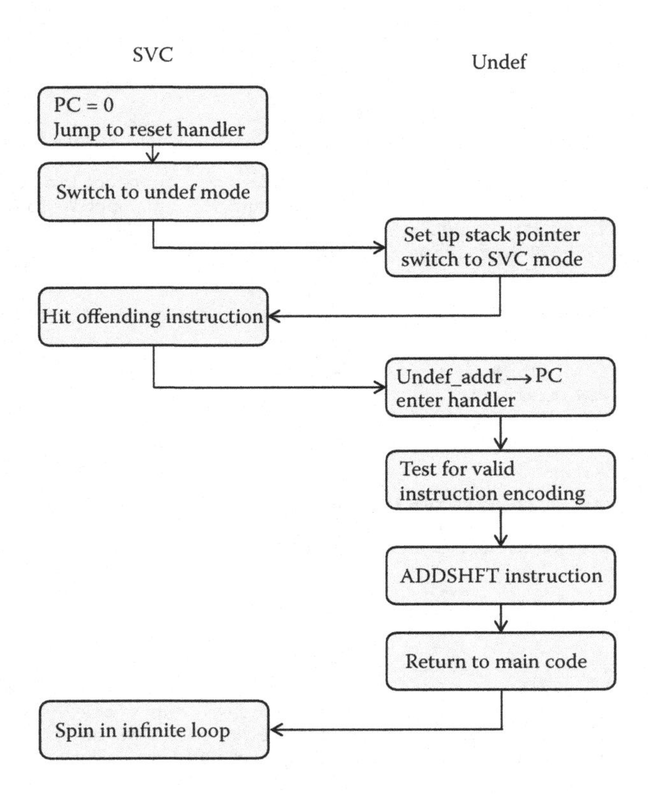

FIGURE 14.4 Exception flow diagram.

it immediately changes the mode to Undef and sets the Program Counter to 0x4, expecting to fetch some type of branch instruction at the exception vector that will ultimately take us to our handler.

Now that the processor has begun exception processing, the instruction at address 0x4 is fetched and executed. This PC-relative load moves the address of our handler into the Program Counter, and fetching begins from there. The first three instructions in the handler save off register and SPSR information to the stack. A comparison is made to determine if the bit pattern the processor rejected was something we wish to support, and if so, the processor branches to the ADDSHFTInstruction routine. When returning from our undefined instruction exception, we restore the SPSR and the register file that was stacked and move the Link Register value into the Program Counter, given as

```
LDMFD  sp!, {r0-r12, PC}^
```

This particular version of LDM does two things: it loads the Link Register value back into the Program Counter, effectively jumping back to where we left off, and it moves the SPSR back into the CPSR. Astute readers will notice that we made no adjustment to the Program Counter or Link Register values before we jumped back to the main code. We mentioned before that in some cases, such as during a branch and link instruction (BL), the Link Register value may be adjusted due to the fact that the Program Counter points two instructions ahead of the instruction being executed. In this case, the Link Register holds the return address of the instruction *following* the offending instruction, so the processor will not re-execute the one that caused the exception. For some exceptions, as we'll see in a moment, you might want to retry an offending instruction. Since we've finished handling the exception, the machine automatically changes back to Supervisor mode.

14.8.3 Interrupts

ARM cores have two interrupt lines—one for a fast interrupt (FIQ) and one for a low-priority interrupt (IRQ). If there are only two interrupts in the entire system, then this works well, as there is already some level of prioritization offered. FIQs have a higher priority than IRQs in two ways: they are serviced first when multiple interrupts arise, and servicing an FIQ disables IRQ interrupts. Once the FIQ handler exits, the IRQ can then be serviced. FIQ interrupts also have the last entry in the vector table, providing a fast method of entering the handler, as well as five extra banked registers to use for exception processing.

But with only two lines, how would an SoC with dozens of potential interrupt sources compete for the processor's time? There are a couple of ways to do this. The first, but not the best way, is to basically wire OR the interrupts coming from peripherals or external devices together. This would then be used to signal an interrupt on one of the two lines going to the processor. However, this would require polling each interrupt source to determine who triggered the interrupt, which would waste thousands of cycles of time (especially if the requesting device is in a hurry and happens to be the last one in the list!) A second way is to use an external interrupt controller, which is a specialized piece of hardware that takes in all of the interrupt lines, assigns priorities to them, and often provides additional information to help the processor, such as a register that can be read for quickly determining who requested the interrupt. When handling interrupts, the processor must first change the Program Counter to either address 0x18 or 0x1C in memory, fetch the instruction that will change the Program Counter, e.g., either a MOV or LDR instruction, then jump to a new address, which is the start of the interrupt handler. The first column of Figure 14.5 shows what happens when you have multiple sources of interrupts. If the incoming lines are wired together or connected to an external interrupt controller, after the processor jumps to the start of the interrupt handler, the handler itself still has to determine which device caused the interrupt, and only after doing so can the processor branch to the correct interrupt handler.

The second column of Figure 14.5 shows the general flow for a better way to handle interrupts. Suppose all of the interrupting peripherals in a system are connected through a controller so that when a device, such as a timer, needs attention (let's say the timer expired), the controller itself pulls on an interrupt line going to the processor. It also has the ability to give the processor the address of the interrupt handler so that all the processor needs to do is load the address into the Program Counter. The instruction to do so would still sit in the vector table as it did before. Programmers would not be absolved of all duties, however, as a few registers would

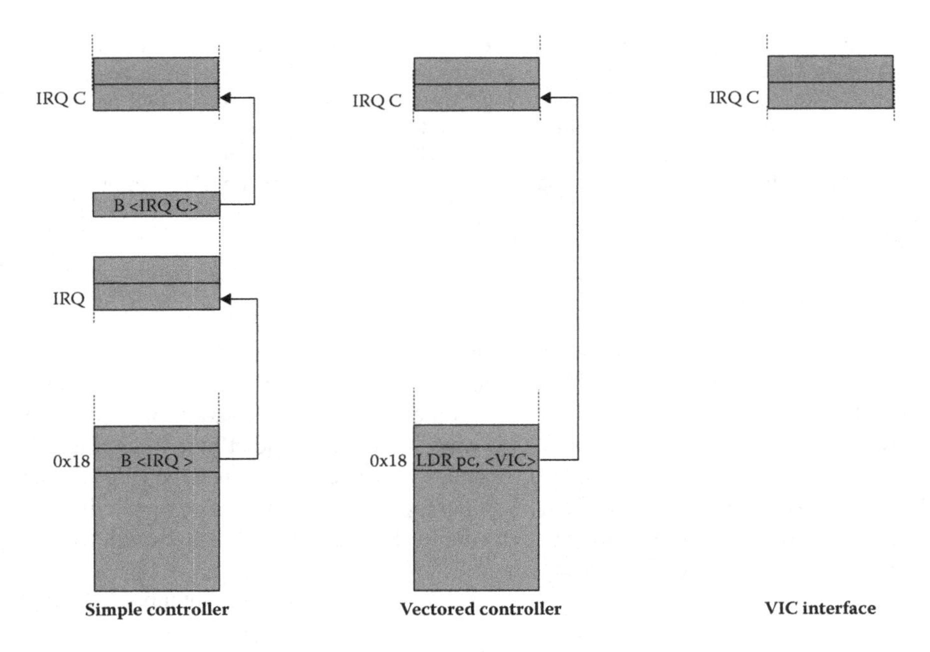

FIGURE 14.5 Three methods of handling interrupts.

need to be configured before using such a controller. The modern solution to handle multiple interrupt sources, therefore, is to use what's known as a vectored interrupt controller (VIC), and given that so many popular microcontrollers have them now, it's worth a closer look.

14.8.3.1 Vectored Interrupt Controllers

Vectored Interrupt Controllers require a bit of thought and effort but make dealing with interrupts less taxing. For hardware engineers, it makes designs more straightforward, since all the logic needed to build complicated interrupt schemes is already there. Software engineers appreciate the fact that everything is spelled out, but it still requires some work to get registers configured, interrupts enabled and defined, and memory locations initialized. Like the other microcontrollers that we've examined so far, the STR910FAM32 contains an ARM core (although this one is an ARM9E-based microcontroller), along with the two AMBA busses (AHB and APB) for interfacing to the memory and peripherals. You can see from Figure 14.6 that the VIC sits off the AHB bus, so it appears as a memory-mapped device. VIC registers exist at memory locations rather than within the processor, a topic we'll examine in much more detail when we look at memory-mapped peripherals in the Chapter 16. For a complete description of the VIC in the STR910FAM32 microcontroller, consult the STR91xFAxxx Reference Manual (STMicroelectronics 2006).

The basic principle behind the VIC is to provide enough information to the processor so that it doesn't have to go searching through all of the possible interrupts to

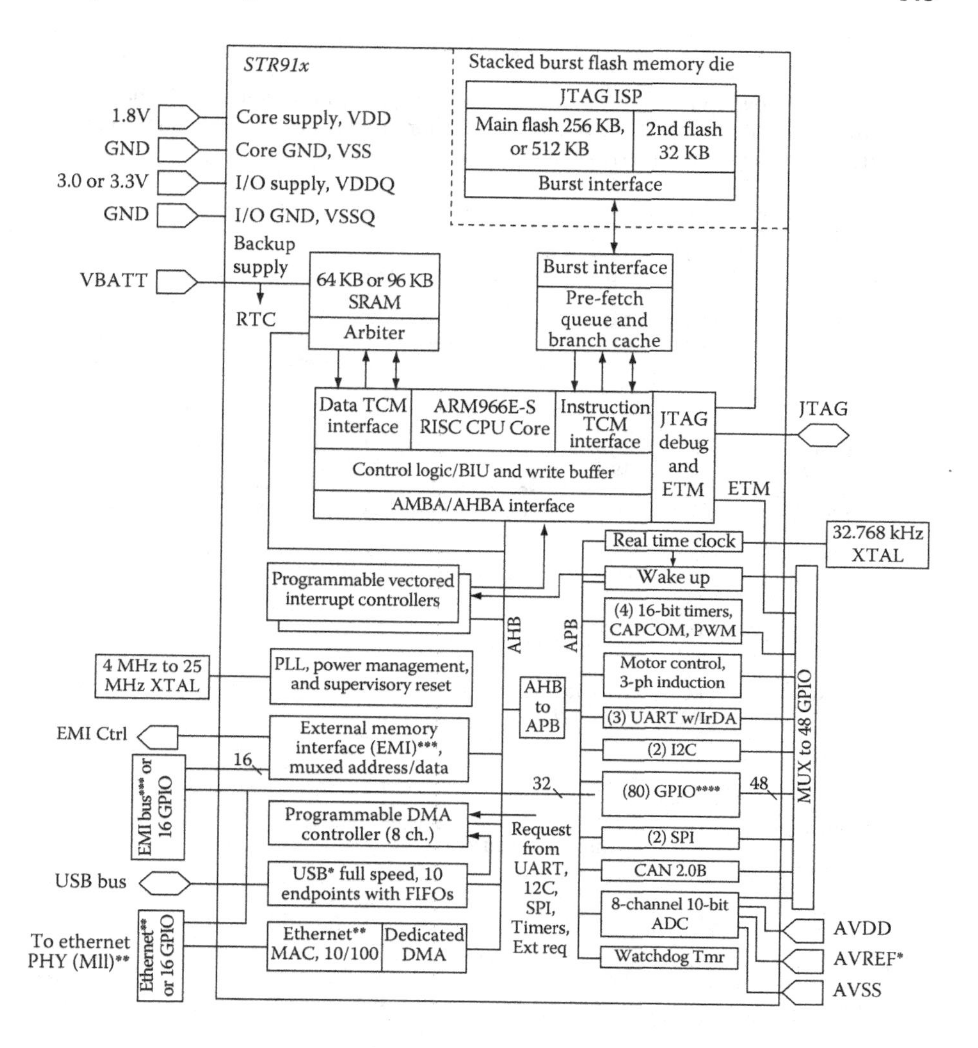

FIGURE 14.6 STR910FAM32 microcontroller. (From STMicroelectronics, *STR91xF data sheet* [Rev4], STMicroelectronics, Geneva, Switzerland. With permission.)

determine the requester. It has multiple inputs and only two outputs for most cores—the FIQ and IRQ interrupt lines. It also provides the most important component—the address of the interrupt handler. Rather than polling every possible interrupt source, the VIC can simply give the processor the address of the handler in a register. The processor then loads this value into the Program Counter by way of an LDR instruction in the IRQ exception vector.

EXAMPLE 14.2

To illustrate how a VIC works, on the following pages is some actual code that can be run on the Keil tools. The code itself is a shortened version of the initialization code available for the STR910FAM32 microcontroller.

```
; Standard definitions of Mode bits and Interrupt (I & F)
; flags in PSRs
SRAM_BASE          EQU     0x04000000
VectorAddr         EQU     0xFFFFF030   ; VIC Vector Address Register
Mode_USR           EQU     0x10
Mode_IRQ           EQU     0x12
I_Bit              EQU     0x80         ; when I bit is set, IRQ is disabled
F_Bit              EQU     0x40         ; when F bit is set, FIQ is disabled

; System Control Unit (SCU) definitions

SCU_BASE           EQU     0x5C002000   ; SCU Base Address (non-buffered)
SCU_CLKCNTR_OFS    EQU     0x00   ; Clock Control register Offset
SCU_PCGR0_OFS      EQU     0x14   ; Peripheral Clock Gating Register 0 Offset
SCU_PCGR1_OFS      EQU     0x18   ; Peripheral Clock Gating Register 1 Offset
SCU_PRR0_OFS       EQU     0x1C   ; Peripheral Reset Register 0 Offset
SCU_PRR1_OFS       EQU     0x20   ; Peripheral Reset Register 1 Offset
SCU_SCR0_OFS       EQU     0x34   ; System Configuration Register 0 Offset

SCU_CLKCNTR_Val    EQU     0x00020000
SCU_PLLCONF_Val    EQU     0x000BC019
SCU_PCGR0_Val      EQU     0x000000FB
SCU_PCGR1_Val      EQU     0x00EC0803
SCU_PRR0_Val       EQU     0x00001073
SCU_PRR1_Val       EQU     0x00EC0803

      PRESERVE8

; Area Definition and Entry Point
; Startup Code must be linked first at Address at which it expects to run.
      AREA       Reset, CODE, READONLY
      ENTRY

      ARM

; Exception Vectors Mapped to Address 0.
; Absolute addressing mode must be used.
; Dummy Handlers are implemented as infinite loops which can be modified.
Vectors
      LDR      pc, Reset_Addr
      LDR      pc, Undef_Addr
      LDR      pc, SVC_Addr
      LDR      pc, PAbt_Addr
      LDR      pc, DAbt_Addr
      NOP                              ; Reserved Vector
      LDR      pc, [pc, #-0x0FF0]
      LDR      pc, FIQ_Addr

Reset_Addr     DCD    Reset_Handler
Undef_Addr     DCD    UndefHandler
SVC_Addr       DCD    SVCHandler
PAbt_Addr      DCD    PAbtHandler
DAbt_Addr      DCD    DAbtHandler
               DCD    0                ; Reserved Address
IRQ_Addr       DCD    IRQHandler
FIQ_Addr       DCD    FIQHandler
UndefHandler   B      UndefHandler
SVCHandler     B      SVCHandler
```

```
PAbtHandler    B    PAbtHandler
DAbtHandler    B    DAbtHandler
IRQHandler     B    IRQHandler
FIQHandler     B    FIQHandler

Reset_Handler

; Setup Clock

        LDR          r0, =SCU_BASE
        LDR          r1, =0x00020002
        STR          r1, [r0, #SCU_CLKCNTR_OFS]

        ; Select OSC as clk src

        NOP
        ; Wait for OSC stabilization
        NOP
        NOP
        NOP
        NOP
        NOP
        NOP
        NOP
        NOP
        NOP
        NOP
        NOP
        LDR          r1, =SCU_CLKCNTR_Val

        ; Setup clock control

        STR          r1, [r0, #SCU_CLKCNTR_OFS]
        LDR          r1, =SCU_PCGR0_Val

        ; Enable clock gating

        STR          r1, [r0, #SCU_PCGR0_OFS]
        LDR          r1, =SCU_PCGR1_Val
        STR          r1, [r0, #SCU_PCGR1_OFS]

; Setup Peripheral Reset

        LDR          r1, =SCU_PRR0_Val
        STR          r1, [r0, #SCU_PRR0_OFS]
        LDR          r1, =SCU_PRR1_Val
        STR          r1, [r0, #SCU_PRR1_OFS]

; Enter IRQ Mode and set its Stack Pointer

        MSR          CPSR_c, #Mode_IRQ:OR:I_Bit:OR:F_Bit
        LDR          sp, =SRAM_BASE+100

; Enter User Mode

        MSR          CPSR_c, #Mode_USR

; VIC registers
```

```
VIC0_VA7R   EQU    0xFFFFF11C    ; Vector Address Register for TIM3 IRQ
VIC0_VC7R   EQU    0xFFFFF21C    ; Vector Control Register for TIM3 IRQ
VIC0_INTER  EQU    0xFFFFF010    ; Interrupt Enable Register


; TIM3 registers
TIM3_CR2    EQU    0x58005018    ; TIM3 Control Register 2
TIM3_CR1    EQU    0x58005014    ; TIM3 Control Register 1

      LDR          r4, =VIC0_VA7R
      LDR          r5, =IRQ_Handler
      STR          r5, [r4]

      ; Setup TIM3 IRQ Handler addr

      LDR          r4, =VIC0_VC7R
      LDR          r5, [r4]
      ORR          r5, r5, #0x27
      STR          r5, [r4]

      ; Enable the vector interrupt and specify interrupt number
      LDR          r4, =VIC0_INTER
      LDR          r5, [r4]
      ORR          r5, r5, #0x80
      STR          r5, [r4]        ; Enable TIM3 interrupt

      ; Timer 3 Configuration (TIM3)

      LDR          r4, =TIM3_CR2
      LDR          r5, =0xFF00
      LDR          r6, =0x200F
      LDR          r8, [r4]
      AND          r8, r8, r5      ; Clear prescaler value
      ORR          r8, r8, r6

      ; Setup TIM3 prescaler and enable TIM3 timer overflow interrupt
      STR          r8, [r4]
      LDR          r4, =TIM3_CR1
      LDR          r5, =0x8000
      LDR          r6, [r4]
      ORR          r6, r6, r5
      STR          r6, [r4]        ; TIM3 counter enable

; main loop

      LDR          r9, =0xFFFFFFFF

Loop  B       Loop

IRQ_Handler

    SUB            lr, lr, #4          ; Update Link Register
    SUB            r9, r9, #1
    STMFD          sp!, {r0-r12, lr}   ; Save Workspace & LR to Stack
    LDR            r4, =0x5800501C     ; r4 = address of TIM3_SR
    LDR            r5, =~ 0x2000
    LDR            r6, [r4]
    AND            r6, r6, r5
    STR            r6, [r4]            ; Clear Timer Overflow interrupt flag
    LDR            r0, =VectorAddr     ; Write to the VectorAddress
    LDR            r1, =0x0            ; to clear
```

```
STR             r1, [r0]            ; the respective Interrupt
LDMFD           sp!, {r0-r12, PC}^  ; Return to program, restoring state

END
```

Nearly all of the code sets up registers, initializes clocks, or sets up stack pointers. Note that this example removes some parts that are not critical to demonstrating how interrupts work. You can see all of the EQU directives that assign names to numeric values—this is purely for convenience. Reading code becomes difficult otherwise. The modes are translated into bit patterns, e.g., Mode_USR is equated to 0x10, which is what the lower 5 bits of the CPSR would look like in User mode.

The code actually starts after the first AREA directive, and you can see the exception vector table being built with the label Vectors starting the table. While we normally use LDR instructions to load the PC when a handler is not close enough to use a B (branch) instruction, the method used here is the most general and copes with any memory map. In fact, take a look at the vector table, as shown in Table 14.2. The IRQ vector now contains an instruction that tells the processor to load the PC with a value from memory. The address of that value in memory is calculated using the difference between the Program Counter (which would be 0x20 when this instruction reaches the execute stage of the ARM7TDMI's pipeline) and the value 0xFF0, giving 0xFFFFF030, which is a strange address and not at all intuitive. It turns out that the VIC has an address register for the processor, called VIC0_VAR, that just happens to sit at address 0xFFFFF030 in memory (STMicroelectronics defined the address—it could have been anything). This register holds the address of our IRQ exception handler, and the address is matched to a particular interrupt source. For example, suppose a timer and a USB interface can both generate interrupts. Inside of the VIC, handler addresses are stored in memory-mapped registers for each interrupt source. So if the USB interface generates an interrupt, *its* exception handler address is placed in the VIC0_VAR register. If the timer generates an interrupt, then the handler address belonging to the *timer* is placed in the VIC0_VAR register. Instead of a generic exception handler for interrupts, which would have to spend time figuring out who triggered the IRQ interrupt, the programmer can write a special handler for each type of interrupt and the processor will jump immediately to that unique handler.

In the example code, TIMER3 is used to generate an interrupt when the counter increments from 0x0000 to 0xFFFF. TIMER3 sits on channel 7 by default and its

TABLE 14.2
Vector Table Showing IRQ Branch Instruction

Exception Vector	Instruction
Reset	LDR pc, Reset_Addr
Undefined Instruction	LDR pc, Undef_Addr
SVC	LDR pc, SVC_Addr
Prefetch Abort	LDR pc, PAbt_Addr
Data Abort	LDR pc, DAbt_Addr
Reserved	NOP
IRQ	**LDR pc, [pc, -0x0FF0]**
FIQ	LDR pc, FIQ_Addr

TABLE 14.3
VIC0 Registers

Address	Register Name	Function
0xFFFFF010	VIC0_INTER	Interrupt enable register
0xFFFFF11C	VIC0_VA7R	Vector address register
0xFFFFF21C	VIC0_VC7R	Control register

interrupt line goes through VIC0. There are three registers that also need to be set up for VIC0 as shown in Table 14.3. You can see on page 316 where the code equates addresses with the names of the VIC registers. Immediately afterward, the address of the timer's interrupt handler, called IRQ_Handler, is stored in the VIC0_VA7R register. Remember that if an interrupt is triggered, the VIC will know it was TIMER3 requesting the interrupt, and then it will move the handler's address from VIC0_VA7R into VIC0_VAR. The remaining code enables and configures the timer.

The handler itself is at the end of the code. It adjusts the Link Register value first so that we can exit the handler with a single instruction (to be discussed in a moment). The second instruction in the handler begins stacking off registers into memory, including the Link Register. The rest of the handler clears the timer overflow flag in the timer peripheral, and it disables the interrupt request by writing to the VIC0_VAR register. An interrupt handler usually contains the code that clears the source of the interrupt.

Returning from an interrupt is not difficult, but it does require a little explanation. The timing diagram in Figure 14.7 shows an example sequence of events in the ARM7TDMI processor's pipeline. Cycle 1 shows the point at which the processor acknowledges that the IRQ line has been asserted. The ADD instruction is currently in the execute stage and must complete, since the processor will allow all instructions to finish before beginning an interrupt exception sequence. In Cycle 2, the processor has now begun handling the IRQ, but notice that the Program

Cycle				1 (IRQ)	2	3	4	5	6	7	8
Address	**Operation**										
0x8000	ADD	F	D	E							
0x8004	SUB		F	Decode IRQ	Execute IRQ	Linkret	Adjust				
0x8008	MOV			F							
0x800C	X				F						
0x0018	B (to 0xAF00)					F	D	E			
0x001C	XX						F	D			
0x0020	XXX							F			
0xAF00	STMFD								F	D	E
0xAF04	MOV									F	D
0xAF08	LDR										F

FIGURE 14.7 Interrupt processing in the ARM7 pipeline.

Counter has already progressed, i.e., the processor has fetched an instruction from this address, and the PC points to the instruction at 0x800C. It is this value that is loaded into the Link Register in Cycle 3. The exception vector 0x18 becomes the new Program Counter value, and the instruction at this address is fetched, which is a branch instruction. In Cycle 4, the Link Register can be adjusted in the same way that it is for BL instructions, but this makes the address in the Link Register 0x8008, which is four bytes off if we decide to use this address when we're done with the interrupt handler.

In our example code, the handler adjusts the Link Register value straight away before stacking it. Notice the first instruction in the handler is

```
SUB    lr, lr, #4
```

This allows us to exit the handler using only a single instruction:

```
LDMFD  sp!, {r0-r12, pc}^
```

which is a special construct. The LDM instruction restores the contents of the registers from the stack, in addition to loading the PC with the value of the Link Register. The caret (^) at the end of the mnemonic forces the processor to transfer the SPSR into the CPSR at the same time, saving us an instruction. This is the recommended way to exit an interrupt handler.

14.8.3.2 More Advanced VICs

Believe it or not, there is an even faster way of handling interrupts. Referring back to Figure 14.5, we've described two methods already, which are shown in the first two columns. The first requires the processor to branch to an address—the start of your interrupt handler. The handler then determines who requested the interrupt, branching to yet another location for the handler. The second method uses a VIC so that the processor still goes to the IRQ exception vector, but instead of branching to a generic handler, it branches to a handler address that is given to it by the VIC.

If the VIC is coupled ever more tightly to the processor, it's possible to forgo an exception vector completely; ergo, a bus is created on the processor that talks directly to the VIC. As shown in the third column of Figure 14.5, when an interrupt occurs, the processor knows to take the address from the dedicated bus. Recall from the previous example that the VIC has memory-mapped registers that are attached to the AHB bus. When an interrupt occurs, the processor gets its interrupt service routine address from the VIC0_VAR register, which is also on the AHB bus. The third method allows the interrupt service routine's address to be given to the core on a dedicated address bus, along with handshake lines to signal that the address is stable and that the core has received it. The processor doesn't even have to go to the exception vector at address 0x18. Since the processor core must be modified to accept a more advanced vectored interrupt controller, this feature is not found on all ARM processors.

14.8.4 Aborts

Aborts have something of a negative connotation to them, but not all of them are bad. Certainly, the processor should be prepared to deal with any that happen to appear, and those that do should be the type that are handled with a bit of extra hardware, since the really awful cases are rare and don't really provide many graceful exits. The topic of caches, physical and virtual addresses, and memory management units is best left for an advanced text, such as Sloss et al. (2004), Furber (2000), or Patterson and Hennessy (2007), but the general idea is that normally you limit the amount of physical memory in a processor system to keep the cost of the device low. Since the address bus coming out of the processor is 32-bits wide, it can physically address up to 4 GB of memory. A common trick to play on a processor is to allow it to address this much memory while physically only having a much smaller amount, say 128 MB. The hardware that does this is called a memory management unit (MMU). An MMU can swap out "pages" of memory so that the processor thinks it's talking to memory that isn't really there. If the processor requests a page that doesn't exist in memory, the MMU provides an abort signal back to the processor, and the processor takes a data or prefetch abort exception. It's what happens in the abort handler that determines whether or not the processor actually begins processing an abort exception. Normally, the MMU can try to swap out pages of memory and the processor will try again to load or store a value to a memory location (or fetch an instruction from memory). If it can't or there is a privilege violation, e.g., a user trying to address memory marked for supervisor use only, then an abort exception may be taken. There are two types of aborts: prefetch aborts on the instruction side and data aborts on the data side.

14.8.4.1 Prefetch Aborts

A prefetch abort can occur if the processor attempts to fetch an instruction from an invalid address. The way the processor reacts depends on the memory management strategy in use. If you are working with a processor with an MMU, for example, an ARM926EJ-S or ARM1136JF-S, then the processor will attempt to load the correct memory page and execute the instruction at the specified address. Without an MMU, this is indicative of a fatal error, since the processor cannot continue without code. If it's possible, the system should report the error and quit, or possibly just reset the system.

The prefetch abort sequence resembles those of other exceptions. When an abort is acknowledged by the processor, it will fetch the instruction in the exception vector table at address 0xC. The abort handler will then be responsible for either trying to fix the problem or die a graceful death. The offending instruction can be found at address ea < LR-4> if the processor was in ARM state, or ea <LR-2> if the processor was in Thumb state at the time of the exception. You should return from a prefetch abort using the instruction

```
SUBS    PC, LR, #4
```

which retries the instruction by putting its address into the Program Counter. The suffix "S" on the subtract instruction and the PC as the destination restores the SPSR into the CPSR automatically.

14.8.4.2 Data Aborts

Loads and stores can generate data aborts if the address doesn't exist or if an area of memory is privileged. Like prefetch aborts, the action taken by the processor depends on the memory management strategy in place. If an MMU is being used, the programmer can find the offending address in the MMU's Fault Address Register. The MMU can attempt to fix the problem by swapping in the correct page of memory, and the processor can reattempt the access. If there is no MMU in the system, and the processor takes the abort exception, this type of error is fatal, since the processor cannot fix the error and should report it (if possible) before exiting.

There is a subtle difference in the way that certain ARM cores handle data aborts. Consider the instruction

```
LDR     r0, [r1,#8]!
```

which would cause the base register r1 to be updated after the load completes. Suppose, though, that the memory access results in a data abort. Should the base register be modified? It turns out the effect on the base register is dependent on the particular ARM core in use. Cores such as StrongARM, ARM9, and ARM10 families use what's known as a "base restored" abort model, meaning the base register is automatically returned to its original value if an abort occurs on the instruction. The ARM7 family of cores uses a "base updated" abort model, meaning the abort handler will need to restore the base register before the instruction can be reattempted. The handler should exit using the instruction

```
SUBS    PC,LR,#8
```

if the processor wants to retry the load or store instruction again.

14.8.5 SVCs

Software interrupts, or SVCs as they are now known, are generated by using the ARM instruction SVC (although you might still see SWI being used in legacy code). This causes an exception to be taken, and forces the processor into Supervisor mode, which is privileged. A user program can request services from an operating system by encoding a request in the 24 bits of the mnemonic for ARM instructions or the 8 bits for Thumb instructions, as shown in Figure 14.8.

Suppose that an embedded system with a keypad and an LCD display are being designed, but only parts of the system are ready—the processor and inputs are ready, but it has no display. The software engineer may choose to use a technique called

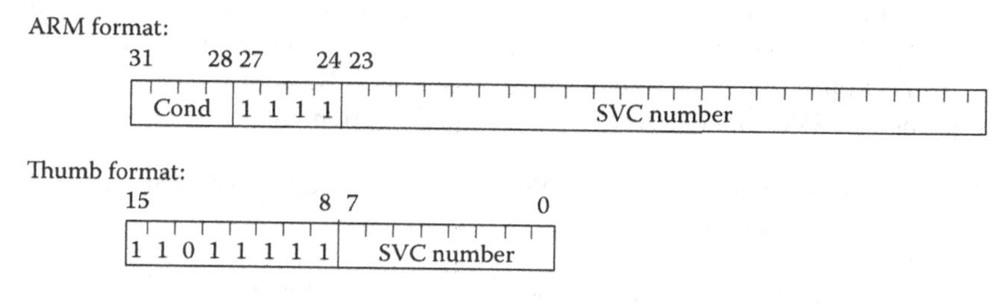

FIGURE 14.8 Opcodes for SVC instructions.

semihosting to write his or her code, where C primitives such as printf and scanf are simulated by a debugger in the development tools, such as the RealView Developer Suite (RVDS). The printf statements force the output onto a console window in the tools instead of to a nonexistent LCD panel. Semihosting uses SVCs to provide system calls for the user, but these are done in the background without the user noticing. Note that the μVision tools do not support semihosting. Consult the RealView ICE and RealView Trace User Guide from ARM (ARM 2008b) for more information on semihosting and its use.

When an SVC instruction is encountered in the instruction stream, the processor fetches the instruction at the exception vector address 0x8 after changing to Supervisor mode. The core provides no mechanism for passing the SVC number directly to the handler, so the SVC handler must locate the instruction and extract the comment field embedded in the SVC instruction itself. To do this, the SVC handler must determine from which state (ARM or Thumb) the SVC was called by checking the T bit in the SPSR. If the processor was in ARM state, the SVC instruction can be found at ea <LR-4>; for Thumb state, the address is at ea <LR-2>. The SVC number in the instruction can then be used to perform whatever tasks are allowed by the handler.

To return from an SVC handler, use the instruction

```
MOVS    PC, LR
```

since you wouldn't want to rerun the instruction after having taken the exception already.

14.9 EXERCISES

1. Name three ways in which FIQ interrupts are handled more quickly than IRQ interrupts.

2. Describe the types of operations normally performed by a reset handler.

3. Why can you only return from exceptions in ARM state on the ARM7TDMI?

4. How many external interrupt lines does the ARM7TDMI have? If you have eight interrupting devices, how would you handle this?

5. Write an SVC handler that accepts the number 0x1234. When the handler sees this value, it should reverse the bits in register r9 (see Exercise 2 in Chapter 8). The SVC exception handler should examine the actual SVC instruction in memory to determine its number. Be sure to set up a stack pointer in SVC mode before handling the exception.

6. Explain why you can't have an SVC and an undefined instruction exception occur at the same time.

7. Build an Undefined exception handler that tests for and handles a new instruction called FRACM. This instruction takes two Q15 numbers,

multiplies them together, shifts the result left one bit to align the binary point, then stores the upper 16 bits to register r7 as a Q15 number. Be sure to test the routine with two values.

8. As a sneak peek into Chapter 16, we'll find out that interrupting devices are programmable. Using STMicroelectronics' documentation found on its website, give the address range for the following peripherals on the STR910FAM32 microcontroller. They can cause an interrupt to be sent to the ARM processor.
   a. Real Time Clock
   b. Wake-up/Interrupt Unit

9. Explain the steps the ARM7TDMI processor takes when handling an exception.

10. What mode does the processor have to be in to move the contents of the SPSR to the CPSR? What instruction is used to do this?

11. When handling interrupts, why must the Link Register be adjusted before returning from the exception?

12. How many SPSRs are there on the ARM7TDMI?

13. What is a memory management unit (MMU) used for? What is the difference between an MMU and a memory protection unit (MPU)? You may want to consult the ARM documentation or a text like (Patterson, Hennessy 2007) for specific information on both.

15 Exception Handling
v7-M

15.1 INTRODUCTION

With the introduction of the Cortex-M3 in 2006, ARM decided to move its considerable weight into the huge market for microcontrollers, devices that normally get very little attention as they're embedded into everything from printers to industrial meters to dishwashers. Building on the success of the ARM7TDMI, which incidentally was and continues to be used in microcontrollers (particularly Bluetooth devices), the version 7-M cores like the Cortex-M4 support deeply embedded applications requiring fast interrupt response times, low gate counts, and peripherals like timers and pulse width modulation (PWM) signal generators. In some ways, these processors are easier to work with and in some ways, more difficult. Theoretically, one should not be programming a Cortex-M3 or a Cortex-M4 device by writing assembly (but *we* will!). They are designed to be completely accessible using only C, with libraries available to configure vector tables, the MPU, interrupt priorities, etc., which makes the programmer's job easier. Very little assembly code ever has to be written. If, however, you are writing assembly, there are only two modes instead of seven and fewer registers to worry about. What makes these processors slightly more difficult to work with is the sheer number of options available: there are more instructions; priority levels can be set on the different interrupt types; there are subpriorities available; faults must be enabled before they can be handled; the Nested Vectored Interrupt Controller must be configured before using it (and while implementation specific, Cortex-M parts can support up to 496 external interrupt inputs!); and there are power management features to consider. In Chapter 14, we saw the exception model for the ARM7TDMI, which is different than the one for version 7-M processors. Here, we'll examine the basics of handling exceptions for a processor like the Cortex-M4 without covering every single variable, since you are not likely to encounter *every* exception while you learn about programming, and there are quite a few options to consider when you have multiple exceptions arriving at the same time, some with higher priorities than others. For more advanced topics such as embedded operating systems, semaphores, tail-chaining interrupts, and performance considerations, books such as (Yiu 2014) and the *Cortex-M4 Technical Reference Manual* (ARM 2009) can be read for details.

15.2 OPERATION MODES AND PRIVILEGE LEVELS

The Cortex-M3 and Cortex-M4 processors have only two operation modes: Handler mode and Thread mode. This is a significant departure from the earlier ARM models where the mode was determined more or less by what the processor was doing, e.g., handling an interrupt or taking an exception. Rather than having unique modes for the different exception types, the Cortex-M processors use Handler mode for dealing with exceptions and everything else runs in Thread mode. One further distinction is introduced, and this has to do with privilege levels. Obviously, you would not want a user application to be able to modify critical parts of a system like configuration registers or the MPU, and it is important that an operating system has the ability to access all memory ranges and registers. There are, then, two privilege levels, aptly named privileged and user. You can see from Figure 15.1 that when the processor comes out of reset, it immediately runs in privileged Thread mode. Once the system is configured, the processor can be put into non-privileged Thread mode by changing the least significant bit of the CONTROL register, shown in Figure 15.2. When the processor takes an exception, it switches to Handler mode, which is always privileged, allowing the system to deal with any issues that may require

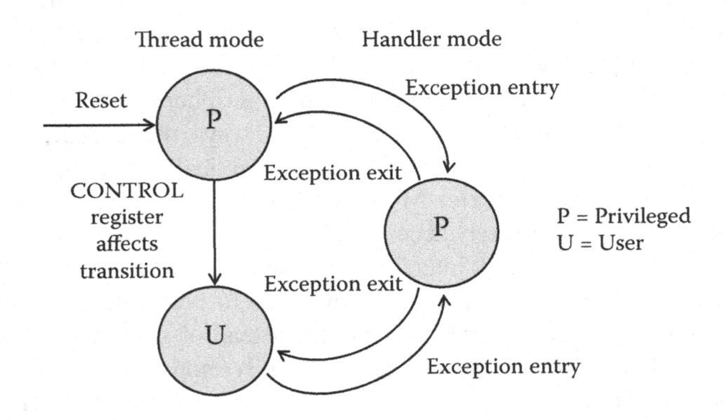

FIGURE 15.1 Cortex-M4 operation modes.

31 … 3	2	1	0
Reserved	FPCA	ASP	TMPL

FPCA – Floating-point context active

1 – Preserve floating-point state when processing exception

0 – No floating-point context active

ASP - Active stack pointer

1 – PSP

0 – MSP

TMPL - Thread mode privilege level

1 – Unprivileged

0 – Privileged

FIGURE 15.2 CONTROL Register on the Cortex-M4.

restricted access to resources. Upon returning from the exception, the processor will revert back to the state from which it left, so there is no way for a user program to change the privilege level by simply changing a bit. It must use an exception handler (which forces the processor into a privileged level) that controls the value in the CONTROL register.

EXAMPLE 15.1

We'll return to this example later in the chapter, with some modifications along the way, as it demonstrates the various aspects of exception handling in the Cortex-M4. Let's begin by building a quick-and-dirty routine that forces the processor into privileged Handler mode from privileged Thread mode. In Chapter 7, the idea of trapping division by zero was only mentioned, leaving an actual case study until now. If you type the following example into the Keil tools, using a Tiva TM4C1233H6PM as the target processor, and then single-step through the code, just out of reset the processor will begin executing the instructions after the label Reset_Handler. Note that many of the registers are memory mapped. For the full list of registers, see the Tiva TM4C1233H6PM Microcontroller Data Sheet (Texas Instruments 2013b).

```
Stack         EQU     0x00000100
DivbyZ        EQU     0xD14
SYSHNDCTRL    EQU     0xD24
Usagefault    EQU     0xD2A
NVICBase      EQU     0xE000E000

      AREA    STACK, NOINIT, READWRITE, ALIGN = 3
StackMem
      SPACE   Stack
      PRESERVE8

      AREA RESET, CODE, READONLY
      THUMB

; The vector table sits here
; We'll define just a few of them and leave the rest at 0 for now

      DCD     StackMem + Stack          ; Top of Stack
      DCD     Reset_Handler             ; Reset Handler
      DCD     NmiISR                    ; NMI Handler
      DCD     FaultISR                  ; Hard Fault Handler
      DCD     IntDefaultHandler         ; MPU Fault Handler
      DCD     IntDefaultHandler         ; Bus Fault Handler
      DCD     IntDefaultHandler         ; Usage Fault Handler

      EXPORT Reset_Handler
      ENTRY

Reset_Handler
          ; enable the divide-by-zero trap
          ; located in the NVIC
          ; base: 0xE000E000
          ; offset: 0xD14
```

```
            ; bit: 4
            LDR         r6, =NVICBase
            LDR         r7, =DivbyZ
            LDR         r1, [r6, r7]
            ORR         r1, #0x10            ; enable bit 4
            STR         r1, [r6, r7]

            ; now turn on the usage fault exception
            LDR         r7, =SYSHNDCTRL (p. 163)
            LDR         r1, [r6, r7]
            ORR         r1, #0x40000
            STR         r1, [r6, r7]

            ; try out a divide by 2 then a divide by 0!
            MOV         r0, #0
            MOV         r1, #0x11111111
            MOV         r2, #0x22222222
            MOV         r3, #0x33333333

            ; this divide works just fine
            UDIV        r4, r2, r1
            ; this divide takes an exception
            UDIV        r5, r3, r0

Exit        B       Exit

NmiISR      B       NmiISR
FaultISR    B       FaultISR
IntDefaultHandler

            ; let's read the Usage Fault Status Register

            LDR         r7, =Usagefault
            LDRH        r1, [r6, r7]
            TEQ         r1, #0x200
            IT          NE
            LDRNE       r9, =0xDEADDEAD
            ; r1 should have bit 9 set indicating
            ; a divide-by-zero has taken place
done        B           done
            ALIGN

            END
```

Continue single-stepping through the MOV and LDR instructions until you come to the first of the two UDIV (unsigned divide) operations. If you examine the registers and the state information using the Keil tools, you see that the first divide instruction is perfectly legal, and it will produce a value in register r2. More importantly, the machine is operating in Thread mode and it is privileged, shown in Figure 15.3. If you try to execute the next divide instruction, one which tries to divide a number by zero, you should see the machine change modes to Handler mode. The program has enabled a particular type of exception (usage faults, which we'll cover in Section 15.6) and enabled divide-by-zero traps so that we can watch the

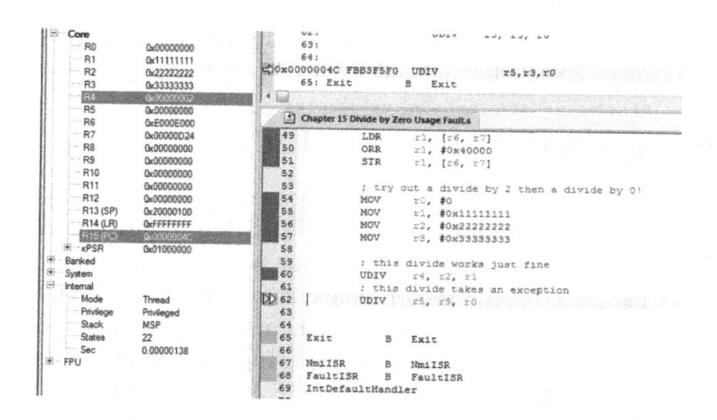

FIGURE 15.3 Cortex-M4 operating in privileged Thread mode.

processor begin working on the exception. At this point, the exception routine does not return back to the main code, but in the next example, we'll add an instruction to effect the return.

EXAMPLE 15.2

To switch privilege levels, the CONTROL register must be used, and this can only be written in a privileged level, so either the processor must be in Handler mode or privileged Thread mode. If we change the exception handler instructions, we can switch the privilege level of the processor. Additionally, we'll add a branch instruction (BX) that will allow the processor to exit exception handling and restore the values placed on the stack. You will notice that the original divide-by-zero exception remains, so that when we return to the main code, the processor will attempt to re-execute the offending instruction. For now, stop your simulation at that point. Our clumsy handler code should then read as:

```
IntDefaultHandler

        ; let's read the Usage Fault Status Register

        LDR             r7, =Usagefault
        LDRH            r1, [r6, r7]
        TEQ             r1, #0x200
        IT              NE
        LDRNE           r9, =0xDEADDEAD
        ; r1 should have bit 9 set indicating
        ; a divide-by-zero has taken place

        ; switch to user Thread mode
        MRS             r8, CONTROL
        ORR             r8, r8, #1
        MSR             CONTROL, r8
        BX              LR

        ALIGN
```

Run the code again and single-step through each instruction, noting the processor mode and privilege level before and after entering the exception handler.

15.3 THE VECTOR TABLE

In Chapter 14, we saw that the ARM7TDMI processor had a unique address associated with each exception type for handling the various interrupts and exceptions that come along. The Cortex-M3/M4 processor has a similar table; however, we pointed out in Chapter 2 that the vector table consists of *addresses*, not instructions like the more traditional ARM processors. When an exception occurs, the processor will push information to the stack, also reading the address at the appropriate vector in the vector table to start handling the exception. Fetching then begins from this new address and the processor will begin executing the exception handler code.

Table 15.1 lists the different exceptions along with their respective vector addresses. Note that the vector table can in fact be moved to another location in memory; however, this is infrequently done. One other point to notice is that the address

TABLE 15.1
Exception Types and Vector Table

Exception Type	Exception Number	Priority	Vector Address	Caused by...
—	—	—	0x00000000	Top of stack
Reset	1	– 3 (highest)	0x00000004	Reset
NMI	2	– 2	0x00000008	Non-maskable interrupt
Hard fault	3	– 1	0x0000000C	All fault conditions if the corresponding fault is not enabled
Mem mgmt fault	4	Programmable	0x00000010	MPU violation or attempted access to illegal locations
Bus fault	5	Programmable	0x00000014	Bus error, which occurs during AHB transactions when fetching instructions or data
Usage fault	6	Programmable	0x00000018	Undefined instructions, invalid state on instruction execution, and errors on exception return
—	7–10	—		Reserved
SVcall	11	Programmable	0x0000002C	Supervisor Call
Debug monitor	12	Programmable	0x00000030	Debug monitor requests such as watchpoints or breakpoints
—	13	—		Reserved
PendSV	14	Programmable	0x00000038	Pendable Service Call
SysTick	15	Programmable	0x0000003C	System Tick Timer
Interrupts	16 and above	Programmable	0x00000040 and above	Interrupts

0x0 is not the reset vector as it is for other ARM processors. On the Cortex-M3/M4 processor, the stack pointer address sits at address 0x0 (holding the value loaded into the Main Stack Pointer, or MSP register, covered in the next section). The reset vector is located at address 0x4.

15.4 STACK POINTERS

There are two stack pointers available to programmers, the Main Stack Pointer (MSP) and the Process Stack Pointer (PSP), both of which are called register r13; the choice of pointer depends on the mode of the processor and the value of CONTROL[1]. If you happen to have an operating system running, then the kernel should use the MSP. Exception handlers and any code requiring privileged access *must* use the MSP. Application code that runs in Thread mode should use the PSP and create a process stack, preventing any corruption of the system stack used by the operating system. Simpler systems, however, such as those without any operating system may choose to use the MSP alone, as we'll see in the examples in this chapter. The topic of the inner working of operating systems literally fills textbooks, but a good working knowledge of the subject can be gleaned from (Doeppner 2011).

15.5 PROCESSOR EXCEPTION SEQUENCE

Aside from the vector table containing addresses, the entry and exit sequences of exception handling differ more than any other aspect of the programmer's model. The overriding idea in the design of the v7-M model is that high-level software and standardized libraries will be controlling everything—the programmer merely calls the appropriate handler functions. Writing these device driver libraries must be done in accordance with the CMSIS standard written by ARM, so a knowledge of assembly will be necessary here. Having said that, someone trying to write or debug code will need a working knowledge of what exactly happens during exceptions. The first step is to look at the fundamentals of exception entry and exiting.

15.5.1 ENTRY

When a processor such as the Cortex-M4 first begins exception processing, eight data words are automatically pushed onto the current stack. This stack frame, as it is called, consists of registers r0 through r3, register r12, the Link Register, the PC, and the contents of the xPSR, shown in Figure 15.4. If a floating-point unit is present and enabled, the Cortex-M4 will also stack the floating-point state. Recall from Section 15.4 that there is an option that controls which stack pointer is used, either the MSP or the PSP, but we'll continue to use the MSP for our next example.

EXAMPLE 15.3

Let's rerun the code from our last example, which trapped the division by zero. Single-step through the code, up to the point where the processor tries to execute the second (faulting) division. Open a memory window to examine the contents of

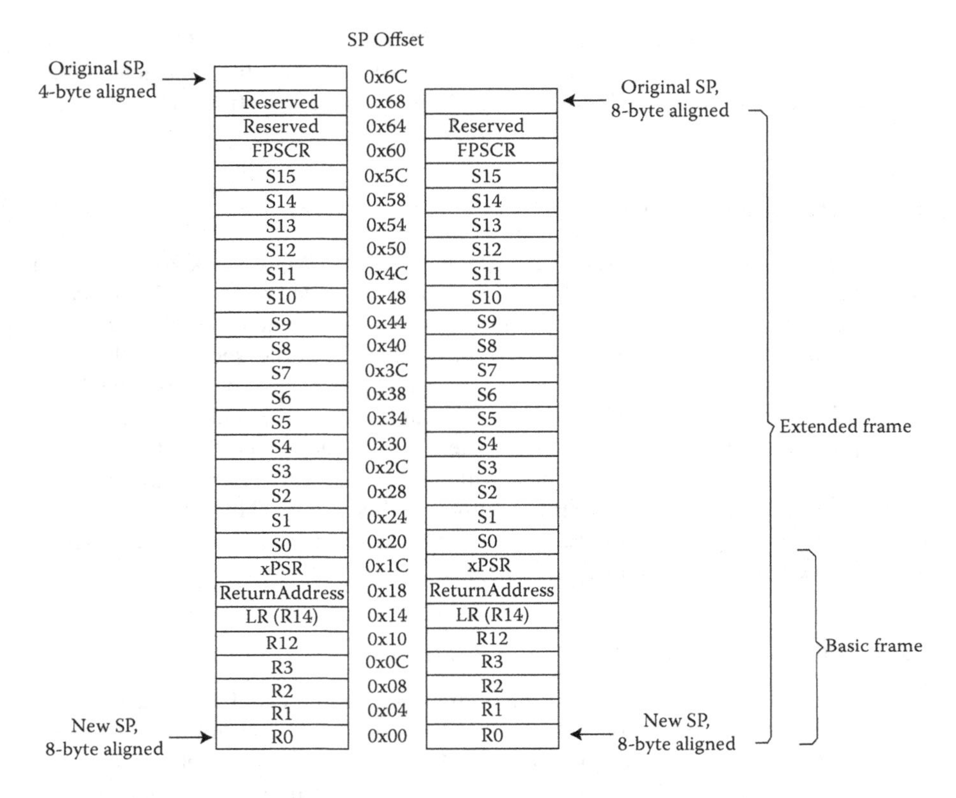

FIGURE 15.4 Exception stack frames.

memory. On the Tiva TM4C1233H6PM microcontroller, SRAM begins at address 0x20000000 and the stack has been defined to be 256 bytes (0x100) at the top of our program. If you look at memory starting just below 0x20000100, you will notice that the contents of registers r0 through r3, register r12, the Link Register, the PC, and the contents of the xPSR have been moved onto the stack, shown in Figure 15.5. Recall that the stack pointer indicates the address of the last full entry, so stacking would begin at address 0x200000FC.

```
Memory 1
Address: 0x20000000
0x200000B0: 00 00 00 00 00 00 00 00 00 00 00 00 00 00 00 00 00 00 00 00 00 00
0x200000C6: 00 00 00 00 00 00 00 00 00 00 00 00 00 00 00 00 00 00 00 00 00 00
0x200000DC: 00 00 00 00 00 00 00 00 11 11 11 11 22 22 22 22 33 33 33 33 00 00
0x200000F2: 00 00 FF FF FF FF 4C 00 00 00 00 00 00 01 00 00 00 00 00 00 00 00
0x20000108: 00 00 00 00 00 00 00 00 00 00 00 00 00 00 00 00 00 00 00 00 00 00
0x2000011E: 00 00 00 00 00 00 00 00 00 00 00 00 00 00 00 00 00 00 00 00 00 00
0x20000134: 00 00 00 00 00 00 00 00 00 00 00 00 00 00 00 00 00 00 00 00 00 00
0x2000014A: 00 00 00 00 00 00 00 00 00 00 00 00 00 00 00 00 00 00 00 00 00 00
Call Stack + Locals | Memory 1
Simulation    t1: 0.00000356 sec    L:62 C:1    CAP NUM SCRL OVR R/W
```

FIGURE 15.5 Exception stack frame in memory.

TABLE 15.2
EXC_RETURN Value for the Cortex-M4 with Floating-Point Hardware

EXC_RETURN[31:0]	State	Return to	Using Stack Pointer
0xFFFFFFE1	Floating-point	Handler mode	MSP
0xFFFFFFE9	Floating-point	Thread mode	MSP
0xFFFFFFED	Floating-point	Thread mode	PSP
0xFFFFFFF1	Non-floating-point	Handler mode	MSP
0xFFFFFFF9	Non-floating-point	Thread mode	MSP
0xFFFFFFFD	Non-floating-point	Thread mode	PSP

While the processor is storing critical information on the stack, it also reads the address of the exception handler in the vector table. In our previous example, the processor is about to take a usage fault exception, so the address found at memory location 0x00000018 would be used. The processor will also store one more value for us, called EXC_RETURN, in the Link Register. This 32-bit value describes which stack to use upon exception return, as well as the mode from which the processor left before the exception occurred. Table 15.2 shows all the values currently used on the Cortex-M4—most are reserved. Notice also from our previous example that the EXC_RETURN value was 0xFFFFFFF9, since the floating-point unit was not enabled at the time we took the exception, and we wish to return to Thread mode.

15.5.2 Exit

Returning from exceptions might be one of the few processes that is easier to do on a Cortex-M4 than on the ARM7TDMI, since the processor does most of the work for us. If we are in Handler mode and we wish to return to the main program, one of the following instructions can be used to load the EXC_RETURN value into the Program Counter:

- A LDR or LDM instruction with the PC as the destination
- A POP instruction that loads the PC
- A BX instruction using any register

As a point of interest, some processors use a dedicated instruction to indicate that an exception is complete, and ARM could have done the same thing given the architectural model of the Cortex-M4. However, the idea is to have a device that you can program entirely in C, so a conventional return instruction is used to allow C subroutines to handle exceptions. Most of the return information is held in the EXC_RETURN value.

15.6 EXCEPTION TYPES

In Chapter 14, we saw the different types of exceptions that ARM processors are asked to handle, and we noted that exceptions require the processor to take some time from normal processing to service a peripheral, deal with an interrupt, or handle a fault of some type. There are even more exception types on the Cortex-M4, some of which are common, some of which are not. In fact, in any given microcontroller application, testing and product development cycles have hopefully removed all of the unexpected conditions so that the processor sees only requests which can be handled easily—interrupts, or possibly a debugger poking around. On the ARM7TDMI, the priorities of the exception types are fixed, so that data aborts overrule interrupts. On version 7-M processors, the exception types are mostly programmable, with a few types being fixed (refer back to Table 15.1): reset (-3 or the highest), non-maskable interrupt (-2), and hard fault (-1). Interrupts will be covered in more detail in Section 15.7.

The following types of exceptions are present on the Cortex-M4:

- Reset
- NMI
- Hard fault
- Memory management fault
- Bus fault
- Usage fault
- SVCall
- Debug monitor
- PendSV
- SysTick
- Interrupt

When the Cortex-M4 processor is reset, it will fetch the value at address 0x0 and address 0x4 (usually located in either Flash memory or some kind of ROM), reading both the initial stack pointer and the reset vector, respectively. As it turns out, there are different ways to reset a system, either parts of it or the entire thing. Depending on what's needed, a reset handler can be very simple, or it may need to perform tasks such as:

- Enable a floating-point unit
- Initialize the memory system (e.g., if a memory protection unit [MPU] is present)
- Initialize the two stack pointers and all of the registers
- Initialize any critical I/O devices
- Initialize any peripheral registers, control registers, or clocks, such as a phase-locked loop (PLL)
- Enable certain exception types

A non-maskable interrupt (NMI) has the second highest priority among the exceptions, which means that in most cases, when the processor sees the request, it will be handled immediately. There are conditions that might prevent this, such as the processor being halted by the debugger or an NMI handler already running, but otherwise, this exception is permanently enabled and cannot be masked. On the

Tiva TM4C1233H6PM, for example, an NMI can be triggered by both hardware and software (there is an Interrupt Control and State Register to do this).

A hard fault can occur when the processor sees an error during exception processing, or when another fault such as a usage fault is disabled. In our example code, if we disable usage faults and then rerun the code, you will notice that the processor takes a hard fault when the UDIV instruction is attempted, rather than a usage fault. You can also see hard faults when there is an attempt to access the System Control Space in an unprivileged mode; for example, if you attempt to write a value to one of the NVIC registers in Thread mode, the processor will take an exception.

Memory management faults occur when the processor attempts to access areas of memory that are inaccessible to the current mode and privilege level (e.g., privileged access only or read only) or that are not defined by the MPU. Generally, the reason for the fault and the faulting address can be found in the Memory Management Fault Status Register, shown in Table 15.3. Like usage and bus faults, memory management faults must also be enabled before the processor can use them.

One of the busses that commonly runs through a conventional SoC is the AMBA High-Performance Bus (AHB), connecting memory and peripherals to the main processor. Bus faults occur when an error response returns from an access on the AHB bus, either for instruction or data accesses. There is a Fault Status Register for these errors as well, so that an exception handler can determine the offending instruction and possibly recover. Since both precise bus faults (where the fault occurs on the last completed operation) and imprecise bus faults (where the fault is triggered by an instruction that may have already completed) can generate an error, recovery is possible in some cases, but it is certainly not easy to do.

Usage faults occur for a number of reasons. If you have enabled usage faults, then the processor will take an exception for the following:

- Trying to divide by zero, assuming that the processor has been told to trap on this event (i.e., setting the DIV_0_TRP bit in the NVIC as we did in Example 15.1)
- Trying to switch the processor into ARM state. Recall that the least significant bit of branch targets, exception vectors, and PC values popped from

TABLE 15.3
Memory Management Fault Status Register (Offset 0xD28)

Bit	Name	Reset Value	Description
7	MMARVALID	0	Indicates the Memory Management Address register is valid
6:5	—	—	—
4	MSTKERR	0	Stacking error
3	MUNSTKERR	0	Unstacking error
2	—	—	—
1	DACCVIOL	0	Data access violation
0	IACCVIOL	0	Instruction access violation

TABLE 15.4
Usage Fault Status Register (Offset 0xD2A)

Bit	Name	Reset Value	Description
9	DIVBYZERO	0	Indicates a divide by zero has occurred (only if DIV_0_TRP is also set)
8	UNALIGNED	0	An unaligned access fault has occurred
7:4	—	—	—
3	NOCP	0	Indicates a coprocessor instruction was attempted
2	INVPC	0	An invalid EXC_RETURN value was used in an exception
1	INVSTATE	0	An attempt was made to switch to an invalid state
0	UNDEFINSTR	0	Processor tried to execute an undefined instruction

the stack must be a 1, since the Cortex-M4 always operates in Thumb state. The INVSTATE bit will be set in the Usage Fault Status Register shown in Table 15.4

- Using an undefined instruction
- Attempting an illegal unaligned access
- Attempting to execute a coprocessor instruction
- Returning from an exception with an invalid EXC_RETURN value

Table 15.4 shows all of the bits in the Usage Fault Status Register that can be examined by a fault handler.

At this point, we might be tempted to clean up our usage fault handler from earlier examples so that the divide-by-zero error is no longer a problem. In an actual system, a warning, an error, or possibly a symbol such as "#DIV/0!" could be printed to a screen, but in an embedded system using only integer math, a division by zero is often catastrophic. There is no recovery that makes sense—what value could you return that either represents infinity or represents a number that could guarantee an algorithm would not exceed certain bounds? Unlike our short examples, a proper usage fault handler should follow the AAPCS guidelines, stacking the appropriate registers before proceeding, and it might even determine the destination register in the offending instruction to make a partial recovery. If an operating system were running, one course of action would be to terminate the thread that generated this exception and perhaps indicate the error to the user.

Supervisor calls (SVCall) and Pendable Service Calls (PendSV) are similar in spirit to the SWI (now SVC) exceptions on the ARM7TDMI, where user-level code can access certain resources in a system, say a particular piece of hardware, by forcing an exception. A handler running in a privileged mode then examines the actual SVC instruction to determine specifically what is being requested. This way, hardware can be controlled by an operating system, and something like an API can be provided to programmers, leaving device drivers to take care of the details. Pending Service Calls can be used in conjunction with SVC instructions to provide efficient handling of context switching for operating systems. See (Yiu 2014) for details on working with service calls and interrupt handling.

The SYSTICK exception is generated by a 24-bit internal timer that is controlled by four registers. When this system timer reaches zero, it can generate an exception with its own vector number (refer back to Table 15.1). Operating systems use this type of timer for task management, that is, to ensure that no single task is allowed to run more than any other. For an excellent reference on operating systems, particularly for embedded systems, see (http://processors.wiki.ti.com/index.php/TI-RTOS_Workshop#Intro_to_TI-RTOS_Kernel_Workshop_Online_Video_Tutorials, 2013).

The last two types of exceptions are almost polar opposites of each other in terms of attention. One, the Debug Monitor exception, is generated by a debug monitor running in a system, and consequently, is of interest to writers of debug monitors and few others (if you are really curious, consult the *RealView ICE User Guide* (ARM 2008b) for details on working with debug components). The second, interrupts, are used by nearly everyone, and therefore deserves a section of its own.

15.7 INTERRUPTS

In Section 14.2 we saw that not all exceptions are unwanted, particularly interrupts, since peripherals can generate them. The Cortex-M4 supports up to 240 interrupts of the 496 allowed by the Cortex-M specification, although most silicon vendors do not implement all of them. In fact, if we carefully examine the TM4C1233H6PM microcontroller from TI, you will notice that it supports only 65—still, quite a few. The interrupt priorities are also programmable, so that the various interrupts coming from different peripherals can either have the same weighting or unique priorities which are assigned to each one. There are so many variations on the interrupts, in fact, that all of the details are best left for a book like (Yiu 2014) or the *Cortex-M4 Technical Reference Manual* (ARM 2009). Interrupts can be masked; interrupts can be held pending; interrupts can have subpriorities; and you can disable only interrupts with a priority below a certain level. All of these options are not critical to our understanding of how they work and what is necessary to configure a peripheral to generate one.

Reading through the partial list of peripherals in Table 15.5, you can see that the various peripherals can generate an interrupt, and all of these interrupts are handled and prioritized by the NVIC (also covered in Chapter 14) once it is configured. A full listing can be found in the Data Sheet (Texas Instruments 2013a).

Configuration is probably the *least* trivial aspect of working with a microcontroller the size of the TM4C1233H6PM. There are dozens of registers than may need to be configured in any given system. Consequently, silicon vendors provide their own APIs to use when programming the controllers in a language like C. The libraries are based around the Cortex Microcontroller Software Interface Standard (CMSIS) from ARM. TivaWare™ from Texas Instruments and the LPCOpen Platform from NXP are examples of libraries that allow peripherals to be enabled and configured using only standard access functions. To fully appreciate a statement like

```
SysCtlPeripheralEnable(SYSCTL_PERIPH_TIMER0);
```

TABLE 15.5
Partial Vector Table for Interrupts on the Tiva TM4C1233H6PM Microcontroller

Vector Number	Interrupt Number (Bit in Interrupt Registers)	Vector Address or Offset	Description
0–15	—	0x00000000–0x0000003C	Processor Exceptions
.	.	.	
.	.	.	
.	.	.	
30	14	0x00000078	ADC0 Sequence 0
31	15	0x0000007C	ADC0 Sequence 1
32	16	0x00000080	ADC0 Sequence 2
33	17	0x00000084	ADC0 Sequence 3
34	18	0x00000088	Watchdog Timers 0 and 1
35	**19**	**0x0000008C**	**16/32-Bit Timer 0A**
36	20	0x00000090	16/32-Bit Timer 0B
37	21	0x00000094	16/32-Bit Timer 1A
38	22	0x00000098	16/32-Bit Timer 1B
37	21	0x0000009C	16/32-Bit Timer 2A
38	22	0x000000A0	16/32-Bit Timer 2B
.	.	.	
.	.	.	
.	.	.	
51	35	0x000000CC	16/32-Bit Timer 3A
52	36	0x000000D0	16/32-Bit Timer 3B
.	.	.	

you need to program the same operation in assembly at least once! That's where we begin.

EXAMPLE 15.4

Let's look at an example of a relatively simple interrupt being caused by a timer counting down to zero. You can see from Table 15.5 that the twelve timers are all given their own vector number, interrupt number, and vector address. We'll set up one 16-bit timer, Timer 0A, to count down from 0xFFFF to 0, sending an interrupt to the NVIC, alerting the core that the timer expired. The processor will acknowledge the interrupt and jump to a handler routine. Once inside the interrupt handler, we'll put a value into a core register and then spin in an infinite loop so that we can see the process happen using a debugger. In this example, we'll use the Tiva Launchpad as a target. As a necessity, the timer must be configured as a memory-mapped peripheral, but this gives us a preview of Chapter 16.

In order to configure the interrupts, the following must take place:

- The system clocks on processor must be set up, similar to the code we'll use in Chapter 16 for the GPIO example.

- The clocks must be enabled to the interrupt block, specifically, using the RCGCTIMER register. Now that the timer is enabled, we can actually write to the memory-mapped registers within it. If the interrupt block is not enabled, any attempts to write to the memory-mapped registers results in a hard fault.
- Rather than having a periodic timer, we will configure it to be a one-shot timer. To do this, we must configure the GPTMTnMR register.
- The timer will be set up as a 16-bit timer, so the initial count will be (by a reset) set to 0xFFFF. By default, the timer will count down to zero, rather than up.
- The interrupt from the timer needs to be enabled. There is a GPTMIMR register, or General-Purpose Timer Interrupt Mask Register, than needs to be configured. Writing a 1 to the appropriate bit enables the interrupt.
- Interrupts needs to be enabled from Timer 0A in the NVIC. Bit 19 of the Interrupt Set Enable register in the NVIC enables Timer 0A.
- The timer needs to be started.

Using the Code Composer Studio tools, you can create a source code file with the following code:

```
MOVW    r0, #0xE000
MOVT    r0, #0x400F
MOVW    r2, #0x60           ; offset 0x060 for this register
MOVW    r1, #0x0540
MOVT    r1, #0x01C0
STR     r1, [r0, r2]        ; write the register's content

MOVW    r7, #0x604          ; enable timer0 - RCGCTIMER
LDR     r1, [r0, r7]        ; p. 321, base 0x400FE000
ORR     r1, #0x1            ; offset - 0x604
STR     r1, [r0, r7]        ; bit 0

NOP
NOP
NOP
NOP
NOP                         ; give myself 5 clocks per spec

MOVW    r8, #0x0000         ; configure timer0 to be
MOVT    r8, #0x4003         ; one-shot, p.698 GPTMTnMR
MOVW    r7, #0x4            ; base 0x40030000
LDR     r1, [r8, r7]        ; offset 0x4
ORR     r1, #0x21           ; bit 5=1, 1:0=0x1
STR     r1, [r8, r7]

LDR     r1, [r8]            ; set as 16-bit timer only
ORR     r1, #0x4            ; base 0x40030000
STR     r1, [r8]            ; offset 0, bit[2:0]=0x4

MOVW    r7, #0x30           ; set the match value at 0
MOV     r1, #0              ; since we're counting down
STR     r1, [r8, r7]        ; offset - 0x30

MOVW    r7, #0x18           ; set bits in the GPTM
LDR     r1, [r8, r7]        ; Interrupt Mask Register
```

```
    ORR     r1, #0x10          ; p. 714 - base: 0x40030000
    STR     r1, [r8, r7]       ; offset - 0x18, bit 5

    MOVW    r6, #0xE000        ; enable interrupt on timer0
    MOVT    r6, #0xE000        ; p. 132, base 0xE000E000
    MOVW    r7, #0x100         ; offset - 0x100, bit 19
    MOV     r1, #(1<<19)       ; enable bit 19 for timer0
    STR     r1, [r6, r7]

    MOVW    r6, #0x0000        ; start the timer
    MOVT    r6, #0x4003
    MOVW    r7, #0xC
    LDR     r1, [r6, r7]
    ORR     r1, #0x1
    STR     r1, [r6, r7]       ; go!!
```

Now that the NVIC, Timer 0A, and all of the control registers are programmed, we can write a very simple handler for the interrupt we are expecting:

```
IntDefaultHandler:
      MOVW    r10, #0xBEEF
      MOVT    r10, #0xDEAD
Spot
      B       Spot
```

This will do nothing more than write a value into register r10 and then spin in a loop.

To run this program on a Tiva Launchpad, you will likely have to reset the system after the code is loaded (as opposed to resetting just the core). Then run the program. Once you hit the stop button, you can see that the processor is in the interrupt handler just executing branch instructions in a loop. The entire program is given in Appendix D.

15.8 EXERCISES

1. How many operation modes does the Cortex-M4 have?

2. What happens if you do *not* enable usage faults in Example 15.1?

3. Which register must be used to switch privilege levels?

4. What are the differences between the ARM7TDMI and Cortex-M4 vector tables?

5. Give the offsets (from the base address 0xE000E000) and register size for the following:
   a. Usage Fault Status Register
   b. Memory Management Fault Status Register

   You may wish to consult the TM4C1233H6PM data sheet (Texas Instruments 2013b).

6. Configure Example 15.4 so that the timer counts up rather than down. Don't forget to configure the appropriate match value!

16 Memory-Mapped Peripherals

16.1 INTRODUCTION

Modern embedded systems generally demand quite a bit from a single piece of silicon. For example, in the 1990s, cell phones were used for making phone calls and little else. Today, their uses range from checking e-mail to watching your favorite movie. The industry also geared up to include GPS on smartphones, so that when you miss that turn while driving down the road (probably because you were arguing with your smartphone), you can also ask it for directions. To build such systems, the hardware has to include more features in silicon, and the software has to learn to talk to those new features. SoC designs are packing ever more devices onto one die. Even off-the-shelf microcontrollers are getting more elaborate, with small, 32-bit processors built to control different types of serial interfaces, e.g., UARTs, I^2C, and CAN; analog devices like temperature sensors and analog comparitors; and motion controllers for motors and servos. How are all of these attached to a single processor? In this chapter, we're going to look at three particular microcontrollers, the LPC2104 and the LPC2132 from NXP, and the TM4C123GH6PM from TI, along with three very useful peripherals, the UART, general-purpose I/O (GPIO), and the digital-to-analog converter (DAC). The UART is a relatively simple serial interface, and we'll program it to send character data to a window in the simulator. The DAC takes a 10-bit value and generates an output relative to a reference voltage. To show off our coding skills, the DAC will generate a sine wave from the sine table we created in Chapter 12. The last example uses an inexpensive evaluation module, the Tiva Launchpad, to continuously change the color of a flashing LED. In writing the three programs, we're going to tie all of the elements from previous chapters into the code, including

- Subroutines and how they are written
- Passing parameters
- The ARM Application Procedure Call Standard (AAPCS)
- Stacks
- Q notation
- Arithmetic
- The ARM and Thumb-2 instruction sets

16.2 THE LPC2104

The best place to start is at the highest level—the SoC, or in our case, the microcontroller. Figure 16.1 shows the block diagram of the LPC2104 from NXP. You can

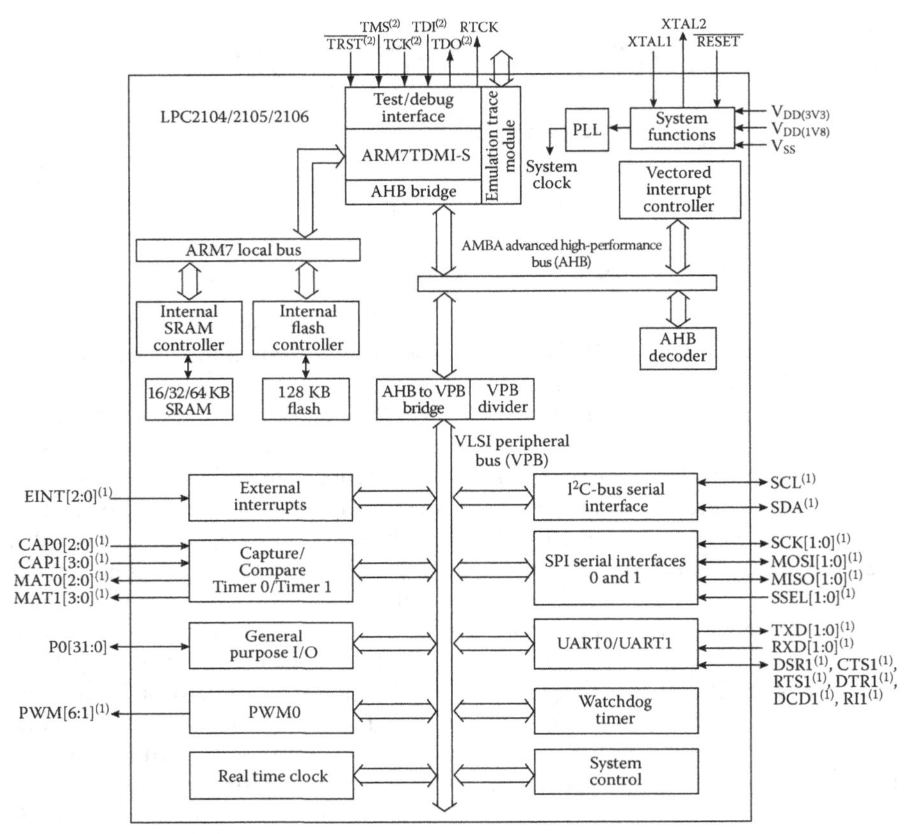

FIGURE 16.1 LPC2104/2105/2106 block diagram. (From Doc. LPC2104–2105–2106–6 Product Data Sheet, NXP Semiconductors, July 2007. With permission.)

see the ARM7TDMI core at the top, and two main busses in the system: the AHB, a high-speed bus designed to have only a few bus masters attached to it, and the VPB, a slower bus designed to have many peripherals attached to it. Between the two busses is a bridge. Fortunately for the programmer, you don't have to focus too much on the internal hardware design, but it is important to know how the peripherals are attached and what type of interface they have (i.e., what pins go with which peripheral). The LPC2104 includes a few different serial interfaces, along with some timers, some general-purpose I/O, two UARTs, and some on-chip memory. Specifically, we're going to use UART0 to write some character data out of the part.

16.2.1 The UART

The Universal Asynchronous Receiver/Transmitter (UART) is probably one of the most ubiquitous peripherals found on microcontrollers. It can be used to implement

TABLE 16.1
UART Configuration Bits in the Control Register

U0LCR	Function	Description	Reset Value
1:0	Word Length Select	00:5-bit character length 01:6-bit character length 10:7-bit character length 11:8-bit character length	0
2	Stop Bit Select	0:1 stop bit 1:2 stop bits (1.5 if U0LCR[1:0] = 00)	0
3	Parity Enable	0: Disable parity generation and checking 1: Enable parity generation and checking	0
5:4	Parity select	00: Odd parity 01: Even parity 10: Forced "1" stick parity 11: Forced "0" stick parity	0
6	Break Control	0: Disable break transmission 1: Enable break transmission. Output pin UART0 TxD is forced to logic 0 when U0LCR6 is actively high	0
7	Divisor Latch Access Bit	0: Disable access to divisor latches 1: Enable access to divisor latches	0

serial transmission standards such as RS-232 or EIA232F, or connect the microcontroller to devices like LCD displays or bar code scanners. While high-speed serial standards such as USB and Firewire have largely replaced the older protocols, UARTs are still used to provide a simple, inexpensive interface to devices that don't necessarily have to transmit and receive data at high speeds.

Asynchronous start-stop communication is done without a clock signal. Rather than using a dedicated clock line, which adds pins, special bits are added to the data being sent to tell the receiver when the data is starting and stopping. Parity bits, like those discussed in Chapter 7, can be added to the transmission. Long ago, these options were all controlled by hardware, through either switches or jumpers. With modern systems, software controls these choices. Table 16.1 shows the options available for character length, stop bits, parity, and break control. You can find more detailed information on UARTs and asynchronous serial ports in (Clements 2000) and (Kane et al. 1981).

16.2.2 The Memory Map

Peripherals on the LPC2104 are memory-mapped, meaning that their configuration registers, receive and transmit buffers, status registers, etc., are each mapped to an address. Accessing peripherals is actually just as easy as accessing a memory block.

You can use LDR and STR instructions just as you would if you were writing a value to memory; although, you should be aware that some peripherals are sensitive to reads and writes. For example, a memory-mapped register may automatically clear its contents after being read.

Looking at Figure 16.2, you can see the memory map for the entire microcontroller. Notice that distinct memory regions are defined. The controller comes with 128 KB of Flash memory for your programs, and in the case of the LPC2104, 16 KB of on-chip RAM for building stacks and holding variables. All of the peripherals lie in the very highest addresses, between addresses 0xE0000000 and 0xFFFFFFFF. If we zoom in a bit more, we will find our UART, called UART0, between addresses 0xE000C000 and 0xE000C01C, as shown in Figure 16.3.

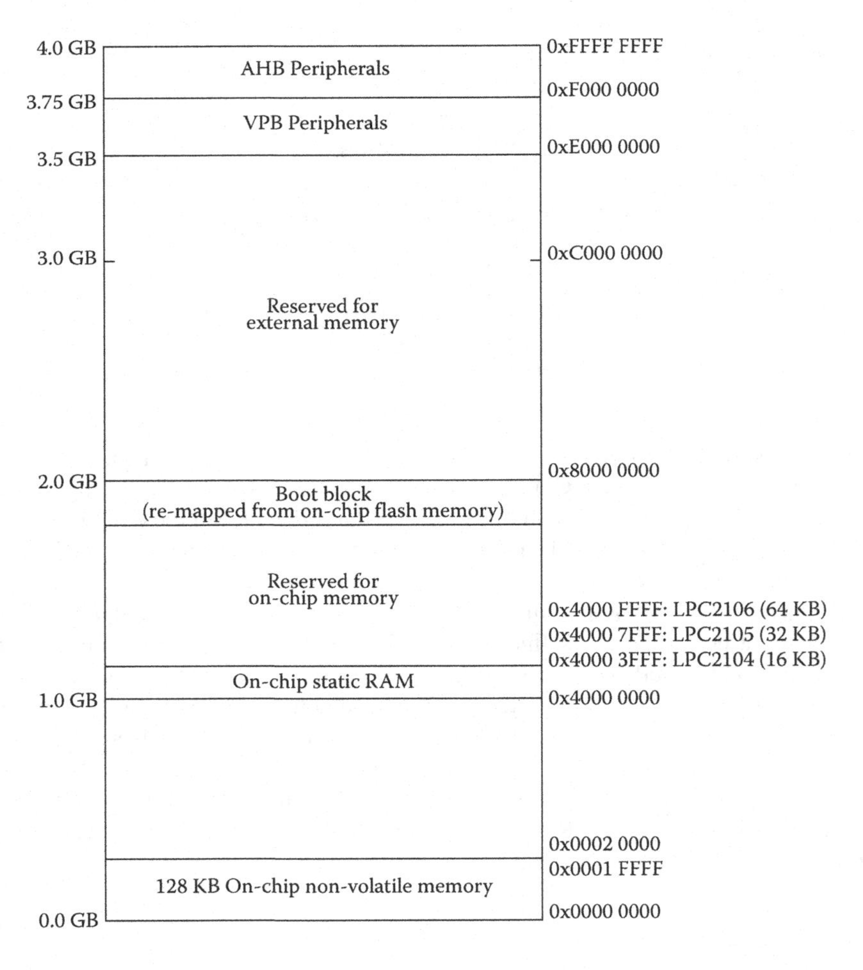

FIGURE 16.2 System memory map. (From LPC2106/2105/2104 User Manual NXP Semiconductors, September 2003. With permission.)

UART 0												
0xE000C000	U0RBR (DLAB=0)	U0 Receiver buffer register	8-bit data								RO	un-defined
	U0THR (DLAB=0)	U0 Transmit holding register	8-bit data								WO	NA
	U0DLL (DLAB=1)	U0 Divisor latch LSB	8-bit data								R/W	0x01
0xE000C004	U0IER (DLAB=0)	U0 Interrupt enable register	0	0	0	0	0	En. Rx Line Status Int.	Enable THRE Int.	En. Rx Data Av.Int.	R/W	0
	U0DLM (DLAB=1)	U0 Divisor latch LSB	8 bit data								R/W	0
0xE000C008	U0IIR	U0 Interrupt ID register	FIFOs Enabled		0	0	IIR3	IIR2	IIR1	IIR0	RO	0x01
	U0FCR	U0 FIFO control register	Rx Trigger		-	-	-	U0 Tx FIFO Reset	U0 Rx FIFO Reset	U0 FIFO Enable	WO	0
0xE000C00C	U0LCR	U0 Line control register	DLAB	Set break	Stick parity	Even parity select	Parity enable	Nm. of stop bits	Word length select		R/W	0
0xE000C014	U0LSR	U0 Line status register	Rx FIFO Error	TEMT	THRE	BI	FE	PE	OE	DR	RO	0x60
0xE000C01C	U0LSR	U0 Scratch pad register	8-bit data								R/W	0

FIGURE 16.3 Memory map of UART0 on the LPC2104. (From LPC2106/2105/2104 User Manual NXP Semiconductors, September 2003. With permission.)

16.2.3 Configuring the UART

To demonstrate how easy it is to talk to a peripheral, we'll write a short block of code that does two things: it calls a subroutine to configure our UART, and it then sends a short message through the UART for the Keil tools to read. If you look at the external pins of the LPC2104, shown in Figure 16.4, you will notice that they are multiplexed pins, meaning that the function of the pin itself is configurable. This allows a package to reduce the pin count, but the programmer must configure the pins to use them. So the first order of business is to set up the LPC2104 so that pins P0.0 and P0.1 become our transmit and receive pins, Tx0 and Rx0, respectively. To do this, we load the address of the pin configuration register, shown in Table 16.2, into a general register, where PINSEL0 is equated to 0xE002C000. Using a read-modify-write sequence (good practice when you don't want to disturb other configuration or status bits), PINSEL0[1:0] and PINSEL0[3:2] are set to 0b01. The assembly would look like the following:

```
LDR     r5, = PINSEL0    ; base address of register
LDR     r6, [r5]         ; get contents
BIC     r6,r6,#0xF       ; clear out lower nibble
ORR     r6,r6,#0x5       ; sets P0.0 to Tx0 and P0.1 to Rx0
STR     r6, [r5]         ; r/modify/w back to register
```

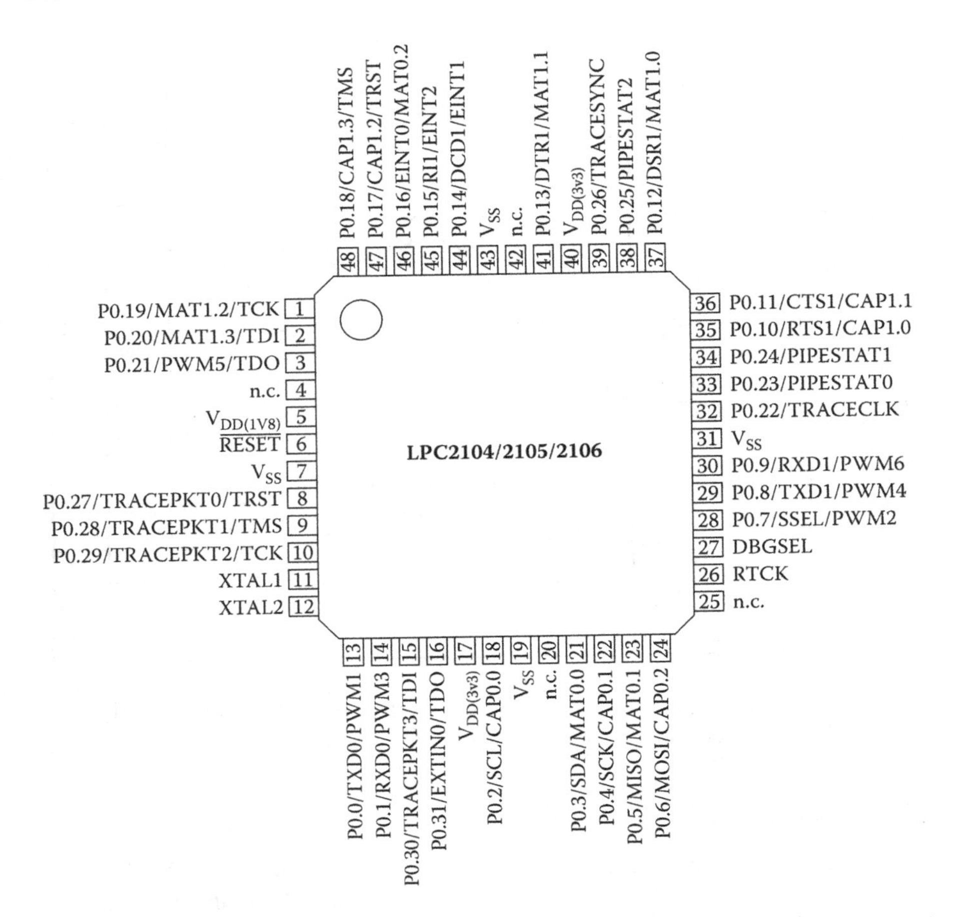

FIGURE 16.4 Pin descriptions for the LPC2104. (From Doc. LPC2104–2105–2106–6 Product Data Sheet, NXP Semiconductors, July 2007. With permission.)

The next step to configuring the UART is to set the number of data bits, the parity, and the number of stop bits. Again, the starting address of the UART0 configuration register, 0xE000C000, is loaded into a general register to be used as a base address. The LCR and LSR registers can be accessed using a pre-indexed addressing scheme, where the offsets are equated to known values at the beginning of the final routine. Here, LCR0 would be equated to 0xC, and for our write routine, LSR0 would be equated to 0x14. Since these are 8-bit registers, they must be accessed using STRB and LDRB instructions. The rest of the configuration code is below.

```
LDR     r5, =U0START
MOV     r6, #0x83          ; set 8 bits, no parity, 1 stop bit
STRB    r6, [r5, #LCR0]    ; write control byte to LCR
MOV     r6, #0x61          ; 9600 baud @15 MHz VPB clock
STRB    r6, [r5]           ; store control byte
```

TABLE 16.2
PINSEL0 Register for Pin Configurations

PINSEL0	Pin Name	Function When 00	Function When 01	Function When 10	Function When 11	Reset Value
1:0	P0.0	GPIO Port 0.0	TxD (UART 0)	PWM1	Reserved	0
3:2	P0.1	GPIO Port 0.1	RxD (UART 0)	PWM3	Reserved	0
5:4	P0.2	GPIO Port 0.2	SCL (I^2C)	Capture 0.0 (Timer 0)	Reserved	0
7:6	P0.3	GPIO Port 0.3	SDA (I^2C)	Match 0.0 (Timer 0)	Reserved	0
9:8	P0.4	GPIO Port 0.4	SCK (SPI)	Capture 0.1 (Timer 0)	Reserved	0
11:10	P0.5	GPIO Port 0.5	MISO (SPI)	Match 0.1 (Timer 0)	Reserved	0
13:12	P0.6	GPIO Port 0.6	MOSI (SPI)	Capture 0.2 (Timer 0)	Reserved	0
15:14	P0.7	GPIO Port 0.7	SSEL (SPI)	PWM2	Reserved	0
17:16	P0.8	GPIO Port 0.8	TxD UART 1	PWM4	Reserved	0
19:18	P0.9	GPIO Port 0.9	RxD (UART 1)	PWM6	Reserved	0
21:20	P0.10	GPIO Port 0.10	RTS (UART 1)	Capture 1.0 (Timer 1)	Reserved	0
23:22	P0.11	GPIO Port 0.11	CTS (UART 1)	Capture 1.1 (Timer 1)	Reserved	0
25:24	P0.12	GPIO Port 0.12	DSR (UART 1)	Match 1.0 (Timer 1)	Reserved	0
27:26	P0.13	GPIO Port 0.13	DTR (UART 1)	Match 1.1 (Timer 1)	Reserved	0
29:28	P0.14	GPIO Port 0.14	CD (UART 1)	EINT1	Reserved	0
31:30	P0.15	GPIO Port 0.15	RI (UART1)	EINT2	Reserved	0

Source: From LPC2106/2105/2104 User Manual NXP Semiconductors, September 2003. With permission.

```
MOV     r6, #3           ; set DLAB = 0
STRB    r6, [r5, #LCR0]  ; Tx and Rx buffers set up
```

16.2.4 Writing the Data to the UART

Now that the UART is configured to send and receive data, we can try writing some data out of the part. In this case, we'll send some character data—the short message "Watson. Come quickly!" The subroutine for this task is written so that the calling routine can send a single character at a time. When the subroutine receives the character, it's placed into the transmit buffer, but only after the processor checks to ensure the previous character has been transmitted. Who's reading this data? In the simulation tools, there is a serial window that can accept data from a UART, driving the necessary handshake lines that are normally attached to the receiver. The assembly code for our transmitter routine looks like the following:

```
        LDR     r5, =U0START
wait    LDRB    r6,[r5,#LSR0]    ; get status of buffer
        CMP     r6,#0x20         ; buffer empty?
        BEQ     wait             ; spin until buffer's empty
        STRB    r0,[r5]
```

16.2.5 Putting the Code Together

Now that we have one subroutine to set up our UART and another to send a character, the remaining code will be responsible for reading a sentence from memory, one character at a time, and calling the subroutine to transmit it. A small loop will read a character from memory and test to see whether it is the null terminator for a string, i.e., the value 0. If so, the loop terminates. Otherwise, the subroutine Transmit is called.

The AAPCS allows registers r0 through r3 to be corruptible, and we've used registers r5 and r6 in our subroutines. While the code could be written without having to stack any registers (left as an exercise), we'll go ahead and set the stack pointer to the start of RAM, which is address 0x40000000. The registers used in our subroutines can then be saved off. The code below is a complete routine.

```
          AREA UARTDEMO, CODE, READONLY
PINSEL0   EQU    0xE002C000      ; controls the function of the pins
U0START   EQU    0xE000C000      ; start of UART0 registers
LCR0      EQU    0xC             ; line control register for UART0
LSR0      EQU    0x14            ; line status register for UART0
RAMSTART  EQU    0x40000000      ; start of onboard RAM for 2104
          ENTRY
start
          LDR    sp, = RAMSTART  ; set up stack pointer
          BL     UARTConfig      ; initialize/configure UART0
          LDR    r1, = CharData  ; starting address of characters
Loop
          LDRB   r0, [r1],#1     ; load character, increment address
          CMP    r0,#0           ; null terminated?
          BLNE   Transmit        ; send character to UART
          BNE    Loop            ; continue if not a '0'
done      B      done            ; otherwise we're done

; Subroutine UARTConfig
; This subroutine configures the I/O pins first. It
; then sets up the UART control register. The
; parameters
; are set to 8 bits, no parity and 1 stop bit.
; Registers used:
; r5 - scratch register
; r6 - scratch register
; inputs: none
; outputs: none

UARTConfig
          STMIA  sp!, {r5,r6,lr}
          LDR    r5, = PINSEL0      ; base address of register
          LDR    r6, [r5]           ; get contents
          BIC    r6,r6,#0xF         ; clear out lower nibble
          ORR    r6,r6,#0x5         ; sets P0.0 to Tx0 and P0.1 to Rx0
          STR    r6, [r5]           ; r/modify/w back to register
          LDR    r5, = U0START
          MOV    r6, #0x83          ; set 8 bits, no parity, 1 stop bit
          STRB   r6, [r5, #LCR0]    ; write control byte to LCR
          MOV    r6, #0x61          ; 9600 baud @15 MHz VPB clock
```

```
            STRB   r6, [r5]            ; store control byte
            MOV    r6, #3              ; set DLAB = 0
            STRB   r6, [r5, #LCR0]     ; Tx and Rx buffers set up
            LDMDB  sp!,{r5,r6,pc}

; Subroutine Transmit
; This routine puts one byte into the UART
; for transmitting.
; Register used:
; r5 - scratch
; r6 - scratch
; inputs: r0- byte to transmit
; outputs: none
;

Transmit

            STMIA  sp!,{r5,r6,lr}
            LDR    r5, =U0START
wait        LDRB   r6,[r5,#LSR0]       ; get status of buffer
            CMP    r6,#0x20            ; buffer empty?
            BEQ    wait                ; spin until buffer's empty
            STRB   r0,[r5]
            LDMDB  sp!,{r5,r6,pc}
CharData
            DCB    "Watson. Come quickly!",0
            END
```

16.2.6 Running the Code

At this point, you should take some time to enter the code and run it in the Keil tools. It should be run in the same manner we have taken with all of the other programs, namely that it starts in memory at address 0x0, and there are no handlers of any kind for exceptions. The tools have additional windows that allow you to view peripherals on the chip. The peripheral we use here, UART0, can be seen by choosing UART0 from the Peripherals menu after the debug session has been started. This will bring up the peripheral window, shown in Figure 16.5.

To see the output from the UART, you can use the Serial Window submenu from the View menu on the toolbar. You should select UART #1, which brings up the window shown in Figure 16.6.

16.3 THE LPC2132

Figure 16.7 shows a block diagram of the LPC2132 microcontroller, which looks very much like the LPC2104. It has the same ARM7TDMI processor, the same AHB and VPB busses, and a very similar set of peripherals, which is fortunate. With a similar structure and memory map, programming our microcontroller should be very straightforward. The peripheral of interest this time is the D/A converter with its associated output pin A_{OUT}. Since we've already covered fractional arithmetic, sine tables, and subroutines, we can tie all of these concepts together by creating a sine wave using the D/A converter. The output A_{OUT} will be monitored on a simulated logic analyzer in the Keil tools so that we can see our sine wave.

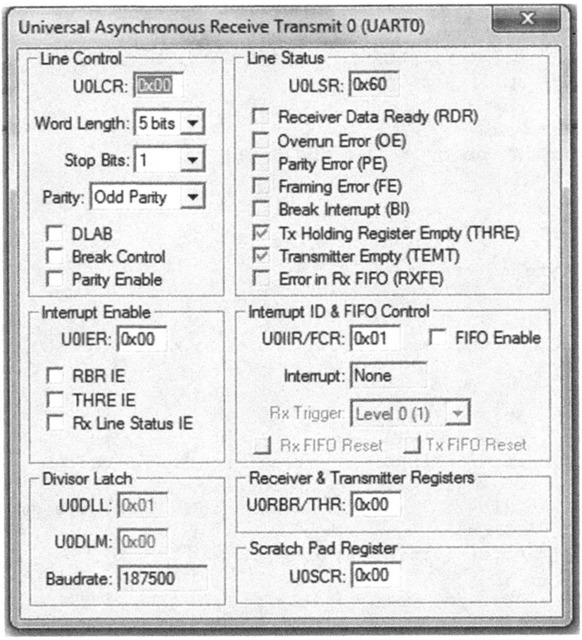

FIGURE 16.5 The UART0 peripheral window.

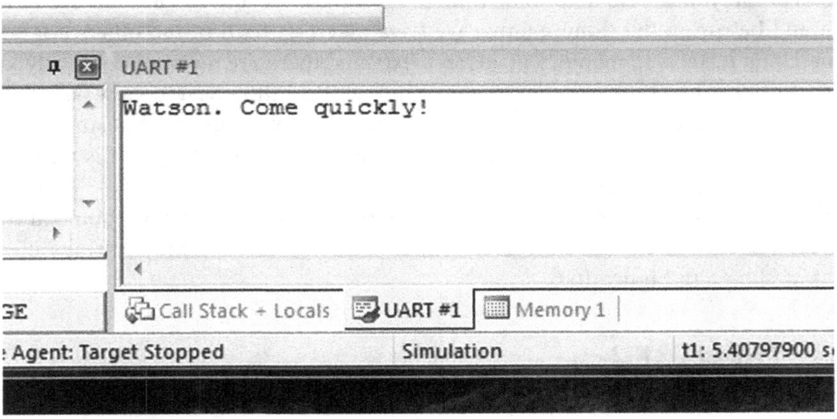

FIGURE 16.6 Serial output window.

16.3.1 The D/A Converter

In many signal processing and control applications, an analog waveform is sampled, processed in some way, e.g., a digital filter, and then converted back into an analog waveform. The process of taking a binary value and generating a voltage based on that value requires a digital-to-analog converter. There are many types, including

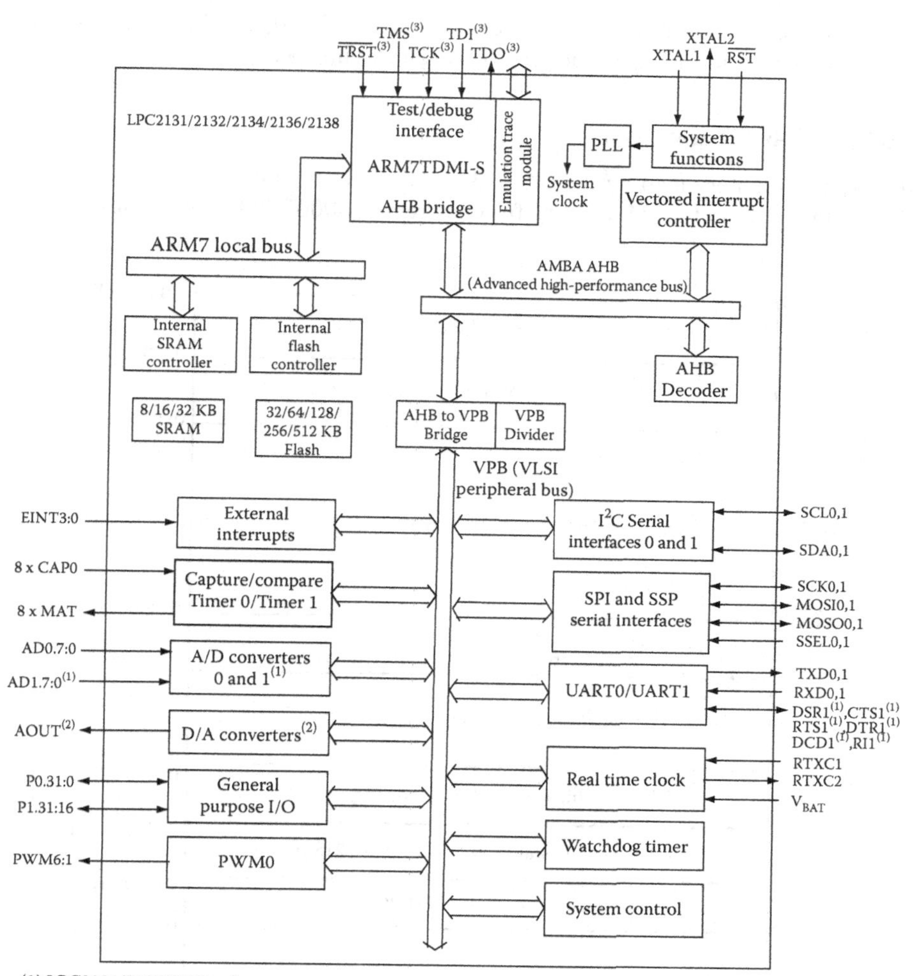

FIGURE 16.7 LPC2132 block diagram. (From UM10120 Vol. 1: LPC213x User Manual, NXP Semiconductors, June 2006. With permission.)

tree networks and R-2R ladders, but their construction lies outside the scope of this book. Fortunately, the electronics can be overlooked for the moment and we can concentrate on using the device.

The basic operation of the D/A converter takes a 10-bit binary value and generates a voltage on A_{OUT} which is proportional to a reference voltage V_{REF}. In other words, if our binary number in base ten is *value*, then the output voltage is

$$A_{OUT} = \frac{value}{1024} \times V_{REF}$$

To use the D/A converter, we will need to set up the pin P0.25 to be our analog output A_{OUT}. Afterward, we can send our 10-bit value to this peripheral to be converted.

16.3.2 The Memory Map

The system memory map for the LPC2132 is shown in Figure 16.8. By now, we recognize the 64 KB of ROM memory from address 0x00000000 to 0x0000FFFF and

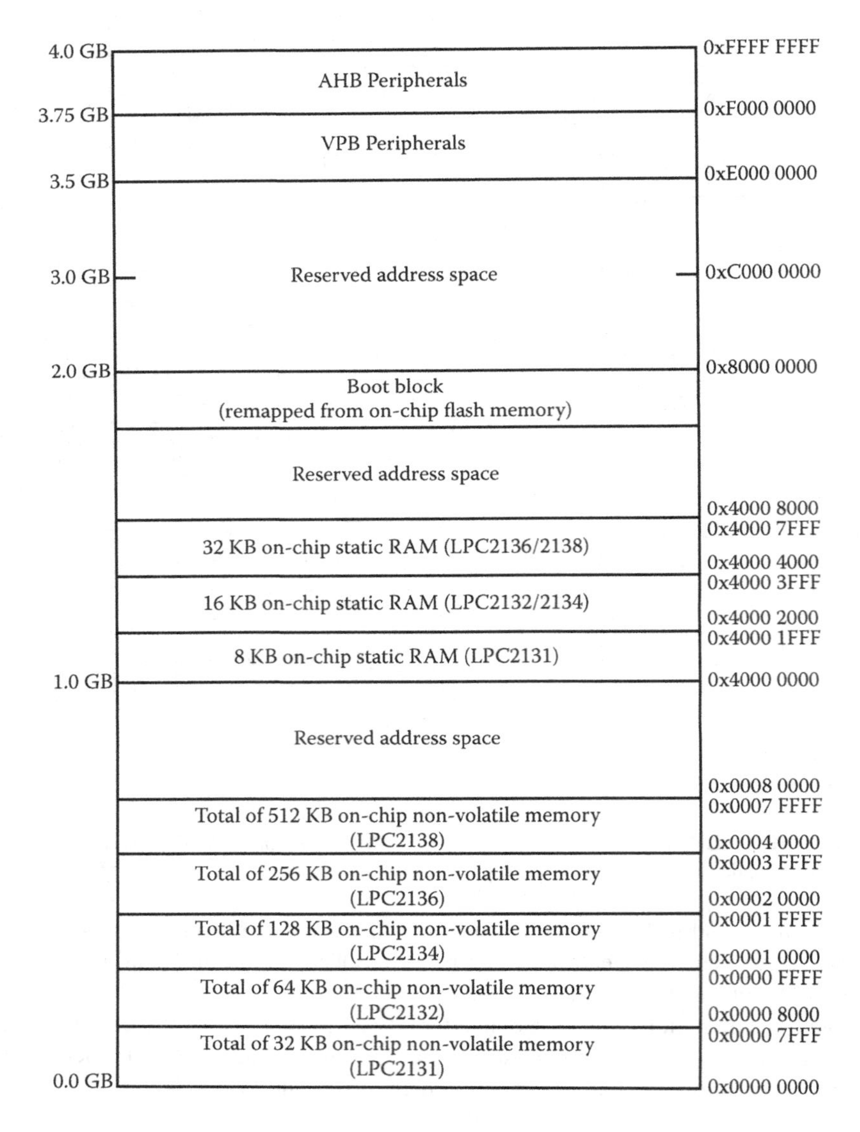

FIGURE 16.8 LPC2132 memory map. (From UM10120 Vol. 1: LPC213x User Manual, NXP Semiconductors, June 2006. With permission.)

TABLE 16.3
DAC Register Bit Description

Bit	Symbol	Value	Description	Reset Value
5:0	—		Reserved, user software should not write ones to reserved bits. The value read from a reserved bit is not defined.	NA
15:6	VALUE		After the selected settling time after this field is written with a new VALUE, the voltage on the A_{OUT} pin (with respect to VSSA) is VALUE/1024 * V_{REF}.	0
16	BIAS	0	The settling time of the DAC is 1 μs max, and the maximum current is 700 μA.	0
		1	The settling time of the DAC is 2.5 μs and the maximum current is 350 μA.	
31:17	—		Reserved, user software should not write ones to reserved bits. The value read from a reserved bit is not defined.	NA

the 16 KB of RAM starting at address 0x40000000. The peripherals are memory-mapped in high memory, starting at address 0xE0000000. Again, zooming in a bit more, our peripheral has a register called DACR, or DAC Register, which is used for configuring the D/A converter as well as giving it a value to convert. The register is located at address 0xE006C000, and bits [15:6] are the 10 bits of digital input, as shown in Table 16.3.

16.3.3 Configuring the D/A Converter

Before using the D/A converter, we need to configure pin P0.25 such that it becomes A_{OUT}, in the same way we changed the UART pins that were also multiplexed. The code below enables the D/A converter and sets the pin by writing 0b10 to bits [19:18] of the Pin Function Select Register called PINSEL1, which is located at address 0xE002C004. A read-modify-write sequence ensures that other bits that are set or clear are not altered:

```
LDR   r6, = PINSEL1          ; PINSEL1 configures pins
LDR   r7, [r6]               ; read/modify write
ORR   r7,r7,#1:SHL:19        ; set bit 19
BIC   r7,r7,#1:SHL:18        ; clear bit 18
STR   r7, [r6]               ; change P0.25 to Aout
```

16.3.4 Generating a Sine Wave

The D/A converter will take an unsigned binary value and generate a voltage that ranges between 0 and V_{REF} (in our simulation, this is 3.3V). To see a complete sine wave, all we have to build is a simple loop that counts from 0 to 359. Say this counter is held in register r1. The sine table we built in Chapter 12 will return a Q31 value given the argument in register r1. However, we will have to scale and shift the output of our sine function, so that the value v sent to the D/A converter is

$$v = 512 \times \sin(r1) + 512$$

since sine returns negative arguments in two's complement. Scaling and shifting the output of our sine table will force v's range between 0 and 1024, which is what the D/A converter understands. To write the 10-bit value to the DAC Register, a half-word store moves the value v that has been shifted 6 bits so that it sits in bits [15:6]. So, the code would be

```
ASR     r0,r0,#16      ; convert Q31 to Q15
LSL     r0,r0,#9       ; x512 now in Q15 notation
ASR     r0,r0,#15      ; keep the integer part only
ADD     r0,r0,#512     ; 512 x sin(r1) +512 to show wave
LSL     r0,r0,#6       ; bits 5:0 of DAC are undefined
STRH    r0, [r8]       ; write to DACR
```

16.3.5 Putting the Code Together

To put everything together, we first convert the sine table to a subroutine, making sure to follow the AAPCS rules for passing arguments, namely to put the argument in register r1 and expect the sine of the argument in register r0. Since registers r4, r5, and r7 were changed in the subroutine, we stack those before using them. While not absolutely necessary, our loop counter counts down rather than up, and we subtract the loop counter from 360 to use as the argument to our sine function. The complete code is shown below.

```
; Sine wave generator using the LPC2132 microcontroller
; This program will generate a sine wave using
; the D/A converter on the controller. The output can be
; viewed using the Logic Analyzer in the Keil tools.

PINSEL1      EQU     0xE002C004
DACREG       EQU     0xE006C000
SRAMBASE     EQU     0x40000000

    AREA    SINEWAVE, CODE
    ENTRY

main
    LDR     sp, =SRAMBASE      ; initialize stack pointer
    LDR     r6, =PINSEL1       ; PINSEL1 configures pins
    LDR     r8, =DACREG        ; DAC Register[15:6] is VALUE
    LDR     r7, [r6]           ; read/modify write
    ORR     r7,r7,#1:SHL:19    ; set bit 19
    BIC     r7,r7,#1:SHL:18    ; clear bit 18
    STR     r7, [r6]           ; change P0.25 to Aout
```

```
outloop
    MOV     r6,#360             ; start counter
inloop
    RSB     r1,r6,#360          ; arg=360 - loop count
    BL      sine                ; get sin(r1)
    ; Now that we have r0=sin(r1), we need to send
    ; this to the DAC converter.
    ; First, we take the Q31 value and make it Q15
    ; and multiply it by 512. Then we offset the result
    ; by 512 to show the full sine wave. Aout is
    ; VALUE/1024*Vref, so our sine wave should swing
    ; between 0 and 3.3 V on the output.

    ASR     r0,r0,#16           ; convert Q31 to Q15
    LSL     r0,r0,#9            ; x512 now in Q15 notation
    ASR     r0,r0,#15           ; keep the integer part only
    ADD     r0,r0,#512          ; 512 x sin(r1)+512 to show wave
    LSL     r0,r0,#6            ; bits 5:0 of DAC are undefined
    STRH    r0,[r8]             ; write to DACR

    SUBS    r6,r6,#1            ; count down to 0
    BNE     inloop
    B       outloop             ; do this forever

; Sine function
; Returns Q31 value for integer arguments from 0 to 360
; Registers used:
;    r0=return value in Q31 notation
;    r1=sin argument (in degrees)
;    r4=starting address of sine table
;    r5=temp
;    r7=copy of argument

sine
    STMIA   sp!,{r4,r5,r7,lr} ; stack used registers
    MOV     r7, r1            ; make a copy
    LDR     r5, =270          ; won't fit into rotation scheme
    ADR     r4, sin_data      ; load address of sin table
    CMP     r1, #90           ; determine quadrant
    BLE     retvalue          ; first quadrant?
    CMP     r1, #180
    RSBLE   r1,r1,#180        ; second quadrant?
    BLE     retvalue
    CMP     r1, r5
    SUBLE   r1, r1, #180      ; third quadrant?
    BLE     retvalue
    RSB     r1, r1, #360      ; otherwise, fourth
retvalue
    LDR     r0,[r4,r1,LSL #2] ; get sin value from table
    CMP     r7, #180          ; do we return a neg value?
    RSBGT   r0, r0, #0        ; negate the value
```

```
        LDMDB  sp!,{r4,r5,r7,pc}     ; restore registers
done    B      done
        ALIGN
sin_data
        DCD 0x00000000,0x023BE164,0x04779630,0x06B2F1D8
        DCD 0x08EDC7B0,0x0B27EB50,0x0D613050,0x0F996A30
        DCD 0x11D06CA0,0x14060B80,0x163A1A80,0x186C6DE0
        DCD 0x1A9CD9C0,0x1CCB3220,0x1EF74C00,0x2120FB80
        DCD 0x234815C0,0x256C6F80,0x278DDE80,0x29AC3780
        DCD 0x2BC750C0,0x2DDF0040,0x2FF31BC0,0x32037A40
        DCD 0x340FF240,0x36185B00,0x381C8BC0,0x3A1C5C80
        DCD 0x3C17A500,0x3E0E3DC0,0x40000000,0x41ECC480
        DCD 0x43D46500,0x45B6BB80,0x4793A200,0x496AF400
        DCD 0x4B3C8C00,0x4D084600,0x4ECDFF00,0x508D9200
        DCD 0x5246DD00,0x53F9BE00,0x55A61280,0x574BB900
        DCD 0x58EA9100,0x5A827980,0x5C135380,0x5D9CFF80
        DCD 0x5F1F5F00,0x609A5280,0x620DBE80,0x63798500
        DCD 0x64DD8900,0x6639B080,0x678DDE80,0x68D9F980
        DCD 0x6A1DE700,0x6B598F00,0x6C8CD700,0x6DB7A880
        DCD 0x6ED9EC00,0x6FF38A00,0x71046D00,0x720C8080
        DCD 0x730BAF00,0x7401E500,0x74EF0F00,0x75D31A80
        DCD 0x76ADF600,0x777F9000,0x7847D900,0x7906C080
        DCD 0x79BC3880,0x7A683200,0x7B0A9F80,0x7BA37500
        DCD 0x7C32A680,0x7CB82880,0x7D33F100,0x7DA5F580
        DCD 0x7E0E2E00,0x7E6C9280,0x7EC11A80,0x7F0BC080
        DCD 0x7F4C7E80,0x7F834F00,0x7FB02E00,0x7FD31780
        DCD 0x7FEC0A00,0x7FFB0280,0x7FFFFFFF

        END
```

16.3.6 Running the Code

A logic analyzer in the MDK tools allows you to place signals in a window for viewing in the same way that you would probe pins on an actual part. For example, you can take the signal A_{OUT} and drag it into the logic analyzer. As the value changes in real time, you can track it, stopping the processor at any point to read values. After you build a project for this code, enter the program and start the debugger. Open the Symbol window found in the View menu. Expand the Virtual Registers listing to show all of the pins. Open the Logic Analyzer window, also found in the View menu under Analysis Windows. Drag the pin called AOUT into the Logic Analyzer window, then start the simulation. You should see the sine wave, shown in Figure 16.9.

16.4 THE TIVA LAUNCHPAD

For a bit of variety, as well as a good illustration of using general purpose input and output lines, we'll turn next to an inexpensive evaluation module from Texas Instruments, shown in Figure 16.10, which contains the TM4C123GH6PM microcontroller. Figure 16.11 shows a block diagram of the microcontroller, which has

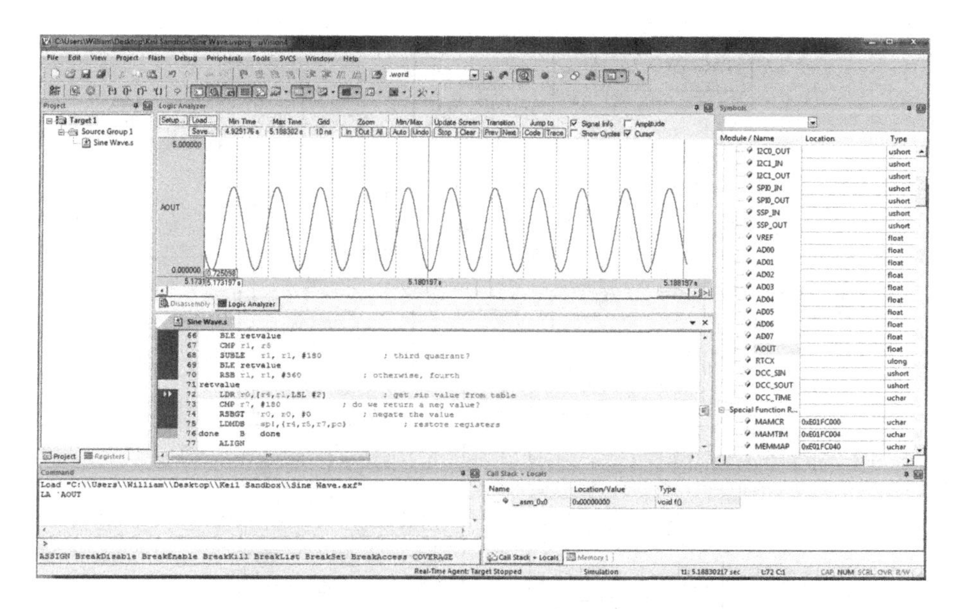

FIGURE 16.9 Simulation window with logic analyzer.

a similar layout as the other microcontrollers we've examined. Rather than an ARM7TDMI core, it uses a Cortex-M4 processor with floating-point hardware, but we will address the peripherals the same way as we did in Sections 16.2 and 16.3. Up until this point, we've used simulation models to run our programs; now we're using real hardware. Using the Code Composer Studio tools, a small block of code

FIGURE 16.10 The Tiva Launchpad Evaluation Module.

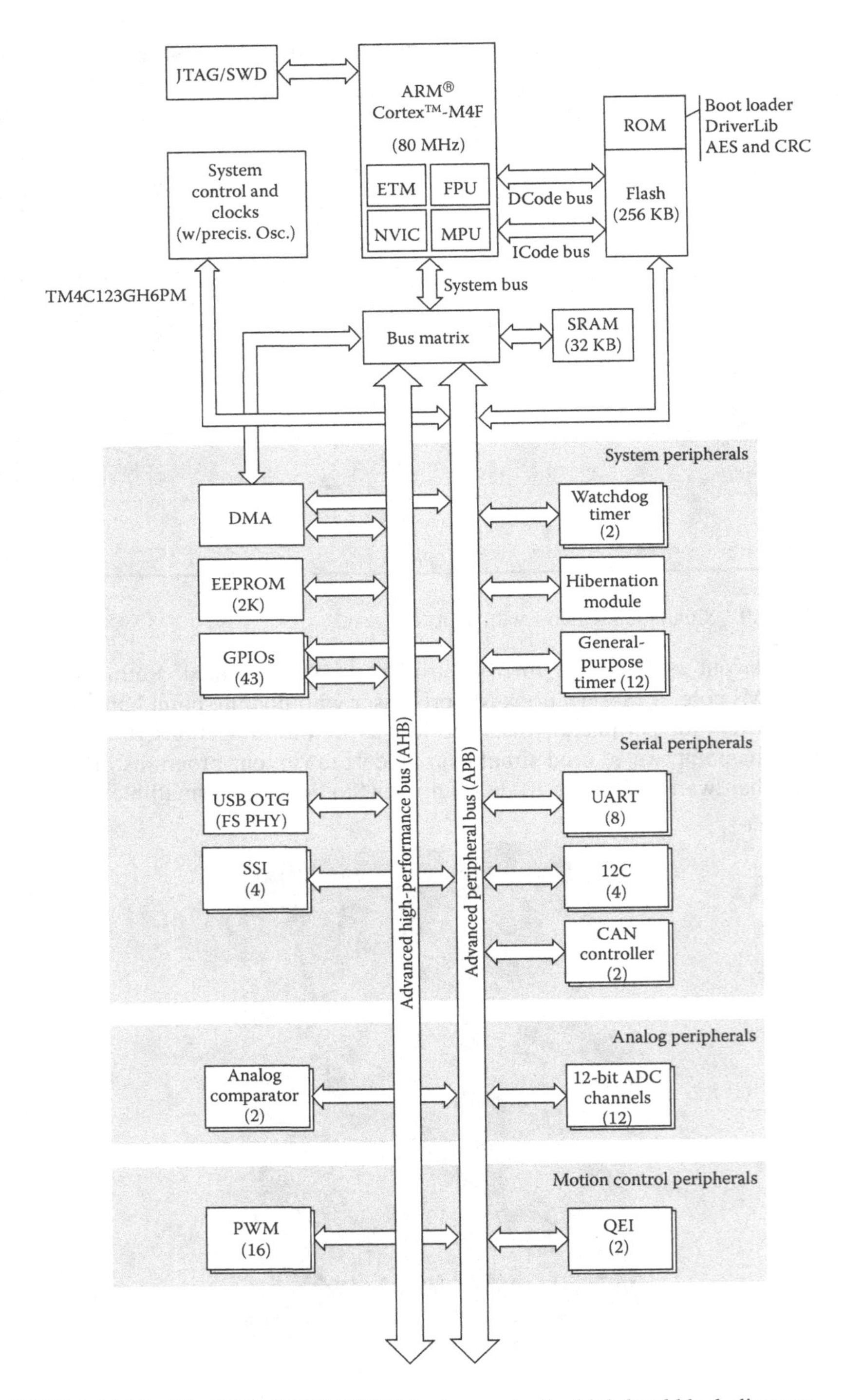

FIGURE 16.11 Tiva™ TM4C123GH6PM microcontroller high-level block diagram.

can be written to turn on the peripheral, set up the clocks, and then change the color of the LEDs on the board. Note that there are no simulation models of the evaluation module, hence the use of actual hardware.

16.4.1 General-Purpose I/O

General purpose I/O lines are probably the most straightforward of all peripherals to understand, and they can be very versatile. Since they can be configured as either input lines or output lines, you can use them for a variety of applications: driving LEDs or other components in a system, accepting parallel data from another device, or using them as a form of interrupt. Their versatility can sometimes prove to be maddening, mostly due to the sheer volume of documentation available on configuring them. Outputs can have different drive strengths, the GPIO lines might be multiplexed with other peripheral lines, some ports are available on both the AHB and the APB bus, etc. For the complete list of registers and options, refer to the Tiva TM4C123GH6PM Data Sheet (Texas Instruments 2013c), but for now, let's examine a (relatively) simple, short block of code to see how everything is configured.

16.4.2 The Memory Map

There are, in fact, so many memory-mapped registers on the TM4C123GH6PM that it's sometimes difficult to know which ones to use. In order to set up the clocks and the PLL so that our evaluation module actually runs code, we will need to configure the Run-Mode Clock Configuration (RCC) Register, which is part of the System Control Registers, which have a base address of 0x400FE000. The entire sequence for setting up the clocks is listed in Section 16.4.5. In order to use the GPIO port, it must be enabled by turning on its clock, which is configured in the General-Purpose Input/Output Run Mode Clock Gating Control (RCGCGPIO) Register. Yes, it's a mouthful. Luckily, it has the same base address as the RCC Register, but its offset is 0x608. The code looks like:

```
; Enable GPIOF
; RCGCGPIO (page 339)
MOVW      r2, #0x608          ; offset for this register
LDR       r1, [r0, r2]        ; grab the register contents
ORR       r1, r1, #0x20       ; enable GPIOF clock
STR       r1, [r0, r2]
```

According to the data sheet, there must be a delay of 3 system clocks after the GPIO module clock is enabled before any GPIO module registers are accessed. There are four instructions between the STR that enables the clock and the STR that sets the direction of the port, satisfying this requirement.

16.4.3 Configuring the GPIO Pins

Moving into a different memory space, the GPIO port itself is configured through dedicated registers. Table 16.4 shows the location of each of the GPIO ports on the

TABLE 16.4
GPIO Port Locations

	Port	Address
APB Bus	GPIO Port A	0x40004000
	GPIO Port B	0x40005000
	GPIO Port C	0x40006000
	GPIO Port D	0x40007000
	GPIO Port E	0x40024000
	GPIO Port F	0x40025000
AHB Bus	GPIO Port A	0x40058000
	GPIO Port B	0x40059000
	GPIO Port C	0x4005A000
	GPIO Port D	0x4005B000
	GPIO Port E	0x4005C000
	GPIO Port F	0x4005D000

TM4C123GH6PM. The LED is actually located on APB bus Port F, so our base address will be 0x40025000. We'll set the direction of GPIO Port F lines 1, 2, and 3 to be outputs:

```
; Set the direction using GPIODIR (page 661)
; Base is 0x40025000
MOVW     r0, #0x5000
MOVT     r0, #0x4002
MOVW     r2, #0x400        ; offset for this register
MOV      r1, #0xE
STR      r1, [r0, r2]      ; set 1 or 2 or 3 for output
```

If you are using the Keil tools, the constant 0x40025000 can be loaded into register r0 with an LDR pseudo-instruction from Chapter 6; however, the Code Composer Studio assembler does not support this, so two separate move instructions will do the trick. There's an additional level of gating on the port—the GPIO Digital Enable Register will need to be configured:

```
; set the GPIODEN lines
MOVW     r2, #0x51c        ; offset for this register
STR      r1, [r0, r2]      ; set 1 and 2 and 3 for I/O
```

16.4.4 Turning on the LEDs

The Tiva Launchpad board has a multi-colored LED that is controlled through three GPIO lines on Port F, one for red, one for green, and one for blue. The red LED is attached to line PF1, the green LED is attached to line PF2, and the blue LED is attached to line PF3. Now that Port F has been enabled and the appropriate lines have been configured as outputs, we can light the LEDs by driving a 1 to the GPIO line of our choice. To showcase all three colors, we can create a loop that selects one color at a time, cycling through all three by changing the value being written to the port.

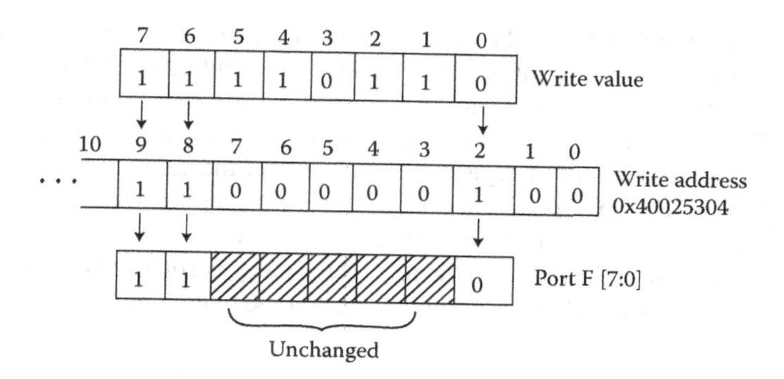

FIGURE 16.12 Masking of the GPIO bits.

In Chapter 5, the concept of bit-banding allowed individual bits of a memory-mapped register to be accessed using a single read or write operation. It turns out the a similar feature is available on the TM4C123GH6PM microcontroller for accessing GPIO lines, where the address is used as a mask, as shown in Figure 16.12. For example, suppose we wish to alter only bits [7:6] and bit [0] of GPIO Port F. We can use bits [9:2] of the address in our write operation to form a bit mask, so that instead of writing to the base address of Port F (0x40025000), we would store our value to address (0x40025304). Now that our mask is in place, no matter what value we write to the port, such as 0xF6, only bits [7:6] and [0] are altered. Specifically for our LED example, we wish to have bits [9:2] of the address be 0b00111000, or 0x38, since we only want to change the LED lines connected to Port F. This value becomes our offset.

The following code shows our loop, complete with a small delay to give the observer a chance to see the individual colors of the LED. The delay value of 0xF40000 is arbitrary, but at 16 MHz, it gives us about 1 second to view a single color.

```
    SUB     r7, r7, r7        ; clear out r7
    MOV     r6, #2            ; start with LED = 0b10
mainloop
    ; turn on the LED
    ; if bits [9:2] affect the writes, then the address
    ; is offset by 0x38
    STR     r6, [r0, #0x38]  ; base + 0x38 so [9:2] = 0b111000
    MOVT    r7, #0xF4         ; set counter to 0xF40000
spin
    SUBS    r7, r7, #1
    BNE     spin
    ; change colors
    CMP     r6, #8
    ITE     LT
    LSLLT   r6, r6, #1        ; LED = LED * 2
    MOVGE   r6, #2            ; reset to 2 otherwise
    B       mainloop
```

The first line of our code sets the value to be written to Port F to 0b0010, and we plan to cycle through the values 0b0100, 0b1000, then back to 0b0010. The STR instruction will use the same base address in register r0 that we used earlier to configure the port, only the offset has now changed to our mask value. By using a MOVT instruction to store a value in the top half of register r7, we can load the value 0xF40000 without shifting any bits. The spin loop then does nothing but subtract one from the counter value until it expires. Once the loop reaches zero, the color is changed by logically shifting left the value to be stored to the port. If the value is equal to 0b1000, then it is reset to 0b0010 and the code branches to the start of the main loop.

16.4.5 Putting the Code Together

The entire program to control the LEDs is listed below. If you follow the suggestions outlined in Appendix A for using the Code Composer Studio tools, this code should assemble without any issues.

```
myStart:
    ; Set sysclk to DIV/4, use PLL, XTAL_16 MHz, OSC_MAIN
    ; system control base is 0x400FE000, offset 0x60
    ; bits[26:23]= 0x3
    ; bit[22] = 0x1
    ; bit[13] = 0x0
    ; bit[11] = 0x0
    ; bits[10:6] = 0x15
    ; bits[5:4] = 0x0
    ; bit[0] = 0x0
    ; This all translates to a value of 0x01C00540
    MOVW      r0, #0xE000
    MOVT      r0, #0x400F
    MOVW      r2, #0x60         ; offset 0x60 for this register
    MOVW      r1, #0x0540
    MOVT      r1, #0x01C0
    STR       r1, [r0, r2]      ; write the register's contents
    ; Enable GPIOF
    ; RCGCGPIO (page 339)
    MOVW      r2, #0x608        ; offset for this register
    LDR       r1, [r0, r2]      ; grab the register contents
    ORR       r1, r1, #0x20     ; enable GPIOF clock
    STR       r1, [r0, r2]
    ; Set the direction using GPIODIR (page 661)
    ; Base is 0x40025000
    MOVW      r0, #0x5000
    MOVT      r0, #0x4002
    MOVW      r2, #0x400        ; offset for this register
    MOV       r1, #0xE
    STR       r1, [r0, r2]      ; set 1 or 2 or 3 for output
    ;set the GPIODEN lines
    MOVW      r2, #0x51c        ; offset for this register
  STR         r1, [r0, r2]      ; set 1 and 2 and 3 for I/O
```

```
    SUB         r7, r7, r7        ; clear out r7
    MOV         r6, #2            ; start with LED=0b10
mainloop
    ; turn on the LED
    ; if bits [9:2] affect the writes, then the address
    ; is offset by 0x38
    STR         r6, [r0, #0x38] ; base+0x38 so [9:2]=0b00111000
    MOVT        r7, #0xF4         ; set counter to 0xF40000
spin
    SUBS        r7, r7, #1
    BNE         spin
    ; change colors
    CMP         r6, #8
    ITE         LT
    LSLLT       r6, r6, #1        ; LED=LED * 2
    MOVGE       r6, #2            ; reset to 2 otherwise
    B           mainloop
```

16.4.6 Running the Code

The Code Composer Studio tools will help you build a project, enter the assembly code into a file, then run the code on the Launchpad evaluation module. The code will be loaded into Flash memory, so that once you build your project, in the future, once power is applied to the board, the same program will immediately execute and you should see the LEDs continue to flash. Take some time to set breakpoints on code segments to see how the register values change, and experiment with different delay values for the loop. Obviously you can restore the evaluation module's default program if you wish using the tools provided by Texas Instruments.

16.5 EXERCISES

1. Write an SVC handler so that when an SVC instruction is executed, the handler prints out the contents of register r8. The program should incorporate assembly routines similar to the ones already built. For example, you will need to convert the binary value in register r8 to ASCII first, then use a UART to display the information in the Keil tools. Have the routine display "Register r8 =" followed by the register's value.

2. Choose a device with general-purpose I/O pins, such as the LPC2103, and write an assembly routine that sequentially walks a 1 back and forth across the I/O pins. In other words, at any given time, only a single pin is set to 1—all others are 0. Set up the Keil tools to display the pins (you might want to compile and run the Blinky example that comes with the tools to see what the interface looks like).

3. Rewrite the two examples described in Section 16.2.5 and Section 16.3.5 using full descending stacks.

4. What is the address range for on-chip SRAM for the LPC2106 microcontroller?

5. Write an assembly program that takes a character entered from a keyboard and echoes it to a display. To do this, you will need to use two UARTs, one for entering data and one for displaying it. The routine that accepts characters should not generate an interrupt when data is available, but merely wait for the character to appear in its buffer. In other words, the UART routine should spin in a loop until a key is pressed, after which it branches back to the main routine. Note that the UART window must have the focus in order to accept data from a keyboard (this is a Windows requirement), so be sure to click on the UART window first when testing your code.

6. Using the D/A converter example as a guide, write an assembly routine that will generate a waveform defined by

$$f(x) = a\sin(0.5x) + b\sin(x) + c$$

where *a*, *b*, and *c* are constants that allow a full period to be displayed. The output waveform should appear on A_{OUT} so that you can view it on the logic analyzer in the Keil tools.

7. What is the address range for the following devices on the LPC2132 microcontroller?
   a. General-purpose input/output ports
   b. Universal asynchronous receiver transmitter 0
   c. Analog-to-digital converter
   d. Digital-to-analog converter

8. What is the address range for the following devices on the STR910FM32 microcontroller?
   a. General-purpose input/output ports
   b. Real time clock
   c. Universal asynchronous receiver transmitter
   d. Analog-to-digital converter

9. The UART example given in this chapter uses registers r5 and r6 to configure the UART and, therefore, must write them to the stack before corrupting them. Rewrite the UART example so that no registers are stacked and the routine is still AAPCS compliant.

10. Modify the routine in Section 16.4.5 so that the LEDs flash alternately between red and blue for about 1 second each.

17 ARM, Thumb and Thumb-2 Instructions

17.1 INTRODUCTION

Throughout the book, we've been using two different instruction sets, ARM and Thumb-2, only mentioning 16-bit Thumb here and there. Recall that the Cortex-M4 executes only Thumb-2 instructions, while the ARM7TDMI executes ARM and 16-bit Thumb instructions. Keeping in mind that a processor's microarchitecture and a processor's instruction set are two different things, Table 17.1 shows how the ARM processor architectures have evolved over the years, along with the instruction sets. They often get developed at the same time, but it is possible for a given microarchitecture to be modified only slightly to support additional instructions, adding more control logic and a bit more datapath, adding registers, etc. Consider the ARM9TDMI which supports the version 4T instruction set and the ARM9E, loosely the same microarchitecture, which supports version 5TE instructions. So when we discuss a processor like the Cortex-A15, we think of pipeline depth, memory management units, cache sizes, and the like, but at the end of the day we're really interested in what instructions the machine supports. Historically for most ARM cores, two instruction sets were supported at the same time—ARM and Thumb—where the processor could switch between them as needed. In 2003, ARM (the company) introduced something called Thumb-2, and well, the water was muddied somewhat, so it's worth a look back to see why there are now effectively *three* different instruction sets for ARM processors and in particular, which processors support any given instruction set.

17.2 ARM AND 16-BIT THUMB INSTRUCTIONS

We've already seen what ARM instructions look like: they're 32-bits long; they contain fields for specifying the operation, the source, and destination operands; they specify whether its execution is predicated upon a condition; etc. This format has also been around since the first ARM1 processor. Interestingly, most ARM processors support them, but not all. Again referring to Table 17.1, you can see that some processors, e.g., the Cortex-M4 and Cortex-M0, do *not* support 32-bit ARM instructions, but we'll come back to this in a moment. In the early 1980s, many processors had either 8- or 16-bit instructions, so the question was eventually raised: can you compress a 32-bit instruction, keeping its code density improvements and features, to take advantage of inexpensive 16-bit memory and improve code density even further if you have 32-bit memory?

TABLE 17.1
Architectures and Instruction Sets

Version	Example Core	ISA
v4T	ARM7TDMI, ARM9TDMI	ARM, Thumb
v5TE	ARM946E-S, ARM966E-S	ARM, Thumb
v5TEJ	ARM926EJ-S, ARM1026EJ-S	ARM, Thumb
v6	ARM1136 J(F)-S	ARM, Thumb
v6T2	ARM1156T2(F)-S	ARM, Thumb-2
v6-M	Cortex-M0, Cortex-M1	Thumb-2 subset
v7-A	Cortex-A5, Cortex-A8,Cortex-A12, Cortex-A15	ARM, Thumb-2
v7-R	Cortex-R4, Cortex-R5, Cortex-R7	ARM, Thumb-2
v7-M	Cortex-M3	Thumb-2
v7E-M	Cortex-M4	Thumb-2

Reducing the size of existing instructions *can* be done by examining the operands and bit fields that are needed, then perhaps coming up with a shorter instruction. Consider the 32-bit pattern for ADD, as shown in Figure 17.1. Normally the required arguments include a destination register Rd, a source register Rn, and a second operand that can either be an immediate value or a register. A simple instruction that adds 1 to register r2, i.e.,

```
ADD r2, r2, #1
```

could be compressed easily, especially since the destination register is the same as the only source register (r2) used in the instruction. The other argument in the addition is 1, a number small enough to fit within an 8-bit field. If we enforce a few restrictions on the new set of instructions, the same operation can be done using only a 16-bit opcode, as shown in Figure 17.2, using Encoding T2, which would make the instruction appear as

```
ADD r2, #1
```

Now the source and destination registers are the same, so they can be encoded in the same field, and the 8-bit immediate value is 1. The other 16-bit format would allow ADD instructions that look like

```
ADDS r2, r3, #3
```

31 28	27 26	25	24 21	20	19 16	15 12	11 0
cond	0 0	1	opcode	S	Rn	Rd	shifter_operand

FIGURE 17.1 ADD instruction format in ARM.

Encoding T1 All versions of the Thumb instruction set.

ADDS <Rd>, <Rn>, #<imm3> Outside IT block.

ADD<c> <Rd>, <Rn>, #<imm3> Inside IT block.

15	14	13	12	11	10	9	8	7	6	5	4	3	2	1	0
0	0	0	1	1	1	0	imm3			Rn			Rd		

Encoding T2 All versions of the Thumb instruction set.

ADDS <Rdn>, #<imm8> Outside IT block.

ADD<c> <Rdn>, #<imm8> Inside IT block.

15	14	13	12	11	10	9	8	7	6	5	4	3	2	1	0
0	0	1	1	0	Rdn			imm8							

Encoding T3 ARMv7-M

ADD{S}<c>.W <Rd>, <Rn>, #<const>

15	14	13	12	11	10	9	8	7	6	5	4	3	2	1	0
1	1	1	1	0	i	0	1	0	0	0	S	Rn			

15	14	13	12	11	10	9	8	7	6	5	4	3	2	1	0
0	imm3			Rd				imm8							

Encoding T4 ARMv7-M

ADDW<c> <Rd>, <Rn>, #<imm12>

15	14	13	12	11	10	9	8	7	6	5	4	3	2	1	0
1	1	1	1	0	i	1	0	0	0	0	0	Rn			

15	14	13	12	11	10	9	8	7	6	5	4	3	2	1	0
0	imm3			Rd				imm8							

FIGURE 17.2 Thumb formats for ADD (immediate).

and

```
ADD r2, r4, #2
```

but again, with restrictions. Note the operand fields have changed from 4 bits to 3 bits, meaning the registers allowed range from r0 to r7, known as the low registers. To access the other registers, known as the high registers, separate instructions exist, including one that adds an immediate value to the Program Counter and two that add a value to the stack pointer. Note that the internal data paths and the registers in the processor would still be 32 bits wide—we're only talking about making the *instruction* smaller.

This first deviation from the more traditional 32-bit ARM instructions is called Thumb, also referred to as 16-bit Thumb, and with it comes its own state (not to be confused with mode) in the processor. The instructions are a subset of the ARM instruction set, meaning that not all of the instructions in ARM are available in Thumb. For example, you cannot access the PSR registers in Thumb state on an ARM7TDMI. There are other restrictions on the use of constants, branches, and registers, but fortunately all of the subtleties in Thumb are left to the compiler, since *you should rarely, if ever, be coding 16-bit Thumb instructions by hand.* It is simply an option to give to the compiler. C or C++ code compiled for Thumb is typically about 65–70% of the size of code compiled for ARM instructions. Fewer instructions necessitates the use of more individual Thumb instructions to do the same thing ARM instructions can do. In practice, code is a mix of ARM and Thumb instructions, allowing programmers to base their use of Thumb instructions on the application at hand, as we'll see in Section 17.5. In some cases, it may be necessary to optimize a specific algorithm in Thumb, such as a signal processing algorithm. Further optimizations would require knowing details of Thumb pretty well, and these can be found in the *ARM Architectural Reference Manual* (ARM 2007c) with the complete list of instructions and their formats.

To demonstrate another advantage of having such an instruction set, an industry benchmark such as Dhrystone can be run on three different types of memory systems: 32-bit memory, 16-bit memory, and a mix of the two. Performance numbers are shown in Figure 17.3, where Dhrystone normally measures the number of iterations of the main code loop per second. For the case where the memory system is made of 32-bit memory only, ARM code clearly performs better than Thumb code, since Thumb must compensate for the loss of some operations by using more than a single instruction. When the system is changed to use 16-bit memory, Thumb code now has the advantage over ARM—it takes two cycles to fetch a complete ARM instruction from 16-bit accesses. Obviously, the performance has decreased for both ARM and Thumb over the original 32-bit configuration. It turns out that if a small amount of 32-bit memory is used for stacks, along with 16-bit memory, the level of performance is nearly comparable to Thumb code running out of 32-bit memory alone. Stack accesses are data accesses, and regaining the ability to fetch 32 bits at a time (even with a 16-bit instruction) shores up the performance numbers. As we'll see shortly, both the ARM7TDMI and the Cortex-M4 execute 16-bit Thumb instructions.

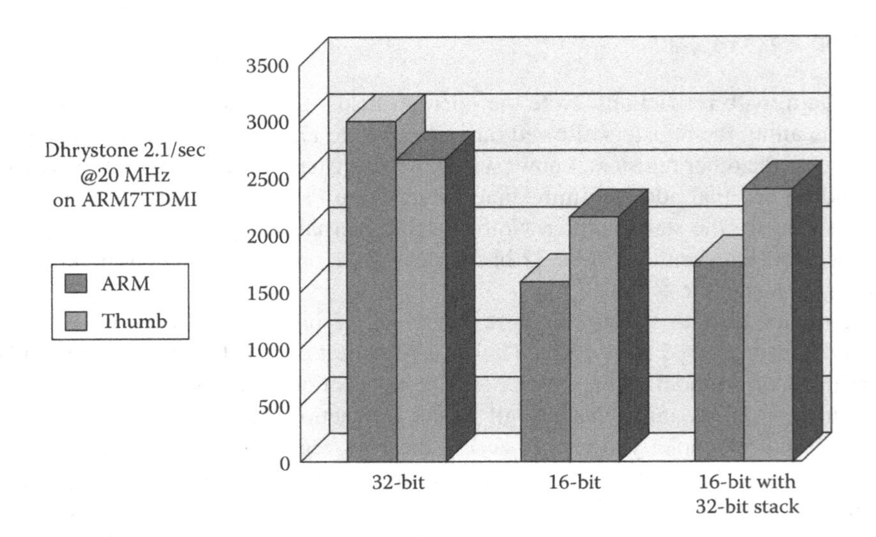

FIGURE 17.3 Dhrystone performance of ARM and 16-bit Thumb code.

17.2.1 Differences between ARM and 16-Bit Thumb

In the creation of a new 16-bit instruction set, we've considered how an ARM instruction can be shortened by restricting the operands, but we still have to take into account other bits in the instruction, such as the S bit that tells it to set the condition codes in the status register, and the conditional field in bits 28 through 31 that allows the instruction to be conditionally executed. To account for the S bit, we could simply say that all ALU instructions set the status flags upon completion. This might be too limiting, so we could further say that depending on the registers used, some instructions will set the flags and some will not. For most data processing instructions, the flags are set by default. However, instructions that use the high registers (except for CMP) leave the flags unaffected, for example:

```
ADD     r12, r4         ; r12 = r12 + r4, flags unaffected
ADD     r10, r11        ; r10 = r10 + r11, flags unaffected
ADD     r0, r5          ; r0 = r0 + r5, flags affected
ADD     r2, r3, r4      ; r2 = r3 + r4, flags affected
```

To account for the conditional field bits, it's necessary to remove conditional execution from 16-bit Thumb code entirely (but don't worry, it comes back with Thumb-2). Branches, however, can still be executed conditionally, since leaving only unconditional branches in the instruction set would be very limiting.

One further restriction that hasn't been mentioned is the lack of inline barrel shifter options available on ARM instructions. There simply isn't enough room in sixteen bits to include an optional shift, so individual instructions have been included in the Thumb instruction set for shifts and rotates (e.g., ASR for Arithmetic Shift Right, ROR for Rotate Right, LSL for Logical Shift Left, and LSR for Logical Shift Right). It wouldn't be realistic to expect a compressed instruction set to include every ARM instruction, so there are noticeable differences between the two sets (some of which we have discussed already). For starters, data processing instructions are unconditionally executed, so loops are not as elegant in Thumb, and the condition code flags will always be updated if you use low registers (r0 through r7). In fact, most of the instructions only act on the low registers, with the notable exceptions being CMP, MOV, and some variants of ADD and SUB.

When it comes to loading and storing data, 16-bit Thumb instructions impose several restrictions. For example, the only addressing modes allowed are

LDR|STR <Rd>, [<Rn>, <offset>]

with the option of two pre-indexed addressing modes: base register + offset register and base register + 5-bit offset (optionally scaled). Even the load and store multiple instructions are different. If you are working with low registers only, then the format

LDMIA|STMIA <Rb>, <low reg list>

can be used. However, for pushing data onto the stack,

PUSH <low reg list, {lr}>

is used instead. Similarly, for popping data off the stack and loading the Program Counter in the process, you can use

POP <low reg list, {pc}>

as we did in Chapter 13.

17.2.2 Thumb Implementation

For the programmer, Thumb shouldn't consist of much more than knowing it's available and knowing how to compile for it. The compiler does all of the work by being told exactly which instruction set to use for any given block of code. However, it's still worth understanding some of the modifications made within the ARM7TDMI architecture to support Thumb. While it's tempting to think that disparate hardware was built specifically to support Thumb instructions, the only affected part of the ARM7TDMI pipeline is the decode stage, as shown in Figure 17.4. After the processor has fetched a Thumb instruction from memory, it goes through a multiplexor first, routing instructions through a programmable logic array (PLA) table, which expands the 16-bit binary pattern into an equivalent ARM instruction before being sent through the decode logic that follows. Afterwards, the encoding is treated exactly as any other instruction would be, necessitating only one decoder in this stage of the pipeline (the decode stage drives the datapath by generating control logic for various blocks within the processor). No penalty results from this extra step of decompression

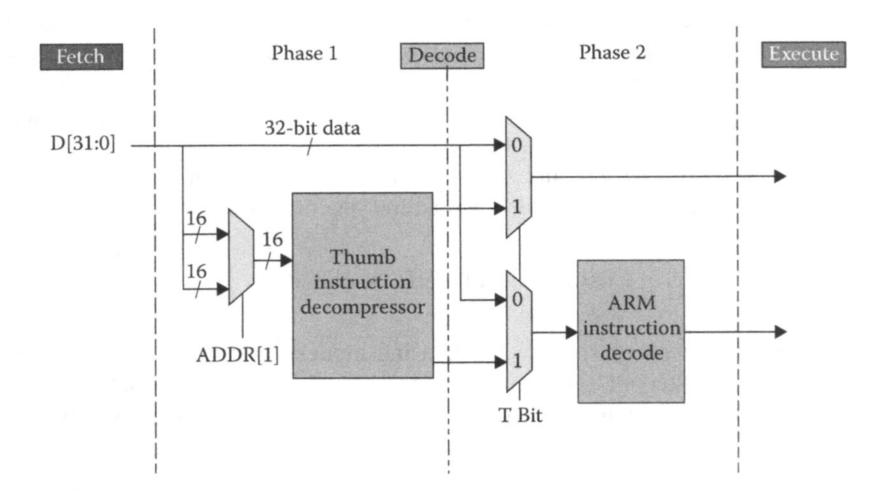

FIGURE 17.4 The ARM7TDMI processor pipeline.

since the first half of the decode stage allows enough time for an instruction to go through the PLA logic (remember, the ARM7TDMI processors generally run at speeds less than 50 MHz). Longer pipelines have less time in each processor stage, so for processors built after the ARM7TDMI, all Thumb and ARM instructions are decoded in parallel—the decision to use only one set of the control signals generated by both decoders is based on the state of the machine. As we'll see shortly, the Cortex-M processors that implement all or part of the version 7-M instruction set need only decode the instructions without worrying about whether they are ARM or Thumb instructions (although there is a certain amount of effort needed to decide whether the machine is looking at a 16-bit Thumb instruction or a 32-bit Thumb instruction). They are *all* Thumb instructions!

17.3 32-BIT THUMB INSTRUCTIONS

With pressure coming from industry standards (e.g., the image compression standard H.264) and applications, support for operating systems, and better handling of interrupts, it was time to reexamine what kinds of processors were possible using ARM, 16-bit Thumb, supersets, and even subsets of these instructions. In 2003, ARM decided to cross the Rubicon and build machines that would support instructions of varying width—some instructions would be 16 bits long and some would be 32 bits long. Part of the reasoning behind doing so was that processors being used in microcontroller applications required features that could not be supported with just a 16-bit instruction set, or even the existing ARM and Thumb instruction sets together. The 16-bit instructions had too many limitations to be used alone, and switching between the two instruction sets added extra cycles that could be better spent handling exceptions and interrupts. However, a compressed instruction set would be beneficial in a microcontroller that only had a limited amount of tightly coupled memory or cache.

Even though the whole idea of instructions with a fixed length is sacrosanct in RISC architectures, with the introduction of 32-bit Thumb instructions, some of the limitations of 16-bit instructions disappear. Consider the two Thumb-2 ADD immediate instructions in Figure 17.5 (shown earlier as Encoding T3 and T4 in

Encoding T3 ARMv7-M

```
ADD{S}<c>.W <Rd>, <Rn>, #<const>
```

15	14	13	12	11	10	9	8	7	6	5	4	3	2	1	0	15	14	13	12	11	10	9	8	7	6	5	4	3	2	1	0
1	1	1	1	0	i	0	1	0	0	0	S	Rn				0	imm3			Rd				imm8							

Encoding T4 ARMv7-M

```
ADDW<c> <Rd>, <Rn>, #<imm12>
```

15	14	13	12	11	10	9	8	7	6	5	4	3	2	1	0	15	14	13	12	11	10	9	8	7	6	5	4	3	2	1	0
1	1	1	1	0	i	1	0	0	0	0	0	Rn				0	imm3			Rd				imm8							

FIGURE 17.5 32-bit Thumb formats for ADD (immediate).

Figure 17.2). Now that the instructions contain 32 bits again, longer constants can be used, condition codes can be optionally set or not set, and the register fields' width allows nearly all the registers to be used with a single instruction (there are minor exceptions). Because there was little room left in the Thumb instruction space, a new encoding would be required to indicate to a processor that a fetched 16-bit instruction was merely the first half of a new 32-bit instruction. It turns out that if you are in Thumb state and the upper three bits of an instruction are all ones and the following two bits are non-zero, the processor can figure out that there are two halves to an encoding. So room does exist in the instruction space for new operations. In other words, the processor can tell by the encoding of the upper five bits whether the instruction is 16 bits long or 32 bits long—if any of the following patterns are seen, it's a 32-bit Thumb instruction:

- 0b11101
- 0b11110
- 0b11111

The newer 32-bit instructions were then combined with older 16-bit Thumb instructions to create something called Thumb-2. This more powerful instruction set eliminated the need to support two instruction sets. Microcontrollers specifically, which often require fast interrupt handling times, did not need to burn cycles switching states from Thumb back to ARM to process an exception. The processor can execute exception handlers and normal code with the same instruction set. If you further examine the *ARM v7-M Architectural Reference Manual*, you'll find that extensions have since been added for DSP operations and floating-point support.

Referring to Figure 17.5, notice that it is now possible to choose whether to set condition codes with a Thumb instruction, and along with this flexibility comes a bit of potential confusion. With the adoption of a Unified Assembly Language (UAL) format, Thumb code and ARM code now look the same, leaving the choice of instruction to the assembler unless you tell it otherwise. For example, if you were to simply say

```
EOR    r0, r0, r1
```

the operation could be performed using either a 16-bit Thumb instruction, an ARM instruction or a 32-bit Thumb instruction. If you happen to be working with a Cortex-M4, for example, then the choice falls to either one of the types of Thumb instructions, but there are two very subtle differences to mind. The first is that an ARM instruction of this type would not set the condition codes, since there is no S appendix in the mnemonic. A 16-bit Thumb instruction *would* set the condition codes. So using 16-bit Thumb instructions, the following two instructions would be equivalent:

```
EOR    r0, r1        ; 16-bit Thumb
EORS   r0, r0, r1    ; 16-bit Thumb using the UAL syntax
```

The second difference to mind is the directives themselves. To identify the more traditional 16-bit Thumb code using ARM tools, then you would use the directive CODE16; if you want to indicate that you are using UAL syntax, then you would use the directive THUMB in your assembly.

EXAMPLE 17.1

Consider the following instruction and directive:

```
THUMB
EOR r0, r0, r1
```

You might be tempted think that an instruction like this (which sets the flags when using these low registers in traditional Thumb) could be done as a 16-bit Thumb instruction. However, since we've indicated to the assembler that we're using UAL syntax by using the directive THUMB, the instruction has no S on the mnemonic and we are therefore telling the assembler *not* to produce an instruction which sets the condition codes. There is no 16-bit instruction in the Thumb instruction set to do this—the assembler is then forced to use a 32-bit Thumb-2 instruction (0xEA800001). This might not be what you want as you attempt to get better code compression. A good general rule is therefore: if you use UAL syntax, then always use an S on those operations that require updating the condition codes.

17.4 SWITCHING BETWEEN ARM AND THUMB STATES

The two processors that we've examined throughout the text, the ARM7TDMI and the Cortex-M4, are perfect examples of the variety that now exists in the ARM product lines. Some cores can execute ARM, Thumb, and Thumb-2 instructions. Some cores only execute Thumb-2. If you do happen to be using a processor that supports both ARM and Thumb instructions, the bulk of C/C++ code in embedded applications might be compiled for Thumb instructions, especially with its performance from narrow memory and its code density. However, there are still times when it will be necessary to switch between ARM and Thumb state. For example, on an ARM926EJ-S, certain operations cannot be done in Thumb state, so if access to the CPSR is needed to enable or disable interrupts, then the core must switch to ARM state. Speed-critical parts of an application may run in ARM state, since it gets better performance in 32-bit memory—a JPEG compression routine, for example, which is common in digital cameras. Processors such as the Cortex-M3 and Cortex-M4 always run in Thumb state, and it was mentioned in Chapter 8 that care must be taken to stay in Thumb state when creating your exception vector table and when doing any branching. Processors such the Cortex-R4, the Cortex-A15, and our venerable ARM7TDMI have more than a single state, so let's examine how to switch between them.

If you recall from Chapter 2, there is a bit in the CPSR, the T bit, that indicates whether the processor is in ARM state or Thumb state. This is only a status bit (meaning it's read-only), and switching between the two states is accomplished by

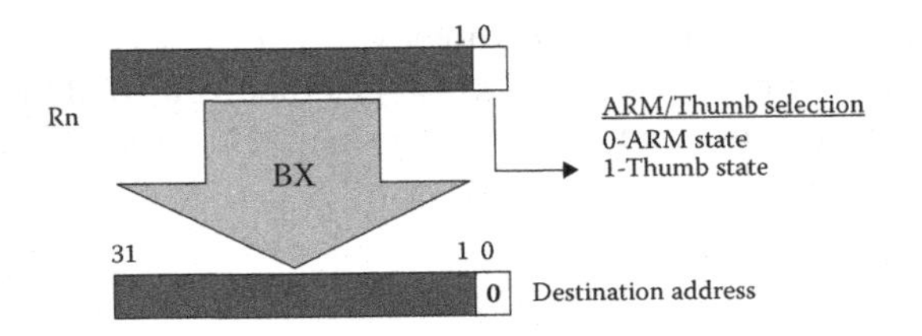

FIGURE 17.6 Changing to Thumb state via BX instruction.

way of a special type of branch instruction—BX, or branch and exchange. The formats for this instruction are

Thumb state:	BX Rn
ARM state:	BX{condition} Rn

where Rn can be any register.

The mechanism used to switch between the states depends on an address held in Rn. Normally, the least significant bit of an address is ignored, since a branch to an unaligned address in both ARM and Thumb states is not allowed. By using bit 0 of Rn and the BX instruction, the state can be changed when the processor jumps to the new address. If bit 0 is a zero, the state is set to ARM; if bit 0 is a one, the state is set to Thumb, as shown in Figure 17.6.

When changing from Thumb to ARM state on the ARM7TDMI, it's important to ensure that bit 1 of the address is also a zero—remember that ARM instructions are always fetched from word-aligned addresses, i.e., addresses that end in 0, 4, 8, or 0xC, so the two least significant bits of the address must be clear. One other important point worth considering is the register used. While the use of the PC as Rn in the BX instruction is valid, it's not recommended, since unexpected results could occur. Depending on how the code is arranged, you can end up jumping to a misaligned address, and from there the system only gets muddled.

EXAMPLE 17.2

Using the Keil tools, the following ARM7TDMI code shows an example of a state change from ARM to Thumb.

```
            GLOBAL Reset_Handler

            AREA Reset, CODE, READONLY
            ENTRY

Reset_Handler

            ARM
start       ADR     r0, into_Thumb + 1
            BX      r0

            CODE16
```

```
into_Thumb
                MOV     r0, #10
                MOV     r1, #20
                ADD     r1, r0
stop            B       stop
        END
```

Notice that the BX instruction is used to jump to an address called into_Thumb, where the least significant bit of the address has been set using the ADR pseudo-instruction. The short section of Thumb code begins with the directive CODE16, indicating the following instructions are Thumb instructions (the newer assembly format uses the directive THUMB). The Thumb code that follows just adds the numbers 10 and 20 together. When you run this example, examine the CPSR and notice that the T bit is set when the branch is made into Thumb code.

Cores that have only one instruction set, such as the Cortex-M3 or the Cortex-M4, do not have to worry about switching states. However, care must be taken to *prevent* switching states. For example, if you were to branch to an address contained in a register on the Cortex-M4, using the BX instruction say, and the least significant bit of that address was a zero, then a Usage fault gets generated since the processor cannot change states. This issue goes away if you are always coding in C or C++, or if your assembly code uses labels and pseudo-instructions to generate branch target addresses, since the tools will compute the correct values for you, even making them odd when necessary. If you generate addresses some other way or enter them by hand in your assembly code, then watch the least significant bit!

17.5 HOW TO COMPILE FOR THUMB

A question that ultimately arises from introducing an instruction such as BX is how certain parts of ARM code might call a Thumb subroutine or vice versa. A section of code can be compiled as either ARM or Thumb code; however, calling and returning from a subroutine might require the ability to switch states. For example, if you were to write an ARM subroutine that called a Thumb subroutine, and these two sections of code were compiled separately without taking some necessary steps, then the ARM subroutine may not be able to switch to Thumb state before jumping, since a BL instruction does not change state. If you're writing all of your own assembly, then obviously you need to mind the state of the machine when putting blocks of code together. However, embedded systems depend heavily on high-level coding, so it's far more likely that you'll be compiling C or C++ to incorporate Thumb code in your application.

Fortunately, compiler options can aid in the use of both ARM and Thumb subroutines in the same program. The process is known as *interworking*, and it can be done through a short bit of code known as a *veneer*. To illustrate how this works, Figure 17.7 shows a subroutine call from `func1` to `func2`, where the subroutine `func1` is to be compiled for ARM and `func2` is to be compiled for Thumb (remember that a function might be called from either state). The BL instruction will not change the

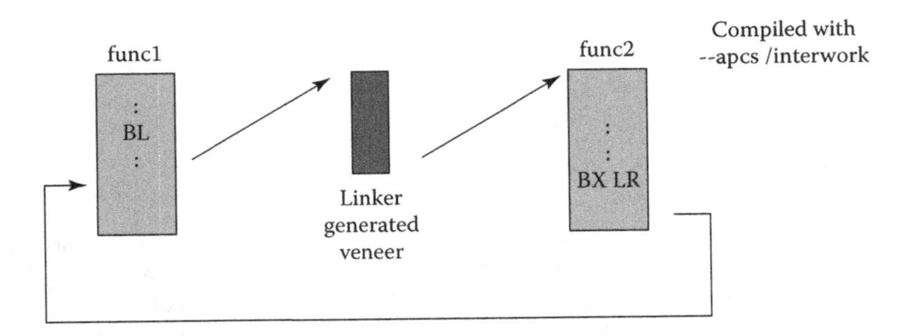

FIGURE 17.7 Linker generated veneers for ARM/Thumb interworking.

state before the jump is made, and the instructions that are normally used to return from a subroutine, i.e.,

```
MOV PC, LR
```

will also not change the state. Therefore, veneers are created to aid in the task of switching between ARM and Thumb. These short blocks of code (usually 8–12 bytes) become the immediate target of the original branch instruction, and include a BX instruction to change the state, e.g.,

```
ADR r12, {PC}+offset
BX r12
```

If the two blocks `func1` and `func2` are compiled with the `--apcs/interwork` option (for the ARM tools), an interwork attribute is set on the code generated, which will in turn be picked up by the linker. The linker will then calculate the offset and insert the veneer in the final code automatically. The called function, `func2`, also returns to the main code via a

```
BX LR
```

instruction instead of using the Thumb POP instruction, which would load the caller's return address from the stack and move it into the PC. In Thumb, the instruction

```
BX LR
```

will always return to the correct state (irrespective of whether a Thumb or ARM function called the subroutine). The BL instruction in Thumb sets the least significant bit of the return address, which would cause problems if the machine was originally in ARM state. Should you decide to mix assembly in C/C++ code, using the`/interwork` option will not change your assembly code, but it will alert the linker that your code is compatible for interworking. You would, however, be expected to use the correct return instruction (`BX LR`) in your own code. Consult the documentation for the tool suite you are using on the rules and usage of ARM/Thumb interworking.

If you are using a processor that does not switch between ARM and Thumb instructions, then by steering the compiler toward the appropriate architecture, the correct assembly will be generated automatically. For example, if you happen to be using the Keil tools to compile code for a Cortex-M4-based microcontroller, you will find the assembler option

```
--cpu Cortex-M4.fp
```

in the command line to tell the assembler that the v7-M instruction set is to be used along with floating-point instructions. No veneers are needed, since the machine never has to change states.

17.6 EXERCISES

1. On the ARM7TDMI, which bit in the CPSR indicates whether you are in ARM state or Thumb state?

2. Give the mnemonic(s) for a 16-bit Thumb instruction(s) that is equivalent to the ARM instruction

```
SUB     r0, r3, r2, LSL #2
```

3. Why might you want to switch to Thumb state in an exception handler?

4. Can you talk to a floating-point coprocessor in Thumb state?

5. Using Figure 17.2 as a guide, convert Program 3 from Chapter 3 into 16-bit Thumb assembly.

6. Describe why veneers might be needed in a program.

7. Convert Example 13.4 into 16-bit Thumb code. Do not convert the entire subroutine—just the four lines of code to perform saturation arithmetic.

8. In which state does the ARM7TDMI processor come out of reset?

9. How do you switch to Thumb state if your processor supports both ARM and Thumb instructions?

10. How do you switch to ARM state on the Cortex-M4?

18 Mixing C and Assembly

18.1 INTRODUCTION

In this last chapter, we're going to examine a few instances where it may make sense to combine your high-level C or C++ code with assembly. Mixing C and assembly is quite common, especially in deeply embedded applications where programmers work nearly at the hardware level. Doing such a thing is not always trivial, and the programmer is forced to be very mindful of variables, pointers, and function arguments. However, a good programmer will need certain tricks in his or her toolbox, and a point was made in the Preface that optimizing code usually requires the ability to recognize what the compiler is doing, and more importantly, the ability to modify code so that a compiler or an assembler generates the best software for the task at hand. There are two ways to add assembly to your high-level source code: the inline assembler and the embedded assembler.

18.2 INLINE ASSEMBLER

Normally, the compiler will try to optimize code as much as possible for you (unless you tell it not to). However, for some applications, algorithms must be optimized by hand, especially in instances where data is manipulated in ways that a compiler would normally not understand. Signal and speech processing algorithms tend to fall into this category. If you're writing an algorithm at a high level, it is possible to give the compiler some assistance by indicating sections of code that should be regarded as important. One way is through a process called inlining, where the __inline keyword is placed in the C or C++ code to notate a function that, when possible, should be placed in the assembly directly, rather than being called as a subroutine. This potentially avoids some of the overhead associated with branching and returning. The compiler will inline as much as possible, given the right optimization settings, but this is an option the user can specify as well. Furthermore, you can even write some functions in your C or C++ code in assembly—this might be placed in a function where you have called for inlining. Using the inline assembler is the easiest way to access instructions that are not supported by the C compiler, for example, saturated math operations, coprocessor instructions, or accessing the PSRs.

EXAMPLE 18.1

To tie a few ideas together, recall from Chapter 7 that Q notation allows us to work with fractional values easily by introducing an assumed binary point somewhere in the number. If we assume a number is Q31, for example, then a 32-bit value would have a sign bit and 31 bits of fractional data behind the binary point. In Chapters 7 and 13, we discussed saturation math, where the result of a signed addition or subtraction could be driven to either the largest positive or negative number, depending on the operation. In the version 5TE instruction set, new instructions were introduced to specifically work with fractional values and saturation math. A new status bit, the Q flag, was added to the CPSR/APSR to indicate that a value had saturated during an operation. The flag is sticky, meaning that it must be specifically written to a zero to clear it once it has been set.

Suppose that we have four Q15 numbers stored in two registers (each register holds two Q15 values). Recall a Q15 value is represented in 16 bits, the leading bit serving as the sign bit, and the remaining 15 bits are fraction bits. So the format of Q15 data is:

$$s.f_{14}f_{13}f_{12}f_{11}f_{10}f_9f_8f_7f_6f_5f_4f_3f_2f_1f_0$$

Further suppose that we need a multiply-accumulate operation to multiply two Q15 values and add the product to a Q31 operand. We can inline an assembly function in our C code to do this. Inside this function, the instruction SMULBB takes two Q15 numbers from the lower half of each source register (the B and B in the mnemonic identify the location of the two operands in the lower half of each source register) and multiplies them together as signed values. The value is now in Q30 notation (represented as two sign bits, one superfluous, and 30 fraction bits), and we must shift the result left by one bit to reformat the result in a Q31 representation. The next instruction, QDADD, performs this function by doubling the operand, checking to see if it requires saturation, then adding the accumulated value to the result, again checking to see if it requires saturation. This whole operation is illustrated in Figure 18.1. If either the shift or the add saturates the result, the Q flag, which is bit 27 in the CPSR of a version 5TE processor and the APSR of a v7-M processor, is set. The code on the following page shows this assembly written in an inline block within the function. Notice that register numbers are not used here—C variables are used inside of the assembly code.

Once we have used saturation math somewhere in our C code, we should check the Q flag (see Exercise 5), take some sort of action if we saturated the

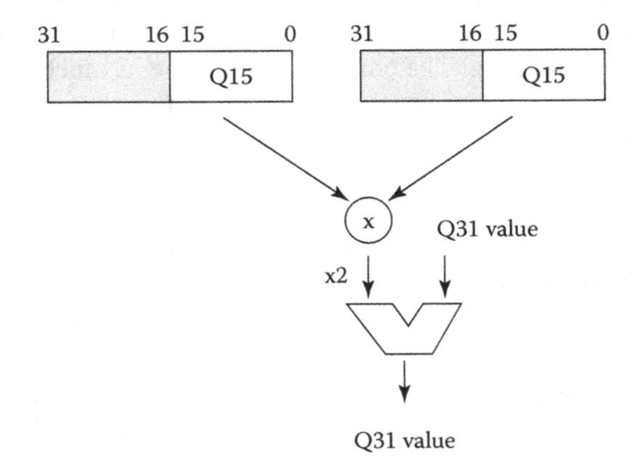

FIGURE 18.1 Multiply-accumulate with two Q15 numbers.

result, and then clear the Q flag. Clearing the flag requires some instructions that the compiler cannot generate, so again, we can write this small function using the inline assembler. In this example, our test code routine takes two numbers, multiplies them together, then adds a number that will produce a saturated result. You should verify running this code sets the Q flag in your simulation. The function Clear_Q_flag is called afterward to clear the Q flag.

```
#define Q_Flag 0x08000000 // Bit 27 of the CPSR

__inline int satmac(int a, int x, int y)
{
     int i;
     __asm
     {
               SMULBB  i, x, y
               QDADD   a, a, i
     }
     return a;
}

__inline void Clear_Q_flag (void)
{
     int temp;

     __asm
     {
               MRS     temp, CPSR
               BIC     temp, temp, #Q_Flag
               MSR     CPSR_f, temp
     }
}

int main(void)
{
//
// Multiply the two Q15 numbers together, then add a Q31
// number to it, which will saturate the result since it
// effectively overflows the precision allowed. This will
// set the Q flag in the CPSR.

//
     unsigned int   b = 0x7000;        // Q15 number
     unsigned int   c = 0x7ABC;        // Q15 number
     unsigned int   a = 0x60000000;    // Q31 number
     unsigned int   r;

     r = satmac(a, b, c);

     // do something with the value here ...

     Clear_Q_flag();

     return 0;

}
```

Using inline assembly code has some advantages, such as allowing you to access C variables directly in your code. You can also use C and C++ expressions as operands in the assembler statements. However, you should be aware of some limitations. First, the compiler will automatically optimize the assembly, so the final instructions may not be exactly what you wrote. Second, you cannot use all of the ARM instruction set in your assembly, e.g., BX and SVC instructions are not supported. In fact, Thumb instructions are not supported at all. Third, if you were to change the mode of the machine, the compiler would not be aware of this and consequently, your code may not behave as you expect. Lastly, be aware that you cannot change the Program Counter, you should not change the stack in any way, and you cannot use pseudo-instructions such as ADR in your inline assembly. In general, the inline assembler should not be used to produce better or more efficient code than the compiler. It should be used to accomplish operations that a compiler cannot, such as accessing coprocessors, performing saturated math operations, changing interrupt status, etc.

While the code in the example has been written for the Keil tools, gnu compilers also support inline assembly. Note, though, that the syntax is significantly different. You should consult the compiler guide for whichever tool you happen to be using. More information on ARM's tools and how to use the inline assembler can be found in the *RealView Compilation Tools Compiler User Guide* (ARM 2010c).

18.2.1 Inline Assembly Syntax

The inline assembler is invoked with the `__asm` keyword, which is followed by a list of assembly instructions inside braces. You can specify inline assembly code using either a single line or multiple lines. For example, single lines would be written as

```
__asm("instruction[;instruction]");// Must be a single string
__asm{instruction[;instruction]}
```

On multiple lines, your code would be written as

```
__asm
{
        ...
        instruction
        ...
}
```

You can use C or C++ comments anywhere in an inline assembly language block, but not the single line structure.

When you use the __asm keyword, be sure to obey the following rules:

- If you include multiple instructions on the same line, you must separate them with a semicolon. If you use double quotes, you must enclose all the instructions within a single set of double quotes.

- If an instruction requires more than one line, you must specify the line continuation with the backslash character (\).
- For the multiple line format, you can use C or C++ comments anywhere in the inline assembly language block. However, you cannot embed comments in a line that contains multiple instructions.
- The comma is used as a separator in assembly language, so C expressions with the comma operator must be enclosed in parentheses to distinguish them, for example,

```
__asm
{
  ADD x, y, (f(), z)
}
```

- Register names in the inline assembler are treated as C or C++ variables. They do not necessarily relate to the physical register of the same name. If you do not declare the register as a C or C++ variable, then the compiler generates a warning.
- Do not save and restore registers in the line assembler. The compiler does this for you. Also, the inline assembler does not provide direct access to the physical registers.
- If registers other than CPSR, APSR, and SPSR are read without being written to, an error message is issue, for example,

```
int f(int x)
{
    __asm
    {
      STMFD sp!, {r0}          // save r0-illegal:read
                               // before write
      ADD r0, x, 1
      EOR x, r0, x
      LDMFD sp!, {r0}          // restore r0 - not needed.
    }
    return x;
}
```

The function must be written as

```
int f(int x)
{
  int r0;
  __asm
  {
     ADD r0, x, 1
     EOR x, r0, x
  }
  return x;
}
```

18.2.2 Restrictions on Inline Assembly Operations

Earlier we mentioned that the inline assembler has some restrictions, but in general, you can still do nearly everything you need to optimize your code. Restrictions mostly apply to the use of registers and the types of instructions allowed. For example, registers r0 through r3, sp, lr, and the NZCV flags in the CPSR/APSR must be used with caution. Other C or C++ expressions might use these as temporary registers, and the flags could be corrupted by the compiler when evaluating those expressions. Additionally, the following instructions are not supported in the inline assembler:

- BKPT, BX, BXJ, BLX, and SVC instructions
- LDR Rn, = expression pseudo-instruction
- LDRT, LDRBT, STRT, and STRBT instructions
- MUL, MLA, UMULL, UMLAL, SMULL, and SMLAL flag setting instructions
- MOV or MVN flag setting instructions where the second operand is a constant
- User mode LDM instructions
- ADR and ADRL pseudo-instructions

All of the restrictions (and even some workarounds) for the inline assembler are detailed in the *RealView Compilation Tools User Guide* (ARM 2010c) and on the Keil Tools website (www.keil.com).

18.3 EMBEDDED ASSEMBLER

If you have a larger routine that requires optimizing by hand, then you can use the embedded assembler rather than the inline assembler. The embedded assembler allows you to declare assembly functions in C and C++ source modules with full function prototypes, including arguments and a return value. Unlike functions written with the inline assembler, these functions cannot be inlined and will always have the overhead associated with function calls. However, you do have access to the full instruction set, so it is possible to insert Thumb assembly functions in an ARM module, for example.

EXAMPLE 18.2

To illustrate how the embedded assembler works, we can write a short routine that copies a string from one memory location and stores it to another. Obviously a compiler would do a pretty good job compiling such a function from C, but it's simple enough to write one just to illustrate the point.

```
#include <stdio.h>

extern void init_serial (void);

__asm void my_strcopy(const char *src, char *dst)
```

```
{
loop
      LDRB    r2, [r0], #1
      STRB    r2, [r1], #1
      CMP     r2, #0
      BNE     loop
      BX      lr
}

int main(void)
{
      const char *a = "Just saying hello!";
      char   b[24];

      init_serial();

      my_strcopy(a,b);

      printf("Original string: '%s'\n", a);
      printf("Copied string: '%s'\n", b);
      return 0;
}
```

The main routine is written under the assumption that standard I/O routines work, i.e., a `printf` function call actually prints to an output device. This is left as an exercise to prove it works. The routine my_strcopy is called with the `main` routine passing the two pointers in memory to our strings. Notice that there is no need to export the function name, but the routine does have to follow AAPCS rules. Arguments will be passed in registers, and values can be pushed and popped to the stack if necessary. The routine has a return sequence (in our case, a simple BX instruction to move the value in the Link Register back to the Program Counter). Be careful when writing embedded assembly routines, as the compiler will not check that your code is AAPCS compliant!

The embedded assembler offers another advantage over the inline assembler in that you can access the C preprocessor directly using the __cpp keyword. This allows access to constant expressions, including the addresses of data or functions with external linkage. Example assembly instructions might look like the following:

```
LDR   r0, =__cpp(&some_variable)
LDR   r1, =__cpp(some_function)
BL    __cpp(some_function)
MOV   r0, #__cpp(some_constant_expr)
```

When using the __cpp keyword, however, be aware of the following differences between expressions in embedded assembly and in C or C++:

- Assembler expressions are always unsigned. The same expression might have different values between assembler and C or C++. For example,

```
MOV   r0, #(-33554432/2)        // result is 0x7f000000
MOV   r0, #__cpp(-33554432/2)   // result is 0xff000000
```

- Assembler numbers with leading zeros are still decimal. For example,

```
MOV   r0, #0700                  // decimal 700
MOV   r0, #__cpp(0700)           // octal 0700 == decimal 448
```

- Assembler operator precedence differs from C and C++. For example,

```
MOV   r0, #(0x23:AND:0xf + 1)     //((0x23 & 0xf) + 1) => 4
MOV   r0, #__cpp(0x23 & 0xf + 1) //(0x23 & (0xf + 1)) => 0
```

- Assembler strings are not null-terminated. For example,

```
DCB   "Hello world!"           //12 bytes (no trailing null)
DCB   __cpp("Hello world!")    //13 bytes (trailing null)
```

18.3.1 Embedded Assembly Syntax

Functions declared with __asm can have arguments and return a type. They are called from C and C++ in the same way as normal C and C++ functions. The syntax of an embedded assembly function is:

```
__asm return-type function-name(parameter-list){
      instruction
      instruction
      etc.
      }
```

The initial state of the embedded assembler (ARM or Thumb) is determined by the initial state of the compiler, as specified on the command line. This means that if the compiler starts in ARM state, the embedded assembler uses __arm. If the compiler starts in Thumb state, the embedded assembler uses __thumb. You can change the state of the embedded assembler within a function by using explicit ARM, THUMB, or CODE16 directives in the embedded assembler function. Such a directive within an __asm function does not affect the ARM or Thumb state of subsequent __asm functions.

Note that argument names are permitted in the parameter list, but they cannot be used in the body of the embedded assembly function. For example, the following function uses integer `i` in the body of the function, but this is not valid in assembly:

```
__asm int f(int i)
{
      ADD i, i, #1//error
}
```

Here, you would use `r0` instead of `i` as both the source and destination to be AAPCS compliant.

18.3.2 Restrictions on Embedded Assembly Operations

No return instructions are generated by the compiler for an __asm function. If you want to return from an __asm function, then you must include the return instructions, in assembly code, in the body of the function. Note that this makes it possible to fall through to the next function, because the embedded assembler guarantees to emit the __asm functions in the order you have defined them. However, inlined and template functions behave differently.

All calls between an __asm function and a normal C or C++ function must adhere to the AAPCS rules, even though there are no restrictions on the assembly code that an __asm function can use (for example, changing state).

All of the restrictions for the embedded assembler are detailed in the *RealView Compilation Tools Compiler User Guide* (ARM 2010c) or at www.keil.com.

18.4 CALLING BETWEEN C AND ASSEMBLY

You may find it more convenient to write functions in either C or assembly and then mix them later. This can also be done. In fact, it's downright easy. Functions can be written in assembly and then called from either C or C++, and vice versa; assembly routines can be called from C or C++ source code. Here, we'll examine mixing C and assembly routines, but refer to the ARM documentation (ARM 2007d) for information on working with C++. When using mixed language programming, you want to ensure that your assembly routines follow the AAPCS standard and your C code uses C calling conventions.

EXAMPLE 18.3

You may have a function defined in C that you want to use in an assembly routine. The code below shows a simple function that is called in the assembly routine with a BL instruction.

C source code appears as

```
int g(int a, int b, int c, int d, int e)
{
    return a+b+c+d+e;
}
```

Assembly source code appears as

```
;int f(int i) {return g(i, 2*i, 3*i, 4*i, 5*i);}
PRESERVE8
EXPORT f
AREA f, CODE, READONLY
IMPORT g                    ; i is in r0
STR    lr, [sp, #4]         ; preserve lr
ADD    r1, r0, r0           ; compute 2*i (2nd param)
ADD    r2, r1, r0           ; compute 3*i (3rd param)
ADD    r3, r1, r2           ; compute 5*i
STR    r3, [sp, #-4]!       ; 5th param on stack
```

```
ADD    r3, r1, r1          ; compute 4*i (4th param)
BL     g                   ; branch to C function
ADD    sp, sp, #4          ; remove 5th param
LDR    pc, [sp], #4        ; return
END
```

EXAMPLE 18.4

The code below shows an example of calling an assembly language function from C code. The program copies one string over the top of another string, and the copying routine is written entirely in assembly.

C source code appears as

```
#include <stdio.h>
extern void strcpy(char *d, const char *s);
extern void init_serial(void);


int main()
{
    const char *srcstr = "First string - source";
    char dststr[] = "Second string - destination";
    /* dststr is an array since we're */
    /* going to change it */
    init_serial();
    printf("Before copying:\n");
    printf("%s\n %s\n",srcstr, dststr);
    strcopy(dststr, srcstr);
    printf("After copying:\n");
    printf("%s\n %s\n",srcstr, dststr);
    return(0);
}
```

Assembly source code appears as

```
    PRESERVE8
    AREA SCopy, CODE, READONLY
    EXPORT strcopy
strcopy
                             ; r0 points to destination string
                             ; r1 points to source string
    LDRB    r2, [r1], #1     ; load byte and update address
    STRB    r2, [r0], #1     ; store byte and update address
    CMP     r2, #0           ; check for zero terminator
    BNE     strcopy          ; keep going if not
    BX      lr               ; return
    END
```

In some cases, features of the processor are not readily available in C and C++. For example, the conversion instructions in the Cortex-M4 for fixed-point and floating-point values we considered in Chapter 9 are not accessible in C and C++. The example below shows how to use the embedded assembly features to create a set of conversion routines for specific formats that can easily be reused.

EXAMPLE 18.5

The code below contains two routines for conversion between signed S16 format values and single-precision floating-point values. Recall that the S16 format specifies a short signed integer of 16 bits. In this example, we are simulating sensor data in the form of a signed fixed-point 16-bit format with 8 fraction bits. The range of input data is {−128, 127 + 255/256}, with a numeric separation of 1/256.

The conversion routine utilizing the VCVT.S16,F32 instruction is shown below. Recall that this instruction operates on two FPU registers, so a move from the input source to an FPU register is required.

```
AREA FixedFloatCvtRoutines, CODE, READONLY
THUMB

EXPORT CvtShorts8x8ToFloat

CvtShorts8x8ToFloat
    ; Use the VCVT instruction to convert a short in
    ; signed 8x8 format to a floating-point single-
    ; precision value and return the float value.
    ; The input short is in register r0.
    ; First move it to a float register - no
    ; format conversion will take place
    VMOV.F32        s0, r0          ; transfer the short to a
                                    ; floating-point register
    VCVT.F32.S16    s0, s0, #8      ; perform the conversion
    BX              lr              ; return
    END
```

A sample C program to use this conversion routine is shown below. The input data is in short integer format representing the signed 8x8 format (check for yourself that these values are correct).

```
//Input data in S16 format with 8 fraction bits.
#include <stdio.h>
extern void EnableFPU(void);
extern float CvtShorts8x8ToFloat(short i);

int main(void)
{
    short Input[10] = {
            1408,    // 5.5 (0x0580)
            384,     // 1.5 (0x180)
            -672,    // -2.625 (0xFD60)
            -256,    // -1.0 (0xFF00)
            641,     // 2.50390625 (2.5+1/256)(0x0281)
            192,     // .75 (0x00C0)
            -32768,  // neg max, -128.0 (0x8000)
            32767,   // pos max, 127+255/256 (0x7FFF)
            -32,     // -0.125 (0xFFE0)
            0
    };
```

```
    int i;
    short InVal;
    float OutVal;

    for (i=0; i<11; i++) {
            OutVal=CvtShorts8x8ToFloat(Input[i]);
            //Operate on the float value
    }
}
```

The conversion routine is stored in a separate file. Multiple routines may be placed in this file and called as needed by the C program. In this way, a library of routines utilizing functions not readily available from the high-level languages may be created to make use of features in the processor.

For further reading, you should consult the ARM documentation about calling C++ functions from assembly and calling assembly from C++. Examples can be found in the *RealView Compilation Tools Developer Guide* (ARM 2007a).

18.5 EXERCISES

1. Example 18.1 gives the program necessary to set the Q flag. Run the code using the Keil tools, with the target being the STR910FM32 from STMicroelectronics. Which registers does the compiler use, and what is the value in those registers just before the QDADD instruction is executed?

2. Example 18.2 demonstrates the embedded assembler. Compile the code and run it. What is the value in the Program Counter just before the BX instruction executes in the function my_strcopy? In order to compile this example, you will need to target the LPC2101 from NXP and include files from the "Inline" example found in the Keil "Examples" directory. Include the source files serial.c and retarget.c in your own project. Also be sure to include the startup file when asked. When you run the code, you can use the UART #2 window to see the output from the printf statements.

3. Write a short C program that declares a variable called TMPTR. Using Example 18.2 as a guide, print out the variable in degrees Celsius, with some initial temperature defined in the main program in degrees Fahrenheit. Write the temperature conversion program as an inline assembly function. You'll want to use fractional arithmetic to avoid division.

4. Using the saturation algorithm discussed in Chapter 13, which performs a logical shift left by *m* bits and saturates when necessary, write a C routine which calls it as an embedded assembly function. The function should have two parameters: the value to be shifted and the shift count. It should return the shifted value. The small C routine should create a variable with the initial value.

5. Modify Example 18.1 so that the function Clear_Q_Flag returns 1 when the function clears a set Q flag; otherwise, if the bit was clear, it returns 0.

6. Run Example 18.4 by creating two separate source files in the Keil tools. Once you have saved these files, you can add them to a new project. The Keil tools will compile the C source file and assemble the assembly language file automatically. When you run the code, you can see the output on UART #2. Refer to Exercise 2 for more details.

7. Run Example 18.5 by creating three separate source files in the Keil tools. Recall that the FPU must be initialized, and this should be one of the three files. Notice the value of OutVal in the variables window and confirm the converted values match the expected inputs from the sensor (see the comments in the array declaration).

8. Expand Example 18.5 by converting the OutVal floating-point value back to S16 8x8 format. Add this routine to the file containing the CvtShorts8x8ToFloat routine and call it CvtFloatToShorts8x8. Verify that the result of the conversion back to S16 8x8 format matches the original value. Experiment with some other formats, such as 9.7 or 7.9, and see what values are produced.

Appendix A: Running Code Composer Studio

A.1 INTRODUCTION

Code Composer Studio (CCS) is freely available from http://www.ti.com and provides a development environment for all of Texas Instruments' ARM-based SoCs and microcontrollers, e.g., Sitara™, Hercules™, and Tiva™. Very inexpensive development platforms and evaluation modules (EVMs) can be found from distributors, so getting a real Cortex-M4 running code takes relatively little effort. A wide variety of microcontrollers are available from Texas Instruments, and their peripherals can be driven with inputs to exercise I/O pins, A/D converters, UARTs, etc. The Tiva Launchpad (shown earlier in Figure 1.6) contains a Cortex-M4-based microcontroller with floating-point hardware, and makes for a quick introduction to using the Code Composer Studio tools. At the time of this writing, there are both Stellaris and Tiva parts available on the Launchpad platform (Tiva is the supported product line now), and while the older LM4F230H5QR microcontroller on the Stellaris Launchpad is nearly identical to the new Tiva TM4C123GH6PM microcontroller, you can use either one. All descriptions in this Appendix will show the newer names.

Code Composer Studio does not generally support building projects in assembly only; however, there are now assembly-only options in the build choices. The tools are based on an Eclipse front-end, so students learning a high-level language like Java may already be familiar with Eclipse-based tools. Just as we'll see with the Keil tools in Appendix B, writing assembly language on a Cortex-M4 will require breaking most industry programming practices. For starters, most of the exception vector table will be omitted to keep things simple. Second, conventional code would require handlers to be in place for dealing with Fault exceptions and interrupts, and we don't really have to consider those just yet. We therefore create our program with handlers that just stay in an infinite loop. This is fine as long as we don't require exception handling. To run a simple program, you first need to specify a particular device for which to assemble your code. You then create a project, write your assembly, and add it to your project. Finally, you build the project and start up the debugger to step through your code.

A.2 RUNNING CODE ON THE CORTEX-M4

The CCS tools do not simulate a Tiva Launchpad, so you'll want to attach the physical hardware to your development tools. You can find information about doing this in the documents provided with the board or on TI's website. There are a couple of things to check before running your code on the Launchpad:

a. When you install CCS for the first time, ensure that the Tiva products are added when the dialog appears asking which processor architectures you wish to support.
b. Make sure you've got the in-circuit debugging interface driver loaded on the host computer.

A.2.1 Creating a Cortex-M4 Project and Selecting a Device

First we'll create a new project file. Start the CCS tools and choose New CCS Project from the Project menu, as shown in Figure A.1. Give your project a name. As an example, you could call it Sample. The project file normally includes source files of code, including C, C++ and assembly, library files, header files, etc., along with a linker command file that tells the linker how to build your executable file. In our case, we will only have an assembly file and a linker command file. When the dialog box appears, select ARM from the Family drop-down menu. Choose Executable for the Output Type. In the Variant box, you should enter 123GH6PM, which will bring up TM4C123GH6PM (the chip on the Tiva Launchpad) in the parts drop-down box. Under Connection, make sure that you have the Stellaris In-Circuit Debug Interface chosen to talk to the Tiva Launchpad (newer versions of CCS may say Tiva In-Circuit Debug Interface). Under Project templates and examples, you will see an option to create an Empty Assembly-only Project, as shown in Figure A.2. Click on this option and then click on Finish. If you go to the Project Explorer pane on the left, and open the project you just created, you will find a list of files in the project. One of those will be a startup file with a ".c" ending. You should delete this file from the project by right-clicking on the name and then choosing Delete.

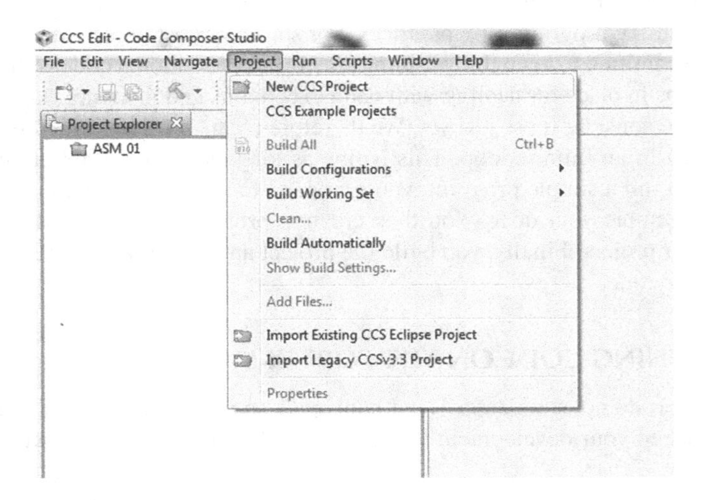

FIGURE A.1 Creating a new project.

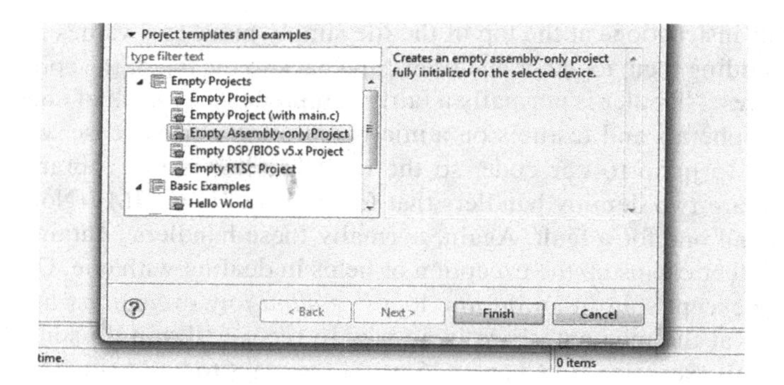

FIGURE A.2 Project options.

A.2.2 Creating Application Code

Now that the project has been created and a device chosen, you will need a source file. The easiest file to create is one that has only a few lines of actual code and a handful of directives to get the tools working. For example, the code below just adds two numbers together. Choose New -> Source File from the File menu, enter a name in the Source file box, such as Sample.asm, then carefully type the following code:

```
       .global myStart, myStack, ResetISR, Vecs, _c_int00, _main
       .sect ".myCode"
myStart:
       MOV    r2, #10
       MOV    r3, #5
       ADD    r1, r2, r3
       B      myStart
       .text
; This is the Reset Handler
_c_int00:
       B      myStart
; This is the dummy NMI handler
NmiSR:
       B      $
; This is the dummy Fault handler
FaultISR:
       B      $
; Here we define the stack
myStack     .usect ".stack", 0x400
; Interrupt vector table (abbreviated)
       .retain ".intvecs"
       .sect ".intvecs"
Vecs: .wordmyStack           ; initial stack pointer
       .word _c_int00        ; the reset handler
       .word NmiSR           ; dummy NMI handler
       .word FaultISR        ; dummy Fault handler
       .word 0               ; we don't care about the rest
```

The four instructions at the top of the file simply move two values into register r2 and r3, adding them together and branching back to the top of the code when finished. The reset handler is normally a fairly comprehensive block of code enabling various peripherals and features on a microcontroller. For our case, we just want the handler to jump to our code, so the reset handler is just a branch instruction. There are two dummy handlers that follow, one for an NMI (Non-Maskable Interrupt) and one for a fault. Again, normally these handlers contain the proper code that either cleans up the exception or helps in dealing with one. Our handlers do nothing except spin in an infinite loop, *so mind any exceptions that occur in your code*—at the moment, there is no way to recover if you do something that causes a Fault exception. In Chapter 15, we'll see ways to build simple handlers to cover these exceptions.

The stack is configured next, with 1,024 bytes reserved for it. The label myStack will eventually be converted into an address that will get stored in the vector table. The section called ".intvecs" is our vector table, and you can see that the individual vectors contain addresses. In the process of creating an executable file, the ELF linker will try to remove any code that isn't actually used, and since it sees the vector table as a set of constants that are not referenced anywhere else in the main code, it will remove them. We therefore add a .retain directive to instruct the linker to keep our vector table. The stack pointer is stored at address 0x0 in memory, and the first value is the address of myStack. The reset handler's address is stored at address 0x4 in memory, and so on. Since we don't need to worry about exceptions yet, we'll only define enough of the vectors to get our code up and running.

A.2.3 Building the Project and Running Code

We're nearly finished. There are a few additional tools issues to deal with before trying to build the project. If you open the linker command file (the file in the project that ends in .cmd), you'll find an equation involving __STACK_TOP. Just comment that out for now. Additionally, we should tell the linker where to put our code, so add an entry under .init_array for the section called .myCode. Your linker command file should now look like this:

```
SECTIONS
{
      .intvecs :          > 0x00000000
      .text :             > FLASH
      .const :            > FLASH
      .cinit :            > FLASH
      .pinit :            > FLASH
      .init_array :       > FLASH
      .myCode :           > FLASH

      .vtable :           > 0x20000000
      .data :             > SRAM
```

```
        .bss :             > SRAM
        .sysmem :          > SRAM
        .stack :           > SRAM
}

// __STACK_TOP = __stack + 512;
```

One last correction is needed. Right click on the project name in the Project Explorer pane, and then choose Properties. In the dialog box, click on the ARM Linker section under Build. You will see a subsection called File Search Path. Ensure that "libc.a" is not included in the File Search Path box on the right (see Figure A.3). If it is there, delete it using the delete button just above the line. If we were writing C code, this is an important library, but since we're only making a small assembly file, this library isn't necessary and the tools will generate a warning.

You can now build the project by either choosing Build Project from the Project menu or you can click on the hammer button in the toolbar.

Launch the debugger by either choosing Debug from the Run menu or hitting the bug button on the toolbar. You should see the four panes shown in Figure A.4—your code is displayed in one pane, along with a disassembly of the code in another. You can open up the Core Registers display to show all of the internal registers when running your code. Use the Assembly Step Into button (the green buttons, which are labeled by mousing over the buttons) to single-step through your assembly code. Examine the contents of the registers for each instruction to ensure things are working well.

Once you've completed a simple example, go back and read the CCS User's Guide, which is available in the Help menu. Many integrated development environments are similar, so if you have already used one from a different vendor, you may find this one very familiar.

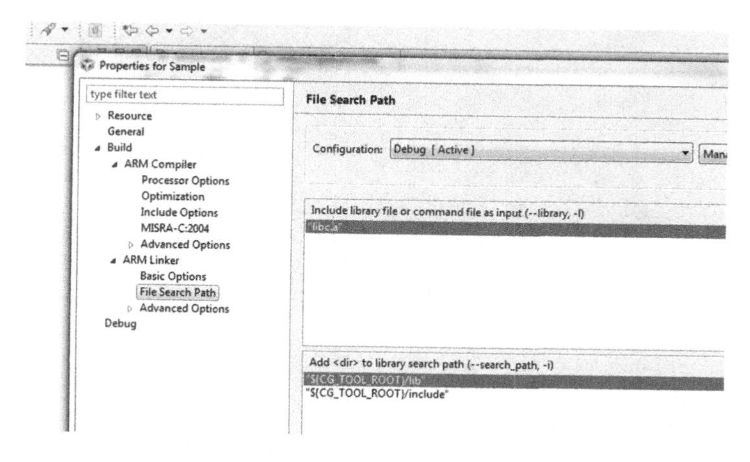

FIGURE A.3 Properties dialog box.

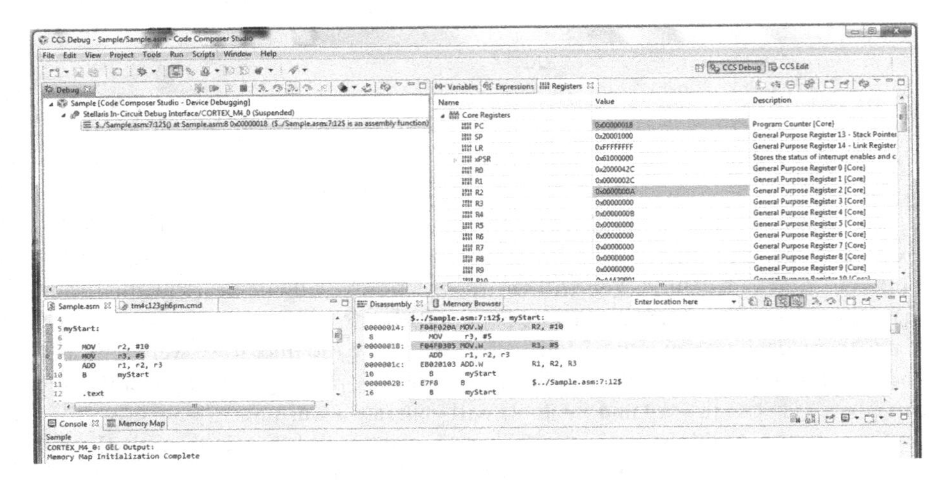

FIGURE A.4 Sample code running in the debugger.

Appendix B: Running Keil Tools

B.1 INTRODUCTION

The RealView Microcontroller Development Kit that is available from http://www.keil.com/demo simulates not only an ARM microprocessor but a complete microcontroller. A wide variety of microcontrollers are available in the tools, and their peripherals can be driven with inputs to exercise I/O pins, A/D converters, UARTs, etc. For the purpose of simulating the microcontrollers described in each chapter, we'll need a way to write assembly code without having to read one hundred pages of manuals, so here's where we bend the rules a bit. While Keil does not formally support building projects in assembly only, it can be done quite easily and the tools provide a nice development environment in which to build your code. There is, however, a rather heretical approach taken when using the RVMDK tools to write only assembly. For starters, code begins at address 0x00000000 on the ARM7TDMI, which is where the exception vector table normally sits, so we can put code there as long as we don't require exception handling. Second, the tools normally expect a default reset handler to be in place, and we don't really want to use that. We therefore create our program in such a way that the tools believe it to be the reset handler. Writing code for the Cortex-M4 works the same way—we will be leaving out exception handlers and assembly code normally created when compiling C code. To run a simple program, you first need to specify a particular device to simulate. You then create a project, write your code, and add it to your project. Finally, you build the project and start up the debugger to step through your code. NB: This appendix uses Version 4.73, so if you are using Version 5.0 or above, be sure to download the software pack that supports the legacy ARM7 and ARM9 microcontrollers.

B.2 WORKING WITH AN ARM7TDMI

This section can be read in conjunction with Chapter 3, since you're likely to look for a way to run your first assembly programs without having to download and build a simulator. If you are learning to program an ARM7- or ARM9-based device, then the procedure for running short blocks of assembly is quite easy. Be aware of how many rules of professional programming are being thrown out the window, but since the goal is to learn to walk before you learn to run, it's permitted to ignore a few things for now.

B.2.1 Creating an ARM7TDMI Project and Selecting a Device

Let's begin by creating a new project file. Start the RVMDK tools and choose **New µVision Project** from the **Project** menu, as shown in Figure B.1.

Give your project a name. As an example, you could call it **My First Program**, as shown in Figure B.2. The project file can include source files of code, including C, C++ and assembly, library files, header files, etc., along with environment options

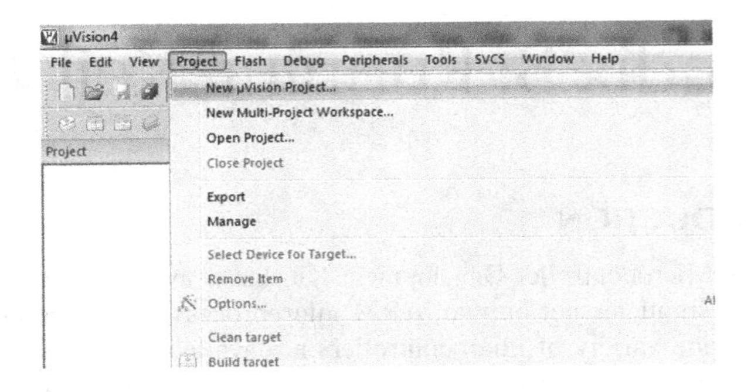

FIGURE B.1 Creating a new project.

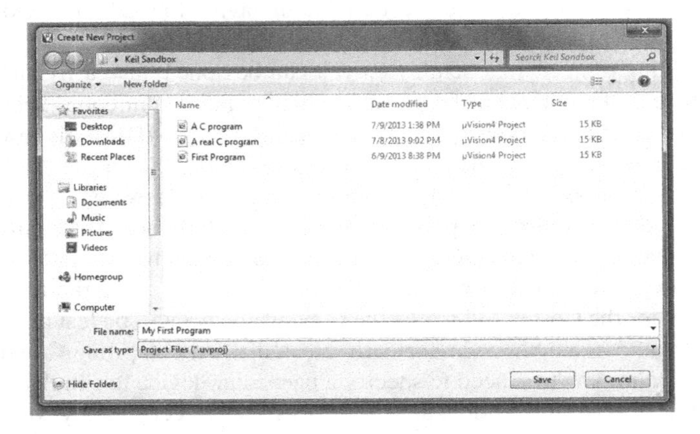

FIGURE B.2 Naming the project.

that you can save (see the µVision IDE User's Guide included with the software for all the different options available). You might wish to create a new folder for each project, just to keep things simple.

At this point the tools will ask you to specify a device to simulate. To continue our example, choose one of the LPC21xx parts from NXP, such as the LPC2104. This is an ARM7-based microcontroller with a few peripherals. You will find all of the available parts in the device database window shown in Figure B.3. Scroll down until you come to the NXP parts and select LPC2104. Notice that the tools detail all of the peripherals and memory options when you choose the device. When you click OK, a dialog box will appear asking if you want to include startup code for this device. Click No, since we are only making a small assembly program and will not need all of the initialization code.

B.2.2 Creating Application Code

Now that the project has been created and a device chosen, you will need to create a source file. From the **File** menu, choose **New** to create your assembly file with the editor. If you like, you can directly copy the small program from Figure B.4 as an

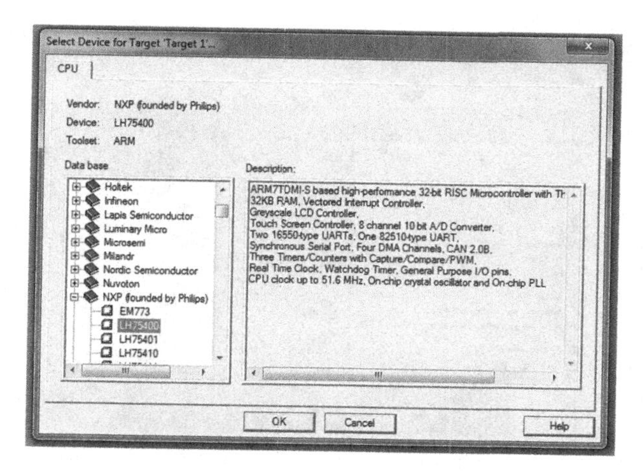

FIGURE B.3 Device database dialog box.

```
My First Program.s
 1
 2          GLOBAL  Reset_Handler
 3          AREA    Reset, CODE, READONLY
 4          ENTRY
 5
 6 Reset_Handler
 7
 8          MOV     r0, #0x11     ; load initial value
 9          LSL     r1, r0, #1    ; shift 1 bit left
10          LSL     r2, r1, #1    ; shift 1 bit left
11
12 stop     B stop                ; stop program
13
14          END
```

FIGURE B.4 Sample code.

example. The latest versions of RVMDK now require a reset handler to be found, and normally this is included in the startup file. However, we have elected not to use this file, so the workaround is to mark your assembly code as the reset handler and declare it globally. Notice the first line of code is the GLOBAL directive. You should also call the block of code reset (or Reset—it is case insensitive) with the AREA directive. After the ENTRY directive, the label `Reset _ Handler` should be placed at the top of your code. You can follow this with another label if you like, say Main or MyCode, but just be sure to include the first label. The remaining code would appear as examples do in the book. Choose **Save As** from the **File** menu, and give it a name, such as **My First Program.s**, being sure to include the ".s" extension on the file. The window should change, showing legal instructions in boldface type, and comments and constants in different colors.

The assembly file must be added to the project. In the Project Workspace window on the left, click on the plus sign to expand the Target 1 folder. Right click on the Source Group 1 folder, then choose **Add Files** to **Group "Source Group 1"** as

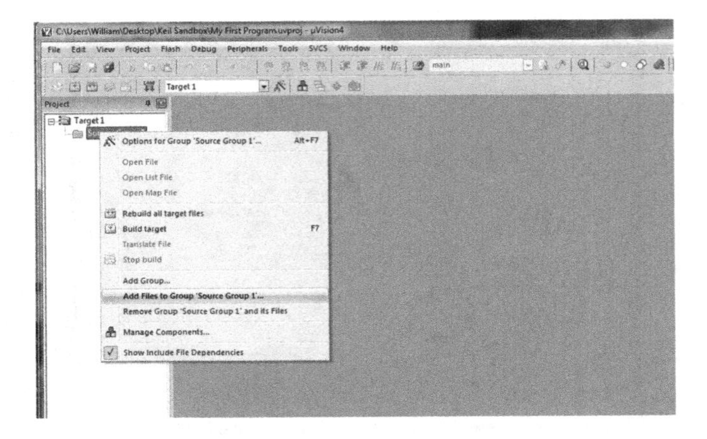

FIGURE B.5 Adding a source file to the project.

shown in Figure B.5. A dialog box will appear. In the dropdown menu called **Files of Type**, choose **Asm Source** file to show all of the assembly files in your folder. Select the file you just created and saved. Click **Add**, and then **Close**.

B.2.3 Building the Project and Running Code

To build the project, select **Build target** or **Rebuild all target files** from the **Project** menu. You will get a warning about the fact that Reset_Handler does not exist, but you can ignore it. Now that the executable has been produced, you can use the debugger for simulation. From the **Debug** menu, choose **Start/Stop Debug Session**.

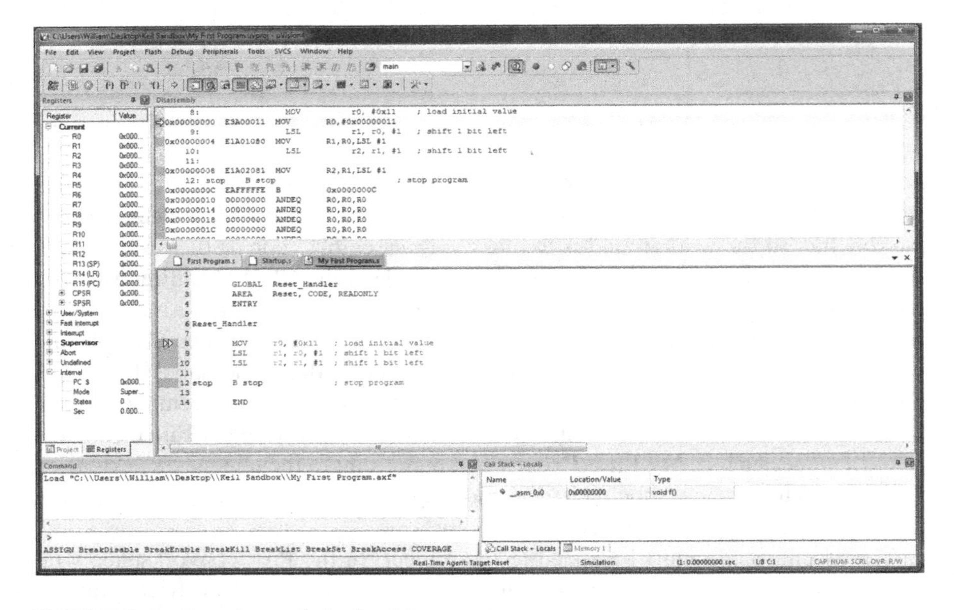

FIGURE B.6 Running code in the debugger.

This puts you into a debug session (shown in Figure B.6) and produces new windows, including the Register window and the Memory window. You can single-step through the code, watching each instruction execute by clicking on the **Step Into** button on the toolbar or choosing **Step** from the **Debug** menu. At this point, you can also view and change the contents of the register file, and view and change memory locations by typing in the address of interest. When you are finished, choose **Start/Stop Debug Session** again from the **Debug** menu. Once you've completed a simple example, go back and read the μVision IDE User's Guide, which is available in the Help menu. Many integrated development environments are similar, so if you have already used one from a different vendor, you may find this one very familiar.

B.3 WORKING WITH A CORTEX-M4

Huge simplifications allow us to make a working environment for Cortex-M4 devices. Unless specific requirements are added for handling exceptions, such as hard faults or interrupts, you must be very careful when writing code since an unexpected condition will send you into the weeds. With the debugging tools available to you, however, most errors can be caught and corrected without too much difficulty. As with the ARM7TDMI projects, you might choose to read this section before reading Chapter 3.

B.3.1 Creating a Cortex-M4 Project and Selecting a Device

First, we'll create a new project file. Start the RVMDK tools and choose **New μVision Project** from the **Project** menu, as shown in Figure B.7. Give your project a name. As an example, you could call it **My First M4 Program**, as shown in Figure B.8. The project file can include source files of code, including C, C++ and assembly, library files, header files, etc., along with environment options that you can save (see the μVision IDE User's Guide included with the software for all the different options available). You might wish to create a new folder for each project, just to keep things simple.

At this point the tools will ask you to specify a device to simulate. To continue our example, choose one of the Tiva parts from TI, such as the LM4F120H5QR (this part number is equivalent to the TM4C1233H6PM). This is a Cortex-M4-based

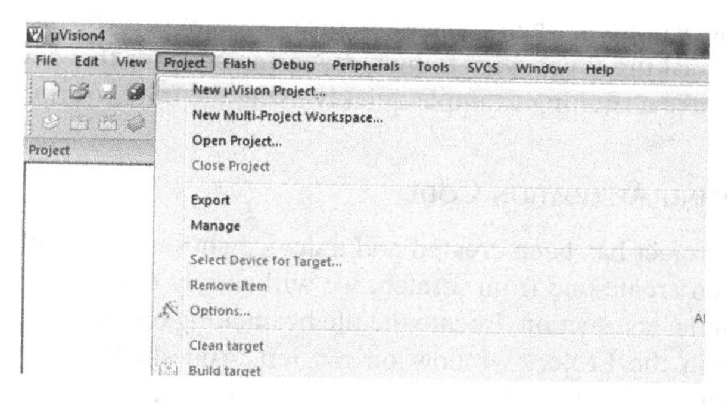

FIGURE B.7 Creating a new project.

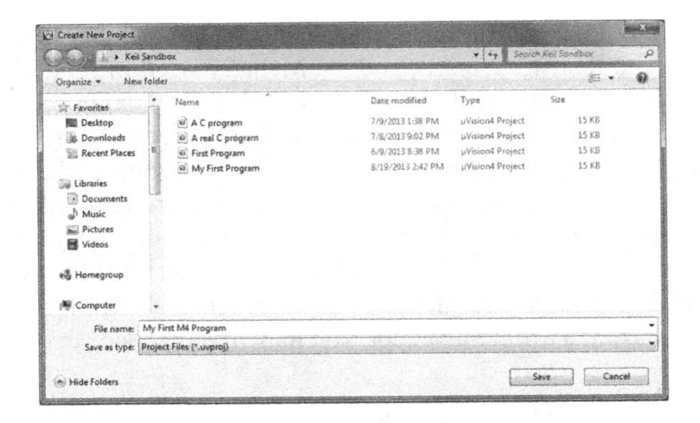

FIGURE B.8 Naming the project.

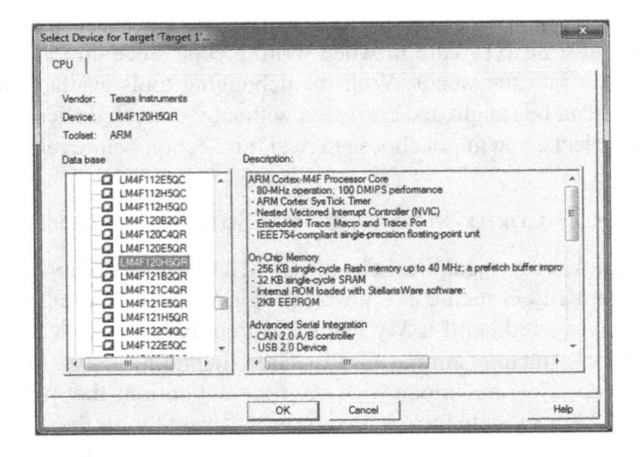

FIGURE B.9 Device database dialog box.

microcontroller with a number of peripherals. You will find all of the available parts in the device database window shown in Figure B.9. Scroll down until you come to the TI parts and select LM4F120H5QR. Notice that the tools detail all of the peripherals and memory options when you choose the device. When you click OK, a dialog box will appear asking if you want to include startup code for this device. Click Yes, since we can make a running example quickly using the initialization code.

B.3.2 Creating Application Code

Now that the project has been created and a device chosen, you will need a source file. Rather than create one from scratch, we will simply edit the Startup.s file that we included in the last section. Locate the file by clicking on the +sign by the Source Group 1 icon in the Project window on the left. You should see one file called Startup.s. Open the file by double-clicking on the name. At this point, you can insert your code. Let's use the example in Chapter 3 for computing a factorial function. You

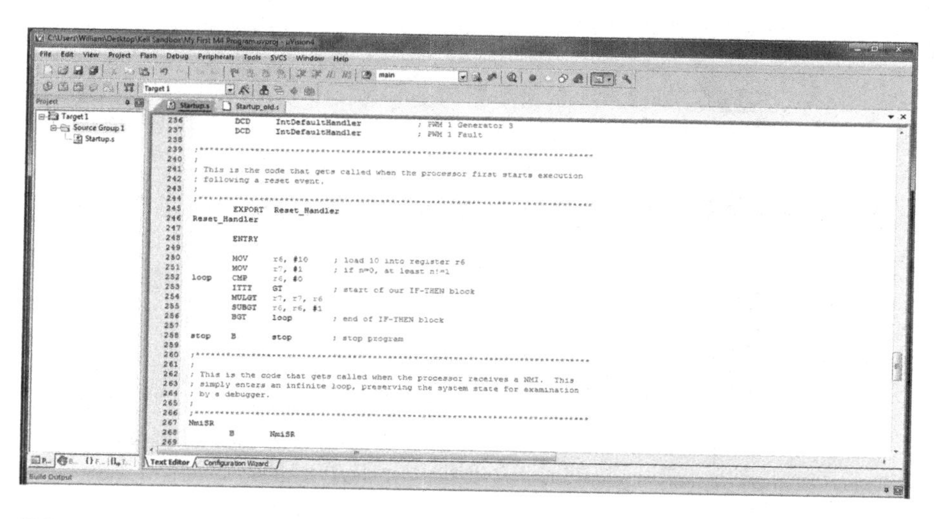

FIGURE B.10 Sample code.

will need to delete the code between the labels Reset_Handler and NmiSR. Add your code after the label Reset_Handler, adding an ENTRY directive so that the code looks like that in Figure B.10.

Comment out the code in the section for handling stack and heap memory locations, since we won't immediately need any of this. Simply add a semicolon to the beginning of each line, so that the code looks like:

```
;**************************************************************
;
; The function expected of the C library startup code for
; defining the stack and heap memory locations. For the C
; library version of the startup code, provide this function
; so that the C library initialization code can find out
; the location of the stack and heap.
;
;**************************************************************
;    IF    :DEF: __MICROLIB
;          EXPORT __initial_sp
;          EXPORT __heap_base
;          EXPORT __heap_limit
;    ELSE
;          IMPORT __use_two_region_memory
;          EXPORT __user_initial_stackheap
;__user_initial_stackheap
;          LDR      R0, =HeapMem
;          LDR      R1, =(StackMem + Stack)
;          LDR      R2, =(HeapMem + Heap)
;          LDR      R3, =StackMem
;          BX       LR
;    ENDIF
```

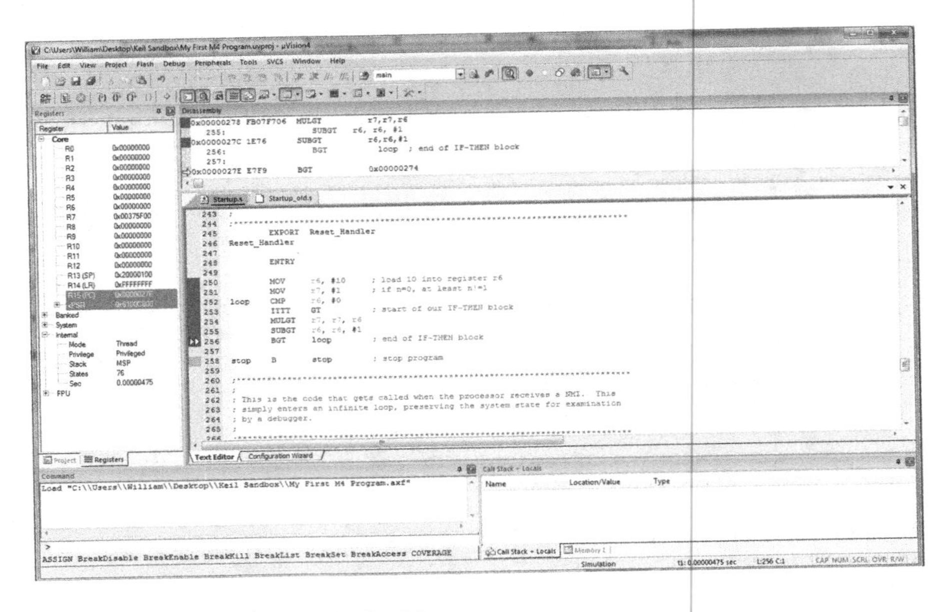

FIGURE B.11 Running code in the debugger.

B.3.3 Building the Project and Running Code

To build the project, select **Build target** or **Rebuild all target files** from the **Project** menu. You will get a warning, but you can ignore it. Now that the executable has been produced, you can use the debugger for simulation. From the **Debug** menu, choose **Start/Stop Debug Session**. This puts you into a debug session (shown in Figure B.11) and produces new windows, including the Register window and the Memory window. You can single-step through the code, watching each instruction execute by clicking on the **Step Into** button on the toolbar or choosing **Step** from the **Debug** menu. At this point, you can also view and change the contents of the register file, and view and change memory locations by typing in the address of interest. When you are finished, choose **Start/Stop Debug Session** again from the **Debug** menu.

Once you've completed a simple example, go back and read the µVision IDE User's Guide, which is available in the Help menu. Many integrated development environments are similar, so if you have already used one from a different vendor, you may find this one very familiar.

Appendix C: ASCII Character Codes

	MOST SIGNIFICANT BITS							
LEAST SIGNIFICANT BITS	0	1	2	3	4	5	6	7
0	Null	Data Link Escape	Space	0	@	P	`	p
1	Start of Heading	Device Control 1	!	1	A	Q	a	q
2	Start of Text	Device Control 2	“	2	B	R	b	r
3	End of Text	Device Control 3	#	3	C	S	c	s
4	End of Transmit	Device Control 4	$	4	D	T	d	t
5	Enquiry	Neg Acknowledge	%	5	E	U	e	u
6	Acknowledge	Synchronous Idle	&	6	F	V	f	v
7	Bell	End of Trans Block	’	7	G	W	g	w
8	Backspace	Cancel	(	8	H	X	h	x
9	Horizontal Tab	End of Medium	)	9	I	Y	i	y
A	Line Feed	Substitute	*	:	J	Z	j	z
B	Vertical Tab	Escape	+	;	K	[	k	{
C	Form Feed	File Separator	,	<	L	\	l	|
D	Carriage Return	Group Separator	-	=	M	]	m	}
E	Shift Out	Record Separator	.	>	N	^	n	~
F	Shift In	Unit Separator	/	?	O	_	o	Delete

The American Standard Code for Information Interchange contains both printable and nonprintable characters, e.g., backspace or line feed. The devices that use ASCII data do not have to implement the entire character set. An LCD panel, for example, probably will not do anything if it receives the control character to ring a bell (0x7). If a device is an ASCII device, then it will only accept ASCII data, and numbers must be sent in their respective ASCII representations. For example, if you want to print a 9, then a printer must receive the value 0x39. Note that the most significant bit of an ASCII value is either zero or a parity bit, depending on how the programmer wants to use it.

Appendix D

```
*****************************************************************
*
* Forward declaration of the default fault handlers.
*
*****************************************************************
     .global myStart, myStack, ResetISR, Vecs, _c_int00, _main
*****************************************************************
* Interrupt vector table
*****************************************************************
      .sect  ".intvecs"

Vecs: .word    myStack + 0x400      ; The initial stack pointer
      .word    _main                ; The reset handler
      .word    NmiSR                ; The NMI handler
      .word    FaultISR             ; The hard fault handler
      .word    IntDefaultHandler    ; The MPU fault handler
      .word    IntDefaultHandler    ; The bus fault handler
      .word    IntDefaultHandler    ; The usage fault handler
      .word    0                    ; Reserved
      .word    0                    ; Reserved
      .word    0                    ; Reserved
      .word    0                    ; Reserved
      .word    IntDefaultHandler    ; SVCall handler
      .word    IntDefaultHandler    ; Debug monitor handler
      .word    0                    ; Reserved
      .word    IntDefaultHandler    ; The PendSV handler
      .word    IntDefaultHandler    ; The SysTick handler
      .word    IntDefaultHandler    ; GPIO Port A
      .word    IntDefaultHandler    ; GPIO Port B
      .word    IntDefaultHandler    ; GPIO Port C
      .word    IntDefaultHandler    ; GPIO Port D
      .word    IntDefaultHandler    ; GPIO Port E
      .word    IntDefaultHandler    ; UART0 Rx and Tx
      .word    IntDefaultHandler    ; UART1 Rx and Tx
      .word    IntDefaultHandler    ; SSI0 Rx and Tx
      .word    IntDefaultHandler    ; I2C0 Master and Slave
      .word    IntDefaultHandler    ; PWM Fault
      .word    IntDefaultHandler    ; PWM Generator 0
      .word    IntDefaultHandler    ; PWM Generator 1
      .word    IntDefaultHandler    ; PWM Generator 2
      .word    IntDefaultHandler    ; Quadrature Encoder 0
      .word    IntDefaultHandler    ; ADC Sequence 0
      .word    IntDefaultHandler    ; ADC Sequence 1
      .word    IntDefaultHandler    ; ADC Sequence 2
      .word    IntDefaultHandler    ; ADC Sequence 3
      .word    IntDefaultHandler    ; Watchdog timer
```

```
.word  IntDefaultHandler    ; Timer 0 subtimer A
.word  IntDefaultHandler    ; Timer 0 subtimer B
.word  IntDefaultHandler    ; Timer 1 subtimer A
.word  IntDefaultHandler    ; Timer 1 subtimer B
.word  IntDefaultHandler    ; Timer 2 subtimer A
.word  IntDefaultHandler    ; Timer 2 subtimer B
.word  IntDefaultHandler    ; Analog Comparator 0
.word  IntDefaultHandler    ; Analog Comparator 1
.word  IntDefaultHandler    ; Analog Comparator 2
.word  IntDefaultHandler    ; System Control (PLL OSC BO)
.word  IntDefaultHandler    ; FLASH Control
.word  IntDefaultHandler    ; GPIO Port F
.word  IntDefaultHandler    ; GPIO Port G
.word  IntDefaultHandler    ; GPIO Port H
.word  IntDefaultHandler    ; UART2 Rx and Tx
.word  IntDefaultHandler    ; SSI1 Rx and Tx
.word  IntDefaultHandler    ; Timer 3 subtimer A
.word  IntDefaultHandler    ; Timer 3 subtimer B
.word  IntDefaultHandler    ; I2C1 Master and Slave
.word  IntDefaultHandler    ; Quadrature Encoder 1
.word  IntDefaultHandler    ; CAN0
.word  IntDefaultHandler    ; CAN1
.word  IntDefaultHandler    ; CAN2
.word  IntDefaultHandler    ; Ethernet
.word  IntDefaultHandler    ; Hibernate
.word  IntDefaultHandler    ; USB0
.word  IntDefaultHandler    ; PWM Generator 3
.word  IntDefaultHandler    ; uDMA Software Transfer
.word  IntDefaultHandler    ; uDMA Error
.word  IntDefaultHandler    ; ADC1 Sequence 0
.word  IntDefaultHandler    ; ADC1 Sequence 1
.word  IntDefaultHandler    ; ADC1 Sequence 2
.word  IntDefaultHandler    ; ADC1 Sequence 3
.word  IntDefaultHandler    ; I2S0
.word  IntDefaultHandler    ; External Bus Interface 0
.word  IntDefaultHandler    ; GPIO Port J
.word  IntDefaultHandler    ; GPIO Port K
.word  IntDefaultHandler    ; GPIO Port L
.word  IntDefaultHandler    ; SSI2 Rx and Tx
.word  IntDefaultHandler    ; SSI3 Rx and Tx
.word  IntDefaultHandler    ; UART3 Rx and Tx
.word  IntDefaultHandler    ; UART4 Rx and Tx
.word  IntDefaultHandler    ; UART5 Rx and Tx
.word  IntDefaultHandler    ; UART6 Rx and Tx
.word  IntDefaultHandler    ; UART7 Rx and Tx
.word  0                    ; Reserved
.word  0                    ; Reserved
.word  0                    ; Reserved
.word  0                    ; Reserved
.word  IntDefaultHandler    ; I2C2 Master and Slave
.word  IntDefaultHandler    ; I2C3 Master and Slave
```

```
.word  IntDefaultHandler    ; Timer 4 subtimer A
.word  IntDefaultHandler    ; Timer 4 subtimer B
.word  0                    ; Reserved
.word  0                    ; Reserved
.word  0                    ; Reserved
.word  0                    ; Reserved
.word  0                    ; Reserved
.word  0                    ; Reserved
.word  0                    ; Reserved
.word  0                    ; Reserved
.word  0                    ; Reserved
.word  0                    ; Reserved
.word  0                    ; Reserved
.word  0                    ; Reserved
.word  0                    ; Reserved
.word  0                    ; Reserved
.word  0                    ; Reserved
.word  0                    ; Reserved
.word  0                    ; Reserved
.word  0                    ; Reserved
.word  0                    ; Reserved
.word  0                    ; Reserved
.word  IntDefaultHandler    ; Timer 5 subtimer A
.word  IntDefaultHandler    ; Timer 5 subtimer B
.word  IntDefaultHandler    ; Wide Timer 0 subtimer A
.word  IntDefaultHandler    ; Wide Timer 0 subtimer B
.word  IntDefaultHandler    ; Wide Timer 1 subtimer A
.word  IntDefaultHandler    ; Wide Timer 1 subtimer B
.word  IntDefaultHandler    ; Wide Timer 2 subtimer A
.word  IntDefaultHandler    ; Wide Timer 2 subtimer B
.word  IntDefaultHandler    ; Wide Timer 3 subtimer A
.word  IntDefaultHandler    ; Wide Timer 3 subtimer B
.word  IntDefaultHandler    ; Wide Timer 4 subtimer A
.word  IntDefaultHandler    ; Wide Timer 4 subtimer B
.word  IntDefaultHandler    ; Wide Timer 5 subtimer A
.word  IntDefaultHandler    ; Wide Timer 5 subtimer B
.word  IntDefaultHandler    ; FPU
.word  IntDefaultHandler    ; PECI 0
.word  IntDefaultHandler    ; LPC 0
.word  IntDefaultHandler    ; I2C4 Master and Slave
.word  IntDefaultHandler    ; I2C5 Master and Slave
.word  IntDefaultHandler    ; GPIO Port M
.word  IntDefaultHandler    ; GPIO Port N
.word  IntDefaultHandler    ; Quadrature Encoder 2
.word  IntDefaultHandler    ; Fan 0
.word  0                    ; Reserved
.word  IntDefaultHandler    ; GPIO Port P (Summary or P0)
.word  IntDefaultHandler    ; GPIO Port P1
.word  IntDefaultHandler    ; GPIO Port P2
.word  IntDefaultHandler    ; GPIO Port P3
.word  IntDefaultHandler    ; GPIO Port P4
```

```
        .word   IntDefaultHandler    ; GPIO Port P5
        .word   IntDefaultHandler    ; GPIO Port P6
        .word   IntDefaultHandler    ; GPIO Port P7
        .word   IntDefaultHandler    ; GPIO Port Q (Summary or Q0)
        .word   IntDefaultHandler    ; GPIO Port Q1
        .word   IntDefaultHandler    ; GPIO Port Q2
        .word   IntDefaultHandler    ; GPIO Port Q3
        .word   IntDefaultHandler    ; GPIO Port Q4
        .word   IntDefaultHandler    ; GPIO Port Q5
        .word   IntDefaultHandler    ; GPIO Port Q6
        .word   IntDefaultHandler    ; GPIO Port Q7
        .word   IntDefaultHandler    ; GPIO Port R
        .word   IntDefaultHandler    ; GPIO Port S
        .word   IntDefaultHandler    ; PWM 1 Generator 0
        .word   IntDefaultHandler    ; PWM 1 Generator 1
        .word   IntDefaultHandler    ; PWM 1 Generator 2
        .word   IntDefaultHandler    ; PWM 1 Generator 3
        .word   IntDefaultHandler    ; PWM 1 Fault

        .sect   ".myCode"

myStart:
        ; Set sysclk to DIV/4, use PLL, XTAL_16 MHz, OSC_MAIN
        ; system control base is 0x400FE000, offset 0x60
        ; bits [26:23] = 0x3
        ; bit  [22]    = 0x1
        ; bit  [13]    = 0x0
        ; bit  [11]    = 0x0
        ; bits [10:6]  = 0x15
        ; bits [5:4]   = 0x0
        ; bit  [0]     = 0x0
        ; All of this translates to 0x01C00540

        MOVW    r0, #0xE000
        MOVT    r0, #0x400F
        MOVW    r2, #0x60       ; offset 0x060 for this register
        MOVW    r1, #0x0540
        MOVT    r1, #0x01C0
        STR     r1, [r0, r2]    ; write the register's contents

;       MOVW    r6, #0xE000
;       MOVT    r6, #0xE000
        MOVW    r7, #0x604      ; enable timer0 - RCGCTIMER
        LDR     r1, [r0, r7]    ; p. 321, base 0x400FE000
        ORR     r1, #0x1        ; offset - 0x604
        STR     r1, [r0, r7]    ; bit 0

        NOP
        NOP
        NOP
        NOP
        NOP                     ; give myself 5 clocks per spec
```

```
        MOVW    r8, #0x0000    ; configure timer0 to be
        MOVT    r8, #0x4003    ; one-shot, p.698 GPTMTnMR
        MOVW    r7, #0x4       ; base 0x40030000
        LDR     r1, [r8, r7]   ; offset 0x4
        ORR     r1, #0x21      ; bit 5=1, 1:0=0x1
        STR     r1, [r8, r7]

        LDR     r1, [r8]       ; set as 16-bit timer only
        ORR     r1, #0x4       ; base 0x40030000
        STR     r1, [r8]       ; offset 0, bit[2:0]=0x4

        MOVW    r7, #0x30      ; set the match value at 0
        MOV     r1, #0         ; since we're counting down
        STR     r1, [r8, r7]   ; offset - 0x30

        MOVW    r7, #0x18      ; set bits in the GPTM
        LDR     r1, [r8, r7]   ; Interrupt Mask Register
        ORR     r1, #0x10      ; p. 714 - base: 0x40030000
        STR     r1, [r8, r7]   ; offset - 0x18, bit 5

        MOVW    r6, #0xE000    ; enable interrupt on timer0
        MOVT    r6, #0xE000    ; p. 132, base 0xE000E000
        MOVW    r7, #0x100     ; offset - 0x100, bit 19
        MOV     r1, #(1<<19)   ; enable bit 19 for timer0
        STR     r1, [r6, r7]

        ;NOP
        ;NOP
        ;NOP
        ;NOP
        ;NOP

        MOVW    r6, #0x0000    ; start the timer
        MOVT    r6, #0x4003
        MOVW    r7, #0xC
        LDR     r1, [r6, r7]
        ORR     r1, #0x1
        STR     r1, [r6, r7]   ; go!!

Spin
        B       Spin           ; sit waiting for the interrupt to occur

*************************************************************
* Interrupt functions
*************************************************************
        .text

;ResetISR:
;_c_int00:
_main
        B       myStart
```

```
NmiSR:
        B       $

FaultISR:
        B       $

IntDefaultHandler:
        MOVW   r10, #0xBEEF
        MOVT   r10, #0xDEAD
        NOP
Spot
        B      Spot

*************************************************************
myStack .usect ".stack", 0x400
```

Glossary

AHB: Advanced High-performance Bus. Part of the AMBA specification for interconnectivity, the AHB is a single-cycle bus to which you normally attach bus masters, such as processor cores, DSP engines, DMA engines, or memory.

AMBA: Advanced Microcontroller Bus Architecture. The AMBA specification provides a bus framework around which systems can be built, and it also defines the manner in which processors and peripherals communicate in a system.

APB: Advanced Peripheral Bus. Part of the AMBA specification, the APB is the bus to which you normally attach peripherals or slower devices in the system.

API: Application Programming Interface. Often APIs come in the form of a library that includes routines for accessing hardware or services at a high level, resembling function calls.

ASIC: Application Specific Integrated Circuit. A description of any integrated circuit which is built for one specific purpose, as opposed to a generic device such as a microprocessor, which can be used in many applications. Examples are anti-lock disc brake circuits for a particular manufacturer or engine controllers for a particular vehicle.

Big-Endian: Byte-ordering scheme in which bytes of decreasing significance in a data word are stored at increasing addresses in memory.

Cache: From the French caché; literally, hidden. This memory, located very near the processor, holds recently used data and allows a processor to find data on the chip before going out to external memory, which is much slower.

CAN: Controller Area Network. Developed by Bosch and Intel, this is a network protocol and bus standard that allows automotive components like transmissions, engine control units, and cruise control to communicate as a subsystem within the car.

CISC: Complex Instruction Set Computer. An older computer architecture which implements a large instruction set, usually microcoded, and can have instructions of varying length.

Die: An individual square produced by cutting a wafer into pieces. Normally a die contains an entire microprocessor or analog device, including contact points (pads).

DSP: Digital Signal Processor. Any device which is specifically designed to transfer large amounts of data while providing arithmetic operations in parallel, particularly multiply and accumulate operations. General-purpose microprocessors can also be used as DSPs.

EEPROM: Electrically Erasable Programmable Read-Only Memory. This type of read-only memory can be programmed via software and erased electrically, rather than through an ultraviolet light source as EPROMs are. Flash memory is a recent form of EEPROM.

Endianness: The scheme that determines the order in which successive bytes of a data word are stored in memory. An aspect of the system's memory mapping.

EFlash Memory: See EEPROM.

Exponent: Eight bits of a single-precision floating-point number or eleven bits of a double-precision floating-point number that follow a sign bit, indicating how the significand is to be scaled. Normally the exponent is biased so that the number is positive.

Fraction: The part of a floating-point number that is in the range [1.0, 2.0), represented in single-precision by 23 bits and in double-precision by 52 bits. Also known as the mantissa.

I²C: Inter-Integrated Circuit. This is a serial bus invented by Philips that allows low-speed peripherals to be attached to motherboards or embedded devices. Common applications of the bus include controlling LCD displays, reading real-time clocks, and accessing low-speed A/D converters.

Intellectual Property (IP): Legally, a term used to describe and protect artistic works, music, inventions, and other creations derived from human intellect. Semiconductor companies can license IP under contract from suppliers such as ARM, and are entitled to certain rights under the contract, e.g., the ability to produce products derived from the design they licensed.

Little-Endian: Byte ordering scheme in which bytes of increasing significance in a data word are stored at increasing addresses in memory.

Mantissa: See Fraction.

MB: Megabyte, or 1,048,576 bytes.

MMU: Memory Management Unit. A hardware option on a microprocessor that allows it to address more memory than physically present.

MPEG: Stands for Moving Picture Experts Group, but generally refers to the different standards for digitally encoding audio and video. Popular formats include MP3 for audio and MPEG-2 and MPEG-4 for video, which include standards for HDTV and high definition DVD players.

PROM: Programmable Read-Only Memory. A type of ROM that is programmed after the device has been built, using fuses that are changed just once.

RAM: Random Access Memory. This type of memory can be written to and read from. Forms of RAM include DRAM (Dynamic RAM) and SRAM (Static RAM).

RISC: Reduced Instruction Set Computer. A computer architecture having a small instruction set, where instructions are of a fixed length. Most RISC instructions execute in a single cycle, and data must be explicitly loaded and stored with separate instructions.

ROM: Read-Only Memory. This type of memory cannot be altered or programmed, and is usually configured at the time of manufacture.

Significand: In IEEE floating-point representations, the significand is the value $1.f$, where f is the fraction.

SoC: System-on-Chip. This term refers to the integration of a processor core or cores, an internal bus, and peripherals on a single die to build a complete system.

Sticky bit: A bit that can only be cleared by explicitly writing a value of zero to it. In floating-point rounding operations, the sticky bit is a bit formed by ORing all bits with lower significance than the guard bit.

TCM: Tightly Coupled Memory. An area of low latency memory that provides predictable instruction execution or data load timing in cases where deterministic performance is required. TCMs are useful for holding important software routines, such as an interrupt handler, or data that is not well suited for caching. TCMs can be either ROM or RAM types.

UAL: Unified Assembly Language. With the introduction of Thumb-2 extensions, the syntax for instructions has been unified to allow the programmer to use a single mnemonic which can be ported to different architectures, especially for newer processors such as the Cortex family.

UART: Universal Asynchronous Receiver/Transmitter. This is a simple buffer which can be used to serially transmit and receive data. UARTs are commonly found on microcontrollers as memory-mapped peripherals.

USB: Universal Serial Bus. Largely a replacement for old serial and parallel connections on computers, the USB specification was developed by the USB Implementers Forum, which included companies such as HP, Apple, Microsoft, Intel, and NEC.

VPB: VLSI Peripheral Bus. A superset of ARM's AMBA Peripheral Bus protocol defined by NXP.

Wafer: A thin, crystalline slice of material (usually silicon) used to make integrated circuits. Wafers are produced by slicing ingots of semiconducting material into thin plates and then polishing them.

Wi-Fi: The trade name for wireless networks based on the IEEE 802.11 standards.

References

ARM Ltd. 2007a. *RealView Compilation Tools Developer Guide*. Doc. no. DUI 0203H, version 3.1. Cambridge: ARM Ltd.

ARM Ltd. 2007b. *Application Binary Interface for the ARM Architecture*. Doc. no. IHI 0036B. Cambridge: ARM Ltd.

ARM Ltd. 2007c. *Architectural Reference Manual*. Doc. no. DUI 0100E. Cambridge: ARM Ltd.

ARM Ltd. 2008a. *RealView Compilation Tools for uVision Assembler User's Guide*. Doc. no. KUI 0100A, Revision B. Cambridge: ARM Ltd.

ARM Ltd. 2008b. *RealView ICE and RealView Trace User Guide*. Doc. no. DUI 0155J, version 3.3. Cambridge: ARM Ltd.

ARM Ltd. 2009. *Cortex-M4 Technical Reference Manual*. Doc. no. DDI 0439C (ID070610). Cambridge: ARM Ltd.

ARM Ltd. 2010a. *ARM v7-M Architectural Reference Manual*. Doc. no. DDI 0403D. Cambridge: ARM Ltd.

ARM Ltd. 2010b. *Cortex-M4 Devices Generic User Guide*. Doc. no. DUI 0553A. Cambridge: ARM Ltd.

ARM Ltd. 2010c. *RealView Compilation Tools Compiler User Guide*. Doc. no. ARM DUI 0205J. Cambridge: ARM Ltd.

ARM Ltd. 2012. *RealView Assembler User Guide* (online), Revision D. Cambridge: ARM Ltd.

Clements, A. 2000. *The Principles of Computer Hardware*. 3rd ed. New York: Oxford University Press.

Cohen, D. 1981. On Holy Wars and a Plea for Peace. *IEEE Computer,* October: 48–54.

Doeppner, T. 2011. *Operating Systems in Depth*. Hoboken: Wiley & Sons.

Ercegovac, M. and T. Lang. 2004. *Digital Arithmetic*. San Francisco: Morgan Kaufmann,

Furber, S. 2000. *ARM System-on-Chip Architecture*. 2nd ed. New York: Addison-Wesley Professional.

Hennessy, J., N. Jouppi, F. Baskett, and J. Gill. 1981. MIPS: A VLSI Processor Architecture. In *Proceedings of the CMU Conference on VLSI Systems and Computations*. Rockville, MD: Computer Science Press.

Hohl, W. and C. Hinds. 2008. A Primer on Fractions. *IEEE Potentials,* March/April:10-14.

Kane, G., D. Hawkins, and L. Leventhal. 1981. *68000 Assembly Language Programming*. Berkeley, CA: McGraw-Hill.

Knuth, D. 1973. *Sorting and Searching*. Vol. 3 of *The Art of Computer Programming*. Reading, MA: Addison-Wesley.

Oshana, R. 2006. *DSP Software Development Techniques for Embedded and Real-Time Systems*. Burlington, MA: Newnes.

Patterson, D. A. and D. R. Ditzel. 1980. The Case for the Reduced Instruction Set Computer. *Computer Architecture News* 8 (6): 25–33.

Patterson, D. A. and J. Hennessy. 2007. *Computer Organization and Design*. 3rd ed. Burlington, MA: Morgan Kaufmann.

Patterson, D. A. and C. H. Sequin. 1982. A VLSI RISC. *IEEE Computer,* September: 8–21.

Roth, R. 2006. *Introduction to Coding Theory*. New York: Cambridge University Press.

Sloss, A., D. Symes, and C. Wright. 2004. *ARM System Developer's Guide*. Burlington, MA: Morgan Kaufmann.

Standards Committee of the IEEE Computer Society, USA. 1985. *IEEE Standard for Binary Floating-Point Arithmetic* (ANSI/IEEE Standard No. 754-1985). New York: IEEE.

Standards Committee of the IEEE Computer Society, USA. 2008. *IEEE Standard for Binary Floating-Point Arithmetic* (ANSI/IEEE Standard No. 754-2008). New York: IEEE.

STMicroelectronics. 2006. STR91xF Reference Manual. Doc. no. UM0216. Geneva, Switzerland: STMicroelectronics.

Texas Instruments. 2013a. *ARM Assembly Language Tools v5.1 User's Guide*. Doc. no. SPNU118L. Dallas, Texas: Texas Instruments.

Texas Instruments. 2013b. Tiva™ TM4C1233H6PM Microcontroller Data Sheet. Doc. no. SPMS351. Dallas, Texas: Texas Instruments.

Texas Instruments. 2013c. Tiva™ TM4C123GH6PM Microcontroller Data Sheet. Doc. no. SPMS376C. Dallas, Texas: Texas Instruments.

Yiu, J. 2014. *The Definitive Guide to the Cortex-M3 and Cortex-M4*. 3rd ed. Massachusetts: Newnes.

Internet

http://babbage.cs.qc.cuny.edu/IEEE-754.old/Decimal.html

http://babbage.cs.qc.cuny.edu/IEEE-754.old/32bit.html

http://processors.wiki.ti.com/index.php/TI-RTOS_Workshop#Intro_to_TI-RTOS_Kernel_Workshop_Online_Video_Tutorials

http://www.h-schmidt.net/FloatConverter

Index

B

C

D

E

F

G

H

I

K

L

M

N

S

W

Z

Printed in the United States
by Baker & Taylor Publisher Services